EXOTIC
ANIMAL CARE
&
MANAGEMENT

SECOND EDITION

EXOTIC
ANIMAL CARE & MANAGEMENT

VICKI JUDAH, AS, CVT (RET.)

KATHY NUTTALL, MS, CVT

CENGAGE
Learning®

Australia • Brazil • Mexico • Singapore • United Kingdom • United States

Exotic Animal Care and Management, Second Edition
Vicki Judah and Kathy Nuttall

SVP, GM Skills & Global Product Management:
Dawn Gerrain

Product Team Manager: Jim Devoe

Associate Product Manager: Daniel Johnson

Senior Director, Development:
Marah Bellegarde

Senior Product Development Manager:
Larry Main

Senior Content Developer: Jennifer Starr

Product Assistant: Jason Kourmourdas

Vice President, Marketing Services: Jennifer
Ann Baker

Marketing Manager: Scott Chrysler

Senior Production Director: Wendy Troeger

Production Director: Andrew Crouth

Content Project Management and Art
Direction: Lumina Datamatics, Ltd.

Cover image(s): ©kimmik/Shutterstock.com;
©iStockPhoto.com/CathyKeifer; ©Denis
Tabler/Shutterstock.com

For product information and technology assistance, contact us at
Cengage Learning Customer & Sales Support, 1-800-354-9706
For permission to use material from this text or product,
submit all requests online at **cengage.com/permissions**
Further permissions questions can be emailed to
permissionrequest@cengage.com

Library of Congress Control Number: 2015954703

ISBN-13: 978-1-285-42508-5

Cengage Learning
20 Channel Center Street
Boston, MA 02210
USA

Cengage Learning is a leading provider of customized learning solutions with office locations around the globe, including Singapore, the United Kingdom, Australia, Mexico, Brazil, and Japan. Locate your local office at:
www.cengage.com/global

Cengage Learning products are represented in Canada by Nelson Education, Ltd.

To learn more about Cengage Learning, visit **www.cengage.com**

Purchase any of our products at your local college store or at our preferred online store **www.cengagebrain.com**

Notice to the Reader
Publisher does not warrant or guarantee any of the products described herein or perform any independent analysis in connection with any of the product information contained herein. Publisher does not assume, and expressly disclaims, any obligation to obtain and include information other than that provided to it by the manufacturer. The reader is expressly warned to consider and adopt all safety precautions that might be indicated by the activities described herein and to avoid all potential hazards. By following the instructions contained herein, the reader willingly assumes all risks in connection with such instructions. The publisher makes no representations or warranties of any kind, including but not limited to, the warranties of fitness for particular purpose or merchantability, nor are any such representations implied with respect to the material set forth herein, and the publisher takes no responsibility with respect to such material. The publisher shall not be liable for any special, consequential, or exemplary damages resulting, in whole or part, from the readers' use of, or reliance upon, this material.

Printed in the United States of America
Print Number: 01 Print Year: 2016

CONTENTS

v

UNIT VI

UNIT VII

UNIT VIII

UNIT IX

E xotic Animal Care & Management was designed to provide information on the ever-growing, ever-changing field of exotic animal husbandry and seeks to develop the technical skills needed for those students who wish to further their education, pursuing careers as veterinary assistants, veterinary technicians, or veterinary technologists. In this same vein, this new *second edition* of *Exotic Animal Care & Management* strives to be as equally informative and timely, providing information to reflect the changing pet population and clearly explaining various techniques required of these professions.

The Intent of This Book

Our human need for companion animals is ingrained and is probably the reason that the first orphaned wolf cub or injured young bird was taken in and cared for by our ancestors. It is a need far different from capturing, killing, and consuming. Over time, our attitudes have changed greatly. Animals are no longer just *pets*, but have become companions and family members. The dogs, cats, and endless hamsters of childhood have been joined by animals that are more unusual, strange, and exotic. Never before have there been such a variety of species and so little accurate information generally available.

The availability and affordability of many species have outpaced the knowledge needed to care for them. This, in turn, has produced countless *experts* whose personal experience is delivered as fact and clouds what real information is available. Pet store shelves are filled with glossy books, and racks of *care sheets*, yet sugar gliders are still hawked in the shopping malls as *cute little pocket pets*, and nonnative species are being released when they become too large, too aggressive, or too noncuddly. The problem is accelerating and there are serious ethical questions regarding many species currently being sold, bred, and traded with little thought given to their very specific needs and welfare. Many states have banned the sale of certain species. USDA permits are required for possession of many others. Laws have been enacted to address irresponsible ownership, the reality of nonnative species introducing disease, and the rising number of incidences of rarely seen zoonotic diseases.

Exotic animals are presented to veterinary staff usually when a major health issue becomes apparent or the home remedy or pet store treatment has failed. Then, an owner expects an immediate cure, whether for an iguana in renal failure ("but he loves cat food"), an aggressive chinchilla that is actually a wood rat (delivered to *the exotic vet* from another small-animal clinic), or the Ball Python presented with an esophageal laceration due to an inept attempt to force feed a tofu sausage concoction ("because eating rats is mean").

Veterinary staff and all animal caregivers play a crucial role in education for a variety of species, many perhaps not seen in the clinics and hospitals before, yet there are no defined educational requirements for credentialed technicians. Many schools offer a course on exotics as an elective, but not as a core requirement. Recommended and essential skills in an accredited program for veterinary technology require only that students be able to perform restraint and nail trims for birds and (unspecified) exotics. There is required curriculum for laboratory species.

It includes biological data and basic husbandry practices, but the approach to the beloved companion rat is appreciably different from meeting the needs and performing the procedures for rat colonies in a laboratory setting.

The majority of reliable information is directed toward the clinician/veterinarian or the very advanced, credentialed technician. This book was written to meet the needs of students. Combined, the authors have more than 42 years of clinical and teaching experience. With this experience has come a simple reality; students learn best when engaged, not when overwhelmed with the pedantic. Appropriate use of terminology is important and has been incorporated. We have made a deliberate choice to avoid the ostentatious, the *journal-speak* of academia, while still introducing students to the terminology required to communicate effectively and professionally, providing a more *learner friendly* text.

The American Veterinary Medical Association (AVMA) provides accreditation for schools offering programs in veterinary technology. The AVMA, in their Model Veterinary Practice Act (AVMA, November 2003), also defines veterinary technicians and veterinary technologists as those students who have attended a two- or three-year accredited program (technicians) and those who have attended a four-year accredited program (technologists). The use of either term implies that a certain level of education has been successfully achieved and that the person has passed the credentialing examination. The Veterinary Technician National Exam (VTNE) is administered at a state organization level. Each state may determine the score required to pass the exam and also to determine the title of successful candidates as Certified, Licensed, or Registered. Each state has autonomy in determining the level of care and procedures performed by a credentialed technician or technologist and whether or not they may be performed under the direct or indirect supervision of a veterinarian. Individual states may also have additional requirements for the performance of certain procedures. However, under no circumstances are credentialed staff allowed to diagnose, prescribe, or incise. A few states have no requirements. Other personnel may act as veterinary assistants but may not perform the tasks of a credentialed technician or technologist, nor should they be referred to as *technicians* or *vet techs*.

How This Book Is Organized

Each unit is broken down into subsets of chapters that provide information on related species. In addition, Unit I starts the book by providing basic information on exotic species as a group:

- **Unit I**—includes Chapters 1 and 2, providing an introduction to the unique needs of the exotic animals, as well as special considerations and concerns of these species.
- **Unit II**—includes Chapters 3–13 and provides species-specific information on exotic small mammals: Introduction to Small Mammals, Ferrets, Rabbits, Guinea Pigs, Chinchillas, Hedgehogs, Degus, Hamsters and Gerbils, Rats and Mice, Short-Tailed Opossums and Sugar Gliders.
- **Unit III**—includes Chapter 14: Avian.
- **Unit IV**—includes Chapter 15: Reptiles.
- **Unit V**—includes Chapter 16: Amphibians.
- **Unit VI**—includes Chapters 17–19: Scorpions, Tarantulas, and Hermit Crabs.
- **Unit VII**—includes Chapter 20: Alpacas and Llamas.
- **Unit VIII**—includes Chapter 21: Miniature Pigs.
- **Unit IX**—includes an all-new Chapter 22: Backyard Chickens.

Each species-specific chapter includes information on the species, its behavior, habitats, and diet. Without this appreciation, it just isn't possible to care for these species, whatever the educational level attained. We have included handling and restraint techniques, veterinary concerns and diseases, parasites, blood sampling, and injection sites. Anesthesia may have a slightly different approach in exotics, and special considerations are addressed.

Features of This Book

This book seeks to provide veterinary technology students with an informative yet engaging approach to the care and management of exotic animals. As such, it includes a variety of thought-provoking features to enhance the learning experience:

- **Tips and Tricks:** We have included many adaptations, devices, *tips*, sometimes *tricks*, that contribute to a successful outcome when dealing with different species. These facts are highlighted in boxes throughout the chapters and can best be categorized as *I wish I'd known that*.
- **Summary and Fast Facts:** A narrative *Summary* reviews the key points presented in each chapter, and *Fast FACTS* in the species-specific chapters provide a quick reference for important statistics, including average weight, life span, reproduction, normal vitals, dental, and zoonotic-related concerns.
- **Review Questions:** Serving as a helpful review, these questions appear at the end of each chapter and evaluate student knowledge of the information presented in the chapter.
- **Case Studies:** Each species-specific chapter includes two to three *Case Studies* applicable to the exotic animal discussed in the chapter. Students are presented with a medical concern and through a series of follow-up questions are encouraged to evaluate and determine the potential ailment of the exotic animal, based on the information presented in the chapter.
- **References:** References are included at the end of each chapter in the *For Further Reference* section and cite journal articles and papers presented during conference sessions, sources not readily available to students. Annual conferences offer great insight, understanding, and knowledge of the issues confronting all of us who have an interest in exotic species. Contributors to exotic animal sessions cannot be appreciated enough. Without them sharing their dedication and experiences, our knowledge would be limited.
- **Appendix—Professional Organizations and Associations:** The information presented here is intended to build on the solid foundation provided in a veterinary technology program, rather than offer, for example, detailed instruction for the operation of a rebreathing system in anesthesia delivery. We have not given specifics of drug therapies or formularies for exotics. Much of the pharmacokinetics of commonly used therapeutics in traditional small animals are unknown in exotics and many therapeutic choices are based on extra-label usage and clinical experience: what works, what doesn't, and what may be a possibility. These decisions should be made only by a licensed veterinarian. There are excellent Exotic Formularies, updated regularly as new information becomes available, and these belong on every bookshelf. Details of these may be found in the *Appendix* along with contact information for exotic specialty diagnostic laboratories and resource organizations.
- **Glossary:** A full glossary, including detailed definitions, offers students a quick review of important terms that encapsulate the unique attributes of exotic animals and their specific needs and requirements in the clinical setting.

New to This Edition

- **Current Concerns and Practices:** Current information is reflected throughout the chapters, including new concerns and practices related to the care of exotic animals as pets in addition to new species-specific information that serves to help further the understanding and acknowledgment of the unique needs of these animals.
- **Clinical Procedures—Expanded:** Additional information related to the skills required in the care of exotic animals, including significant laboratory results and radiological procedures, prepares students for the clinical setting.

- **Case Studies:** Formerly included in the accompanying workbook, the *Case Studies* now appear at the end of the species-specific chapters, encouraging students to directly apply what they have learned related to the exotic animal presented in the chapter. Two new cases are also presented at the end of the all-new chapter on *Backyard Chickens*.
- **Chapter 22—Backyard Chickens:** An all-new chapter on the subject of backyard chickens follows the same format as the other species-specific chapters and reflects the ever-changing exotic pet population.

Learning is a lifelong commitment. We also learn by teaching and hands-on experiences, with the questions students ask and the answers we don't know.

Supplement to This Book

Instructor Companion Site—This site features secure information that is specifically designed for the instructor:

- *Answers to the Review Questions* appearing at the end of each chapter allow instructors to track and validate student learning.
- *Lesson Plans* outline the key points in each chapter and review important skills that students should be learning.
- *PowerPoint®* presentations align with the Lesson Plans and include photos and graphics to visually reinforce the key points in each chapter.
- *Testing powered by Cognero*, a flexible online system, enables instructors to
  - author, edit, and manage test bank content from multiple resources
  - create multiple test versions in an instant
  - deliver tests from instructor/institution-specific LMS or classrooms
- *Image Gallery* offers full-color images from the book to enable instructors to further enhance classroom presentations

For these instructor-specific resources, please visit CengageBrain.com at http://login.cengage.com and follow the prompts for obtaining access to this secure site.

About the Authors

VICKI JUDAH is a Certified Veterinary Technician (Ret.) with more than 30 years' experience in animal care and a special interest in exotic animal medicine and captive care. The former program director of Companion Animal Science at the Jordan Applied Technology Center in Sandy, Utah, Ms. Judah developed the curriculum for courses in Animal Care and Science, Veterinary Assisting and Technology, Equine Science, Agricultural/Livestock Care and Management, and Aquatics. Now retired, Ms. Judah lives in Florida.

KATHY NUTTALL was the developer and instructor of the Veterinary Assistant/Veterinary Technician Program at the Jordan Applied Technology Center, and a Certified Veterinary Technologist. Currently, she is starting a new business educating veterinary professionals, Mountain Edge Veterinary Technology. She belongs to the Colorado Association of Certified Veterinary Technicians, the National Association of Career and Technical Educators, the Utah Association of Career & Technical Educators, the National Educators Association, the Utah Educators Association (AR 03/present), and the North American Veterinary Technicians Association. She has been awarded Teacher of the Year 2004 and CTE Teacher of the Year 2013. Ms. Nuttall obtained her Masters of Agricultural Systems and Education in 2013. She completed the state curriculum for Veterinary Assisting in the state of Utah.

ACKNOWLEDGMENTS

THIS BOOK COULD NOT HAVE BEEN WRITTEN without the willing and more-than-generous assistance of many people. We are very fortunate to have so many doors opened to us and to our students, for people at the end of a phone, for those of you who have spent hours reviewing the manuscript, generously giving of your time, contributing photographs, wisdom, and vast amounts of experience. Thank you for your insight, perspective, and dedication. Thank you for your contributions to learning. Without each and every one of you, our lives would be less enriched.

Daniel S. Schapiro, MD; Cathy A. Johnson-Delany, DVM; Martin G. Orr, DVM; Eric. Klaphake, DVM; Laura Devries, CVT; Susan Kelleher, DVM; Marc Kramer, DVM; Linda R. Harrison; and the Zoological Education Network. Janet and Clark Otterness & Alta Mist Alpacas, Faith Ching, Ching Sanctuary, Sonja Craythorn, CC Rabbitry, Ronaleigh Wheelright, Ronie's For The Love of Birds, Pet Kingdom USA (Las Vegas, NV), Beverly and Dan Ring, Chloe Long, Degutopia, Cindy Burningham and Mountain Top Alpacas. Scott Olsen, director of Jordan Applied Technology Centers (Ret.). The students at JATC Companion Animal Sciences, Veterinary Technology and Veterinary Assistant programs: without you, the questions may have never been asked. Thank you, also, to the Cengage Learning team.

In addition, we would like to express our gratitude to the reviewers of both the first and second editions, since each have contributed to the final form of this book:

Terry D. Canerdy, DVM
Murray State University
Murray, Kentucky

Amy Doherty, CVT
Minnesota School of Business—Rochester
Rochester, Minnesota

Besty Drobnitch, LVT
PIMA Medical Institute
Las Vegas, Nevada

Cynthia Hatch, LVT
PIMA Medical Institute
Las Vegas, Nevada

Elizabeth Kamaka, DVM
PIMA Medical Institute
Seattle, Washington

Susan Kopp, DVM
La Guardia Community College
Long Island City, NY

Angela Lathrop, DVM
Lodi, Wisconsin

Dr. Bill Plummer
Cal Poly San Luis Obispo
San Luis Obispo, California

Stuart L. Porter, DVM
Blue Ridge Community College
Weyers Cave, Virginia

Jennifer Serling, CVT
PIMA Medical Institute
Tucson, Arizona

Leland S. Shapiro, DVM
L.A. Pierce College
Woodland Hills, California

Gary Wilson
Moorpark College
Moorpark, California

FOREWORD

AS A VETERINARY TECHNICIAN EDUCATOR who has worked in an exotic animal practice and as an owner of exotic animals, I was excited to learn of this book. Much of my own knowledge of exotics had to be learned on the job. There was little emphasis on exotics in school and a lack of even the most rudimentary exotic animal textbooks. Experience has shown me how complicated exotic animal medicine and care can be. There is still a great deal of misconception and misinformation. Veterinary staff are often the first to provide crucial information to the owner of an exotic pet. We not only educate but in many instances also have to persuade owners that *what they were told* is often far from accurate.

Most of my students have had little exposure to exotic species. An introduction to this field needs to provide a background of the species, the animal's natural habit, and its situation now as a pet. An understanding of the species is essential to an understanding of the animal's health. Information is included in this book to ensure that proper housing and diets can be provided and behaviors better understood.

Medical problems and diseases, the issues of restraint, and techniques in anesthesia are delivered with clear language, yet the text also develops professional terminology. This book provides a beginner-friendly approach to learning. It is straightforward, ensuring an accurate understanding so the student has a concrete foundation to build a future working with exotic animals in many capacities.

Additionally, the text sets the stage for responsibility, offering both legal and ethical considerations in owning exotic animals and the profound impact that collection of wild specimens for the pet industry has had on many popular species. Zoonotic implications are also addressed in a manner that enables the student to appreciate the seriousness of animal-transmitted disease.

My personal acquaintance with the authors began while I was an undergraduate, working part-time in an exclusive exotic animal practice. With their guidance, I was fortunate *to learn the ropes* of being not only a veterinary technician but one working in a very specialized field. I attribute much of my practical knowledge to them, as opposed to the material taught in the classroom. They are able to turn *book knowledge* into real-life application, teach to a high standard of personal responsibility, and encourage sensitivity and understanding.

As a result of their tireless efforts, they have become top educators in the field of animal care and veterinary technology. Their teaching experience and practical knowledge has now become available to all students, presented at an appropriate level so more students are able to learn.

In writing this book, the authors have provided a much needed text that fills a gap between advanced-level textbooks and simple, often inaccurate, pet owner handbooks.

Laura Devries, CVT

UNIT I

CHAPTER 1

INTRODUCTION

OBJECTIVES

After completing the chapter, the student should be able to

- Be familiar with CITES.
- Know what species are covered under CITES.
- Be familiar with the laws that protect exotic pets.

Introduction to Exotic Animals

The trade in exotic animals has grown enormously in the past decade. There are exclusive exotic pet stores and specialized products ranging from cages to canned diets. There are exotic animal veterinarians, organizations, magazines, and Internet forums. A dictionary might define *exotic* as *foreign* or *unusual and different*, and this, in part, is the appeal of keeping an exotic animal.

With dogs and cats, there are specific laws that require licensing and rabies vaccinations. When dealing with exotic animals, laws become far more complicated. The issues are complicated even further by location; an animal that may be legal in one state is illegal in another and may be subject to confiscation, euthanasia, fines, or imprisonment for the owner. Federal laws and enactments ban certain species that can be kept by private individuals, but these regulations are in place to protect not only the species but also the public, public health, and native species. Before acquiring any exotic animal, regardless of the source, it is the responsibility of the owner to comply with the law and to obtain the permits and/or licenses required. The United States Department of Agriculture (USDA) and the Animal and Plant Health Inspection Service (APHIS) are federal governmental agencies that create and oversee the laws governing possession of exotic animals.

Many species that are privately held have come through black-market suppliers or have been purchased through Internet sources. It is likely that they were wild caught and possible that they bring with them parasites and the potential of zoonotic disease. Some species may be listed as threatened or endangered by the Convention on International Trade in Endangered Species (CITES).

In 1963, members of the World Conservation Union proposed an international voluntary agreement between governments of participating countries. The purpose was to ensure that the international trade in wild (exotic) animals and plants did not threaten their survival in the wild.

In 1975, what is now known as The Convention on International Trade in Exotic Species, (CITES) of Wild Fauna and Flora came into effect. There was worldwide concern that the illegal trade in many exotics was fueling such a demand that many species were being exploited to the point of extinction. The traffic in exotic animals amounted to billions of dollars and was *second only to illegal narcotics traffic*.

Species covered by CITES are listed in appendices according to the degree of protection they need. CITES Appendix I lists species that are threatened with extinction. CITES Appendix II lists species that are not directly threatened with extinction but for which trade must be controlled "to avoid utilization which is not compatible with their survival." CITES Appendix III lists species that are protected in at least one country. The country concerned has asked for assistance from other member countries in controlling trade in specified species.

CITES is one of the world's largest conservation agreements. There are now approximately 179 member countries (parties to the agreement). Not one species protected by CITES has become extinct, and for many species, the agreement prohibits capture in the wild.

Understanding the Concerns

There is a growing problem in the release of unwanted exotic animal species. Animals are released and expected to survive on their own. Many do not survive, but those that do can quickly become established, invading habitats and consuming native species, many of which are already endangered.

The Everglades National Park in Florida has a large and growing population of Burmese pythons, released by owners who no longer had an interest in them or the ability to care for them. Since October 2005, when the problems with this invasive species became evident, 200 or more of these giant *pet* snakes have been captured. Because of their adaption to the environment, they have been able to reproduce to such an extent that the Burmese python can now be found in other Florida counties, many in urban environments. In the spring of 2013, the largest python captured and killed was 18 feet, 8 inches (5.7 meters); the species of the snake and its length were confirmed by the University of Florida. In 2012, the federal government placed a ban on importation and sale of *injurious species* in Florida due to the problem caused by the release of these large snakes. Injurious species of snake include the Burmese python, two species of the African Rock Python, and the yellow anaconda. Federal and state officials are concerned that these species have the potential of spreading to other states. In December 2013, Florida state biologists issued a statement regarding the *menace* of the presence of the African Rock Python. During a hunt organized by wildlife officials, a 12-foot Rock Python was captured and killed. In a survey conducted prior to the hunt, it was determined that in an urban area just west of Miami, as many as 30 of these aggressive snakes had been captured. The area is close to shopping malls and residential areas and in the vicinity of where a Rock Python had killed a Siberian Husky in the dog's yard.

Another example of the release of nonnative species is Gasparilla, Florida, an island that has been populated by spiny-tailed iguanas. The island is overrun with an estimated population of more than 13,000 offspring from three pets released in 1970. To date, it is estimated that there are now 10 iguanas per human inhabitant of the island. There are frequent news reports of large snakes and other reptiles being found or causing injury to unsuspecting people and more traditional pets. This problem is not urban legend, but the consequence of human irresponsibility.

Many people think that zoos will readily take their unwanted exotic animals. This is rarely the case because of the potential for introducing disease and parasites. There are legitimate rescue organizations, but they are already full to capacity and struggling to find the funding required to care for and feed abandoned exotic animals. Responsible people know exactly what is involved in caring for an exotic species and, more importantly, are committed to the welfare of the animal for its lifetime in captivity.

People who are interested in and involved with exotic animals are valued members of veterinary practices, rescue organizations, and conservation efforts. Experience in this growing and specialized field of care often begins with volunteer work. Opportunities to volunteer may be found through veterinarians, zoos, and state and federal wildlife organizations. Working as a volunteer reflects commitment and a desire to learn new skills. These qualities often lead to gainful employment. Many jobs in the field of caring for exotic animals are offered to volunteers who have demonstrated their commitment by working long and sometimes difficult hours, accepting any task and performing it to the best of his or her ability.

Attending a two or four year college that offers programs in veterinary technology, wildlife management, animal science, or zoology greatly increases career opportunities and choices. An associate's or bachelor's degree can provide for more career opportunities in the educational field, research, and private industry.

Summary

The growing popularity of keeping exotic pets has resulted in many laws and regulations regarding the trade in exotic species. These laws and regulations not only protect endangered species but also ban the importation of specific species that have become a threat to native species, the public, and more traditional pets. CITES is one of the world's largest conservation agreements between countries. Not one of the species listed by member countries has become extinct, and for most of the listed species, collection from the wild is prohibited. People who are involved with exotic animals have many opportunities to contribute through volunteer organizations and pursuing two- or four-year degrees in a related field.

Review Questions

1. Research the different categories of animals listed in the CITES appendices and determine the following:
   a. How many species listed in each appendix are available in the pet trade?
   b. What impact has the trade in exotic species had on wild populations?

2. Investigate the instances of unwanted exotics being released into the wild.
   a. What are some of the common species that have been identified as being nonnative?
   b. What problems are created when nonnative species invade a habitat?

3. Suppose you wanted to acquire a skunk for a pet but are unsure if it is legal where you live.
   a. How could you determine if it is permitted and what special requirements would need to be met?
      i. Source(s) used?
      ii. Permits?
      iii. Requirements?
   b. What could be the consequences in your area of keeping an illegal species?
   c. Consequences for the animal?
   d. Consequences for the owner?

4. List some of the reasons there are laws prohibiting the keeping of certain species.

5. Why are some of the laws so complicated? For example, it may be that possession of a certain species is legal in a state, but not legal in a county or municipality.

For Further Reference

"Captive Exotics and Wild Animals as Pets." The Humane Society of the United States. http://www.humanesociety.org (accessed June 6, 2005).

"Conventional International Trade in Endangered Species." http://www.cites.org/ (accessed August 2007).

"The Dangers of Keeping Exotic Pets." Animal Protection Institute. http://www.bornfreeusa.org (accessed June 6, 2005).

"EPA-MAIA. Introduced Species." http://www.epa.gov/ (accessed August 2007).

http://www.aphis.usda.gov

http://www.newswatch.nationalgeographic.com (accessed November 6, 2013).

http://THELEDGER.com (accessed December 22, 2013).

Liebman, M. G. (2004). *Overview of Exotic Pet Laws.* East Lansing, MI: Animal Legal & Historical Center, Michigan State University, College of Law. http://www.animallaw.info.

"Non-native Species Sink Their Claws into Florida Island. Burmese Pythons in Everglades." FOXNews.com. Wednesday October 5, 2005 (AP). (accessed November 6, 2013).

ZOONOTIC DISEASES

OBJECTIVES

After completing the chapter, the student should be able to

- Define zoonotic disease.
- Understand how zoonotic diseases are contracted and spread.
- Know how to incorporate safety procedures in handling and housing exotics.
- Identify exotic pets that have a higher risk of transmitting zoonotic diseases.

Introduction

Zoonotic diseases are transmitted directly from animals to humans. The causative agent may be bacterial, viral, fungal, protozoal, or parasitic. Common routes of transmission include inhalation, direct contact, exposure to urine and fecal material, contaminated bedding and food/water bowls, and handling dead or diseased animals and their body fluids and tissues. Methods of transmission can be as variable as each disease-causing organism.

Diseases may be mild to severe and, in some instances, life-threatening. Because more people are coming into close contact with exotic species, the exposure to disease and zoonotic potential increases. Anyone who works with animals is at risk. Those who have close and frequent contact with exotic species have an increased risk. Children and the elderly have higher risk factors because their immune systems are less able to ward off disease. Anyone who is already immunocompromised has an even greater risk.

Species commonly available, including those sold in pet stores, at reptile and bird expos, and by private breeders, all have the potential to transmit disease. Some bacterial zoonoses (plural for zoonosis), such as Salmonella, are well recognized, while others, such as tularemia, are known to have occurred. Tularemia is a contagious strain of Pasteurella bacteria, *Francisella tularensis*. One of the biggest concerns is new and emerging unknown diseases. The disease may not be new, the pathogens (disease-causing agents) may be known, but the zoonotic implications are sometimes discovered only after the first occurrence is diagnosed in a human.

Pathogens, like other life forms, are continually adapting and responding to environmental changes. Many are opportunistic, invading convenient hosts not considered to be the *normal* reservoir. Viruses mutate regularly with different strains emerging annually. Bacteria are becoming drug resistant, producing **superinfections** with deadly consequences to humans.

Few vaccines have been developed to protect against zoonotic diseases. The human rabies vaccine is one exception and is readily available, yet few people choose to be inoculated. People with a high risk of exposure to rabies include veterinarians, veterinary technicians and assistants, wildlife personnel, and diagnostic laboratory workers.

When different species are housed collectively, the opportunity for pathogens to invade other organisms increases. There may be a direct leap to a new host or links and bridges formed from one host to the next until multiple species are infected and ultimately transmitted to a human host. The **monkey pox** outbreak demonstrates how easily this can occur. Monkey pox is caused by an **orthopox** virus and is closely related to the human smallpox virus, **variola**.

In June 2003, monkey pox was diagnosed in several people in the United States. Prior to this, the disease had been limited to people and animals native to the rainforest regions of central and west Africa. The source of the outbreak was determined to be an exotic pet store in a suburb of Chicago, Illinois. All of the victims had one thing in common; they had close contact with wild-caught prairie dogs that were offered for sale. The infected prairie dogs (native to the western states of America) had been previously exposed through a Texas distributor's facility. In the Texas facility, they were housed in close proximity to Gambian giant rats and other small mammals imported from Africa and destined for the pet market.

In Illinois, the exposed prairie dogs were sold as pets. The premises also housed several other exotic mammals. No one knows how many other species of mammals are susceptible. The CDC strongly suggests caution because the potential is unknown, and all mammals should be considered susceptible to the orthopox virus.

The CDC and the **Food and Drug Administration (FDA)** have issued a legal order that prohibits the importation of all African rodents and also *stops movement, sale, or release of prairie dogs* (CDC, 2003). Also included are six genera of African rodents known to be already in the United States.

Hamsters, mice, rats, and guinea pigs are known reservoirs of the virus responsible for causing **lymphocytic choriomeningitis (LCM)**, a disease transmissible to humans. Although long recognized in pet hamsters as a *wasting disease*, there was little public awareness until May 2005, when three of four recipients of organ transplants died from LCM. The organ donor carried the virus but was asymptomatic at the time of his death. He was infected by a pet hamster. The CDC was able to confirm that the original supplier of the hamster was a breeding facility in Arkansas. Unknowingly, infected animals were shipped to retailers in several states. The hamster was purchased at a store in Rhode Island. Further investigation revealed that not only was the hamster the direct source, but other animals in the store also tested positive for LCM. These included two other hamsters and a guinea pig.

The common house mouse, *Mus musculus*, is the reservoir carrier of the LCM virus. It is likely that the breeding facility was contaminated by mouse droppings and urine. The virus is also shed in contaminated bedding and saliva as well as transmitted from infected animals to their offspring in utero. Once infected, animals shed the virus throughout their lives. It is not fatal in mice and there may be no signs that they are infected.

The CDC has issued an awareness and warning fact sheet, especially for pregnant women. Exposure to the LCM virus during early pregnancy can cause abortion, or the child "may suffer permanent developmental skills." Fact sheets are available from the CDC website (http://www.cdc.gov).

Salmonella bacteria is known to cause **septicemia**, systemic blood infections causing death in rodents, rabbits, guinea pigs, and humans. The bacteria is considered normal flora in reptiles but is pathogenic in other species. Cases of humans contracting the disease from vacuum-packed feeder mice have been documented. In 2011, the **Centers for Disease Control (CDC)** and the **United States Department of Agriculture (USDA)** investigated an increase in the number of incidents of Salmonella in humans who had been exposed to pet hedgehogs.

Campylobacter bacteria has been isolated in hamsters and ferrets. Reported incidents of human Campylobacter cases have been directly linked to pet exposure. Companion animals, however exotic, are brought into homes and are given family member status. People develop strong attachments and spend much time caring for and interacting with them. It is human nature to want to hold, hug, kiss, and share. There is rarely a thought given to potential disease. Iggy the iguana probably drags Salmonella bacteria across the kitchen counter; Buddy the bird always gives and receives kisses and takes food from the owner's mouth and *could* be carrying psittacosis, coccidia, or giardia. It is not only reptiles and birds that can transmit disease. Owners need to be aware (as noted earlier in text) that *potentially* any animal could be a source of human disease.

There are more than 200 known zoonoses. Source animals have been identified, methods of transmission known, and the specific symptoms are recognized and treated in humans. The majority of zoonoses are confined to people with direct contact with a specific host and are not usually transmitted from person to person. However, throughout human history, there have been **pandemic diseases** that have killed millions. Pandemic outbreaks of a disease involve multiple countries and continents.

Black Death, **bubonic plague**, swept through Europe from 1347 to 1350, killing approximately one-third of the population. The source was a bacterium, *Yersinia pestis*, that is found in the gut of fleas and thrives in the blood of rats. Infected rats spread the disease to people. Infected people spread the plague to family members and entire villages. People who fled in an attempt to escape the disease spread it even further, carrying it across much of Europe. People became infected from direct contact with infected people, fleas, and rats, causing two of the three forms of the plague: septicemic and bubonic. The other form, **pneumonic**, was transmitted by airborne droplets, sprayed by the coughing and sneezing of infected people.

While it is easy to think that this distant disease was limited to the Dark Ages, the Natural Bridges National Park (San Juan County, Utah) was temporarily closed to the public in April 2006 because the wild rodent population was infected with the plague. In June 2006, a cat in Arizona was positively diagnosed with bubonic plague due to probable close contact with wild rodents. Even today, there are sporadic outbreaks in wild rodent colonies, such as prairie dogs and chipmunks. These outbreaks are quickly identified and warning signs are posted in affected state and national parks until the concern is resolved, either by humane extermination of infected colonies or through natural population demise.

Ectoparasite infestation is also a concern when considering zoonotic potential. Scabies is caused by a burrowing mite (Sarcoptes) and can be found in most animal species. *Trixacarus caviae*, another type of mange mite specific to guinea pigs, has shown to be zoonotic as well as Cheyletiella (*walking dandruff* mite), a common mite found in rabbits.

Allergic alveolitis (hypersensitization) is an inflammation of the alveoli in the lung. The inflammation decreases lung capacity and makes it difficult to move air through the lungs when breathing. People become hypersensitive to inhaled dusts from feathers and fecal material. This is most commonly seen with exposure to avian species. Today, with air travel, human migration and expansion, encroachment into relatively isolated areas, and destruction of habitats, zoonotic diseases can quickly become global concerns. Humans are coming into contact with pathogens to which they have no immunity, more exotic species, and disease potential that is unknown.

The Children's Rhyme

Ring Around a Rosie is constantly being argued over, interpreted, reinterpreted, and dissected. Is this little verse an oral history of the Black Death?

Ring around a rosie: The plague was characterized by pustules ringed with a rosie color.

Pocket full of posies: Strongly scented flowers and herbs (posies) were carried in an attempt to ward off the disease and cover up the stench of decaying corpses.

Ashes, ashes: Bodies were piled together and burned.

All fall down: Countless, seemingly endless bodies, and all were expecting to die.

Most zoonotic diseases are acquired through casualness, carelessness, and lack of education. Zoonoses from animals kept as companions are real and the potential is known. Responsible ownership includes knowledge of what disease potential there may be and how to prevent possible infections. It is everyone's responsibility to inform and educate. Pet stores, animal handlers, veterinarians, and veterinary staff are all in the forefront of public education and awareness.

Summary

Zoonotic diseases are transmitted directly from animals to humans. The causative agent may be bacterial, viral, fungal, protozoal, or parasitic. All species of animals have the potential to transmit a zoonotic disease. Most zoonotic diseases are acquired through casualness, carelessness, and lack of education. Responsible ownership includes knowledge of what disease potential there is and the best practices to prevent possible infections.

TABLE 2-1

SAFE PRACTICES
• Exotic animals should not be allowed to roam freely. This especially applies to reptiles.
• Reptiles should be provided with dedicated tubs for soaking, not placed in the family bathtub or kitchen sink.
• No animal should be in or near food preparation areas.
• Animals should be fed only from dishes reserved exclusively for them, never from hands, and should not be allowed to take food from a human mouth or to lick dinner plates and other utensils.
• All animal food and water dishes should be disinfected and washed separately, not placed in the kitchen dishwasher.
• Animal food and water bowls should not be exchanged between species (e.g., the snake bowl should not be used to provide water for the family dog).
• All soiled bedding and fecal material should be handled with latex or nitrile gloves and disposed of in tied plastic bags. It should not be used as garden mulch!
• Hands should *always* be thoroughly washed with soap and water after handling any animal, animal equipment, and enclosure contact, even if wearing disposable gloves.
• Disposable gloves should be pulled off so they are inside-out, tied at the wrist, and disposed of in a plastic bag.
• Hands should be washed between handling other animals.
• In cleaning cages that are very dirty, a disposable face mask should be worn.
• When handling a diseased animal, protective eyewear should also be worn.

TABLE 2-2: Short List of Common Zoonotic Infectious Agents and Source Groups

SHORT LIST OF COMMON ZOONOTIC INFECTIOUS AGENTS AND SOURCE GROUPS	
Bacteria	**Source**
Salmonella	All animals
Aeromonas	Reptiles
Mycobacterium (tuberculosis)	Mammals, reptiles, birds
Chlamydophila psittaci	Birds
Listeria	Amphibians
Tularemia	Mammals (especially rabbits)
Leptospirosis	Mammals

TABLE 2-2: Continued

SHORT LIST OF COMMON ZOONOTIC INFECTIOUS AGENTS AND SOURCE GROUPS	
Campylobacter	Reptiles
Enterobacter	Reptiles
Streptobacillus	Mammals
Pasteurella	Mammals, birds
Colibacillus	Birds
Yersinia pestis	Mammals (rodents and fleas)
Staphylococcus	Mammals, birds
Virus	**Source**
LCM (lymphocytic choriomeningitis)	Mammals
Rabies	Mammals (rarely reported in birds of prey)
Monkey pox	Potentially any mammal
Yeast	**Source**
Candida	Reptiles, birds
Fungus	**Source**
Dermatophytes (ringworm)	Mammals (commonly identified in rabbits, guinea pigs)
Cryptococcus	Birds
Protozoa	**Source**
Giardia	All species potential
Coccidia	All species potential
Ectoparasites	**Source**
Sarcoptes	All species potential
Cheyletiella	Mammals

Review Questions

1. Define zoonotic disease.

2. What are some common routes of transmission for zoonotic diseases?

3. List eight safety practices that will reduce the potential for contracting a zoonotic disease.

4. Hamsters are known carriers of a disease called LCM. What is it and how is it contracted?

5. Which government agency tracks and provides information on zoonotic diseases?

6. Discuss the reasons that close contact with exotic animals could potentially cause disease in humans.

7. Which population groups have a greater risk of developing severe complications from a zoonotic disease?

8. Explain why mixed species of animals should not be housed together. Give one example of what could happen.

9. List five different groups of pathogens that have zoonotic potential.

10. Access the website of the Centers for Disease Control (CDC) and
   a. Determine the incidence of zoonotic diseases. How frequently are they reported?
   b. Determine which species of animals are most commonly implicated.

For Further Reference

Centers for Disease Control and Prevention. (November 5, 2003). *Fact Sheet: Monkeypox in Animals: The Basics for People Who Have Contact with Animals.* http://www.cdc.gov (accessed November 6, 2013).

Centers for Disease Control and Prevention. (January 26, 2006). *Fact Sheet: Embargoed African Rodents and Monkeypox Virus.* http://www.cdc.gov (accessed November 6, 2013).

"Foreign Animal Disease Alert: Investigation Uncovers First Outbreak of Monkeypox Infection in Western Hemisphere." (June 23, 2003). American Veterinary Medical Association. http://www.avma.org (accessed March 28, 2006).

Hoff, B., & Smith, C., III. (2000). *Mapping Epidemics: A Historical Atlas of Disease.* London: Franklin Watts.

Kaplan, M. (2014). *Potential Zoonotic Diseases in Exotic Pets.* Herp Care Collection. http:www.anapsid.org (accessed June 12, 2014).

"Monkey Pox Backgrounder." American Veterinary Medical Association. http:www.cfsph.iastate.edu (accessed June 23, 2003)

Nolen, S. *CDC on the Offensive to Stamp Out Rodents Virus.* http://www.avma.org (accessed March 28, 2006).

Veterinary Learning Systems. (2002). *Emerging Vector-Borne and Zoonotic Diseases.* (Supplement to Compendium, Vol. 24, no. 1(A) of Veterinary Learning Systems, Muriel, January 2002).

Williams, B. (2013). *Zoonoses of Small Mammals.* Tulsa, OK: Clinicians Brief, pp. 53–56.

Marina Jay/Shutterstock

UNIT II

3 INTRODUCTION TO SMALL MAMMALS

OBJECTIVES

After completing the chapter, the student should be able to

- Explain the term *eutherian*.
- Describe a metatherian.
- Explain the uniqueness of a monotreme.
- Describe the diet of a herbivore.
- Know what defines a carnivore.
- Describe the diet of an omnivore.

Introduction

Mammals are an enormously diverse group within the animal kingdom. They represent a great variety of species, ranging from the very smallest of the known mammals, Kitti's Hog-nosed Bat, which weighs considerably less than an ounce, to the greatest of all living creatures, the Blue Whale, with an estimated weight of 150 tons. Although the diversity of mammalian species is great, they all share distinguishing characteristics.

All mammals are vertebrates, possessing a spinal column that is part of the endoskeleton, a framework of bones that provides structure and support to the body. Mammals have a highly developed central nervous system and brain. They are all endotherms, meaning that they are able to generate and maintain body temperature, have hair, and produce milk to feed their young.

Mammals are separated into three distinctive groups based primarily on embryonic development. The most familiar species are the eutherians, those mammals with a placenta that nourishes the developing young, connecting them to the uterus until the time of birth. The young of some species are precocial. This means they are fully developed at birth; they have their eyes and ears open, teeth erupted, and a full coat of hair; and they are able to function with a degree of independence. A few of these species include the degu, chinchilla, and guinea pig. More commonly, eutherian young are born altricial; when they are born their eyes and ears are closed, they have little hair growth, and they are entirely dependent on maternal care for survival. Examples of these species include mice and rats, rabbits, and ferrets.

The second group is metatheria, or marsupials. The young of all marsupials are born in an embryonic state. They must make their way to either a pouch or a teat attachment to complete their development. Two examples of marsupials commonly kept as pets are sugar gliders and Brazilian short-tailed opossums.

The third group, monotremes, are mammals that lay eggs. There are only two living species of monotremes, the duck-billed platypus and the echidna, a type of spiny anteater. The word *monotreme* means having one opening for excretion and reproduction. While they have many features characteristic of other mammals, adult monotremes do not possess teeth. They also lack nipples, but instead secrete milk to feed their young directly through the skin that covers the mammary glands. The echidna has a specialized snout for probing into anthills and a very long tongue, while the platypus has a *bill*, not dissimilar to that of some ducks, which is used to probe underwater in search of food items.

With the exception of the monotremes, all mammals have two sets of teeth. The deciduous (baby teeth) are shed and replaced by permanent, adult teeth. If adult teeth are lost, they are not replaced.

Other features unique to mammals include the structure of the mandible, or lower jaw. The mandible consists of one bone in mammals and many smaller bones in nonmammals. Fossil evidence in the evolution of mammals has determined that these smaller bones of the lower jaw were once present but have evolved to form the bones of the middle ear. They have become the malleus (the hammer), the incus (the anvil), and the stapes (the stirrup).

Additionally, mammals have a diaphragm, a strong muscle that assists in respiration. It is absent in birds and reptiles. Mammals also possess a soft and a hard palate, which separate the esophagus and the trachea.

Mammals consume a variety of diets. They may be classified as herbivores, animals that eat only plant material; carnivores, those that eat only meat; and omnivores, those that eat both plants and other animals. Some mammals may be referred to as insectivores; they feed primarily on insects.

Summary

Mammals are a diverse group within the animal kingdom. All mammals share certain characteristics: they are all vertebrates, have hair, produce milk for their young, and have a highly developed brain and nervous system. Additionally, all mammals are endotherms. They are divided into three groups based on the manner in which the young develop prior to birth. Mammals consume a variety of foods. Mammals may also be described by their diet: carnivore, omnivore, herbivore, or insectivore.

Review Questions

1. What is one reason mammals are divided into groups?

2. Give definitions for the following groups of animals:
   a. eutherians
   b. metatherians
   c. monotremes

3. What is the function of the placenta?

4. How does the structure of the mammalian mandible differ from other classes of animals?

5. What are the names of the bones of the middle ear?

6. What are the deciduous teeth?

7. A diaphragm is unique to mammals. What is its function?

8. Define the following terms:
   a. herbivore
   b. carnivore
   c. omnivore
   d. insectivore

9. The young of mammals are described as being either precocial or altricial. What is the difference between these two?

10. Create reference lists for pet owners to describe the appropriate diets with examples for
    a. a small mammal carnivore.
    b. a small mammal herbivore.
    c. a small mammal omnivore.
    d. For each, determine how each item meets the species' dietary needs.

For Further Reference

Earth-Life Web productions. (April 2006). www.earthlife .net (accessed December 4, 2006).

www.animaldiversity.org.
www.ucpm.berkley.edu (accessed December 4, 2006).

FERRETS

OBJECTIVES

After completing the chapter, the student should be able to

- Understand the husbandry needs for ferrets.
- Provide appropriate client education to new ferret owners.
- Know what vaccines ferrets should be given and the vaccination schedule.
- Provide basic nursing care to a ferret.
- Assist in the anesthesia of a ferret.
- Demonstrate appropriate restraint techniques.
- Understand common ferret disorders.

KEY TERMS

fitch
congenitally
hob
jill
kit
scruff
fixed-formula
free choice
hypoglycemia
anaphylaxis
Mycoplasma
dyspnea
NSAID
melena
inflammatory bowel disease (IBD)
alopecia
pruritus
erythematosis
hyperestrogenism
ecchymosis
insulinoma

Introduction

In Latin, *Mustela putorius furo* means "smelly mouse-eating thief." Ferret owners could agree that ferrets are smelly thieves, but not all are mouse-eaters. What they will all agree on is how funny, playful, mischievous, and delightful ferrets are as companions. The ferret has a firmly established place as an animal companion, rivaling the popularity of dogs, cats, and birds.

The ferret has been domesticated for centuries. In 450 BC, Aristophanes, a Greek playwright, mentioned ferrets in "The Achaeans," referring to certain people and ferrets together as thieves. The presence of ferrets in ancient writing and art, coupled with the uncertainty of their ancestors, has led many to think that the domestic ferret may never have been a wild species.

Domestic ferrets are closely related to the European polecat (*M. putorius*), a wild ferret that ranges across much of the forests of Europe and Great Britain. A polecat is also known as a fitch. Domestic ferrets and polecats can interbreed, and their physical appearance is barely distinguishable, so it is likely that selective breeding developed the domestic ferret. Closely related is the endangered black-footed ferret, which was originally found in the prairies of North America. Captive breeding programs have successfully reintroduced several groups in the hope that the black-footed ferret will be able to reestablish wild breeding populations.

Ferrets were kept as hunting animals and for pest control. They are efficient hunters with the same predatory instinct of all carnivores. They were introduced into Australia and New Zealand in the 1800s in an attempt to rid these countries of an overpopulation of rabbits (also an introduced species), which were digging up the countryside and ruining crops. In Great Britain and many other European countries, ferrets are still kept for sport hunting (see Chapter 5).

The family *Mustelidae* includes ferrets as well as skunks, weasels, mink, badgers, and otters. Common to animals in this family are very strong-smelling scent glands. Ferrets sold in pet stores are de-scented and spayed or neutered. Anal scent glands, located just inside the rectum, are surgically removed; however, ferrets that have been de-scented still have a musky odor. They have small, scent-producing sebaceous glands on the abdomen that cannot be surgically removed.

Ferrets may be given a bath with a shampoo specifically formulated for ferrets or one made for kittens. Dog and human shampoos should not be used because they reduce the natural oils of the coat and result in dry, flaky skin. Ferrets should not be bathed more frequently than monthly because frequent bathing actually increases sebaceous gland secretion in compensation for oils lost. Diet can also contribute to the odor, especially diets that contain blood and blood by-products. The easiest way to reduce odor is by keeping the ferret's cage clean and by changing the bedding often.

The American Ferret Association recognizes 30 color variations, including sable (the most common and natural coat color), silver and a silver mitt (silver with white feet), chocolate, albino, black-eyed whites, and cinnamon. Some of the newer color variations have distinctive white markings on the head and body. Selective breeding for coat color is often genetically linked to other less desirable traits. Because of the inbreeding required to establish new coat colors and variations, many ferrets with white marking on their heads are **congenitally** deaf; that is, they are born without the ability to hear.

Hobs (males) can be twice the size of **jills** (females). This is especially evident if the hobs were not neutered at an early age. Hobs weigh an average 2 to 4.5 pounds (1 to 2 kg). Jills often weigh as little as 1 to 2.5 pounds (0.6 to 0.9 kg).

Twice a year, healthy ferrets shed their coats and their color may change from dark to light, depending on the season. Ferret skin is normally thick and tough, especially around the neck.

In the United States, ferrets are kept as companion animals. Regulations regarding ferrets vary from state to state. The possession of ferrets is illegal in California, while in many other states, counties, and towns there are strict regulations that require ferrets to be spayed, neutered, and licensed. In addition, a special permit may be required for possession of a ferret. Before obtaining a ferret, always check local laws first. If it is illegal to keep a ferret, ordinances have provisions for confiscation and the ferret may be euthanized. Persons illegally housing ferrets may also be subject to animal violation citations and fines. Many of the laws regarding ferrets were enacted before there was an understanding of their behaviors. The concerns were often based on hearsay: that ferrets would attack and kill human infants, that they would decimate other animal populations, and that they were carriers of rabies. These concerns have now been largely discounted through public education, and many areas have repealed laws that prohibited keeping ferrets as pets. However, no one should attempt to smuggle a ferret into an area where their possession is illegal.

Behavior

Ferrets spend a great deal of time sleeping. It is not unusual for a healthy ferret to sleep 75 percent of their day while the remaining 25 percent is spent in active play. The sleep is so deep, especially in **kits** (immature ferrets), that many new owners become alarmed, thinking that their pet might be dead. When awake, ferrets are nonstop bundles of curious energy.

The spine of a ferret is long and very flexible, allowing it to maneuver through small spaces and still be able to turn around. Ferrets will investigate any space and are capable of flattening their bodies to squeeze into and under small areas. For this reason, homes must be "ferret-proofed" as ferrets may disappear into small spaces that the owner may not have even noticed. It is especially important to check under cabinets for small, recessed openings between the floor and the cabinet bottom. All doors, latches, and window screens need to be checked carefully. Ferrets have no trouble climbing up a screen, making a small opening and crawling out through it. They can squeeze under a closed door that is not flush with the floor or go into heating vents or areas that have been cut out for water pipes.

Ferrets are active diggers and can throw out all the soil in a potted plant in a matter of minutes. They will dig at carpets and in the corners of rooms. Soft furnishings are a favorite play area; a ferret will go in, over, and under, pulling itself along on its back by its nails, potentially doing as much damage as a cat clawing at the furniture.

Ferrets Should Not Be Declawed.

Declawing of cats is controversial, but declawing a ferret is crippling, and the ferret will suffer permanent disability. The claws of ferrets are not retractable, and the anatomy of the digit is completely different from that in cats. Declawing a ferret includes not only removal of the claw but also amputation of the last joint of each toe, making it very difficult for the ferret to walk normally.

Owners must be very careful when the ferret is out of its cage and disappears. Many pieces of furniture are perfect hiding places for ferrets and can become death traps. This is particularly true of reclining chairs and sofa beds. Many ferrets have been critically injured or killed by people who did not realize the animal was sound asleep inside the chair, pulled the recliner lever, and crushed the ferret, or sat on the sofa with a ferret under the cushions. Trauma can include dislocation of a joint or fractures (**FIGURE 4-1**).

Bedding and laundry baskets are other places ferrets like to curl up and sleep. More than one distraught owner has found a missing ferret when pulling laundry from the washing machine. All drawers must be opened carefully, as ferrets climb up the backside of chests or cabinets, be they in the bathroom, bedroom, or kitchen.

Ferrets are very playful and enjoy games with their owners. An excited, playful ferret performs an array of movements by jumping, hunch-backed, bouncing

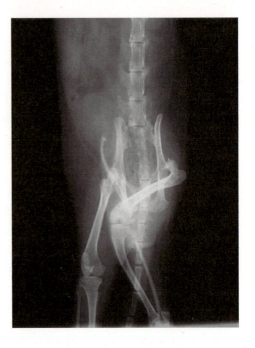

FIGURE 4-1 The owner of this male ferret was unaware that he was sleeping in the back of the couch. When she sat down, the result to the ferret was a luxation of the left femoral head.

up and down, running backward or around in tight circles, shaking its head, and making soft chuckling and hissing sounds. Many ferrets will attack bare toes and ankles, and others may have a strong attraction to socks and slippers. Ferret owners quickly learn to wear shoes when their ferrets are out for playtime.

Play between ferrets can be very rough, especially with kits. Like kittens that learn to stalk, pounce, and bite, kits have no hesitation in being just as rough with people. Ferrets grab and bite each other's **scruff** (the thick skin around the area of the neck) or drag each other around by the ears or any other appendage. They should be taught at a young age that biting and pulling are not so enjoyable for humans. Ferrets are intelligent and responsive to the human voice, quite capable of understanding the meaning of *NO*. Scruffing a kit by grasping the area of skin at the nape of the neck, and saying "No" in a firm voice is effective.

Ferrets should never be given physical punishment. One of the worst recommendations is to flick the ferret on the end of the nose. This is not only very painful for the ferret and completely unjustified but it also teaches the ferret that hands cause pain, and the lesson for the ferret is one of avoidance or aggression. A simple, painless scruff is all that is ever required—a time-out. When ferrets are scruffed with all four feet off the ground, they relax completely and will usually yawn widely and appear to be asleep.

Toys must be chosen carefully. Ferrets that have swallowed pieces of toys often require surgery to remove foreign body obstructions. Soft, wool-covered toys; fur mice; plastic cat balls; feather wands; and squeaky toys are all favorites. Toys need to be checked regularly to make sure the ferret is not chewing on them and swallowing small pieces. The biggest danger is not from toys, however, but from other items around the house. Common causes of intestinal blockage are foam rubber from headphones and furniture cushions, pencil erasers, rubber bands, and cotton balls. Other items that have been surgically removed are small pieces of plastic, wads of thread, and Velcro®, which is another favorite chew toy for ferrets.

Shoes with Velcro® fastenings and all shoes with rubber soles need to be examined for missing pieces.

Thief? Ferrets will take anything that appeals to them and stash the item away. Missing items are found all at the same time and in the same place. For ferret owners, this is the answer to the age-old mystery, Where did the other sock go? In addition to socks, people have found car keys, wrapped sweets, prescription bottles, cash and wallets, ornaments they had not even realized were missing, toothbrushes, and countless other items. Clearly, these are not food items for a ferret. It is not food hoarding, as seen in other species, and the reason for this behavior is not clear. Wise owners know the location of the stash and have learned to look there first for any missing items. Good places to begin a treasure hunt are the inside of a box spring mattress and in the back of the family sofa or upholstered chair.

Ferrets will use a litter box, provided it is in the right place at the right time. During play, they may not actively seek out the litter box but instead run backward to the nearest corner. Plastic sheets or loose floor tiles in the corners of rooms where the ferret is allowed to play make cleanup considerably easier. Ferrets elevate their hind quarters to urinate and defecate, so the protected corners should also cover a small area of the wall. Litter boxes designed specifically for ferrets are triangular in shape and fit into corners. The triangular back of the litter box is higher to accommodate the ferret's elimination behavior. Many ferrets will not use a litter box, no matter how many or how well placed the litter boxes are. Instead, they prefer to go behind the litter box, often shoving it away to get to the corner. Basic unscented cat litter is the best choice. Clumping litters could have the potential for creating an intestinal blockage if ingested.

Routine care includes nail trims and ear cleaning. Each foot has five claws that can grow very long and become entangled in carpet and bedding. These can easily be trimmed with small human nail clippers, being careful not to cut into the quick. Depending on the ferret, it may require one person to scruff and dangle while another performs the trim. What often works well is to place a small amount of cat hairball remedy on a tongue depressor and allow the ferret to lick it while the nails are being trimmed. They are very fond of the malt flavor and will usually be distracted enough not to object to the nail trim.

Ferrets normally produce grainy, reddish-brown exudates from their ears. Ear-cleaning solutions are available from pet stores and veterinarians. A few drops of the solution should be placed in each ear and massaged in. The ear can be wiped out with cotton-tipped applicators, being careful the applicator does not enter the ear canal. If the debris is very dark and has an odor, the ferret may have ear mites, which can be confirmed and treated by a veterinarian.

Housing

One of the things ferrets do best is sleep and elaborate cages are not required. With the dramatic increase in the popularity of ferrets has come a variety of cages and related products, in particular the ferret condo.

The condo is an upright, multilevel wire cage with a solid floor. Ferrets are ground dwelling, and the various levels designed within these cages can contribute to injury. Because the eyes of ferrets are located more laterally, their vision can be limited when items are directly in front of them. Negotiating ladders and ramps can lead to falls. Most condo setups suggest using sleeping sacks or hammocks suspended from the top of the cage. Ferrets do not naturally climb up to sleeping dens but go down and under to a small space. Many ill or sleeping ferrets have fallen from the unguarded edges of platforms, resulting in fractures. Condos, if used, should be equipped with padded ramps rather than open-rung wire ladders. Sleeping sacs are safer and more easily accessed by the ferret when they are placed on the cage floor.

Many ferrets are housed in large, sturdy pet carriers. An adequate-size carrier would be a medium-to-large dog pet transporter. Ferrets that are allowed to play out of their enclosures do not play in their sleeping den. The sleeping den needs to be provided with a water bottle, a full food cup, soft bedding, and a litter box at the opposite end of the carrier. When the ferret is let out for exercise and play time, the door should be left open, making access to food and water uncomplicated.

Cage substrates, including any type of wood shavings, recycled paper products, and corn cob, should not be used for ferrets. The cage floor can be lined with plain newspaper. A pile of soft towels, old clothes, or pieces of blanket are all a ferret needs for comfort and security. Items used should be those that can be easily and frequently laundered without the addition of fabric softeners or drying sheets. Frequent laundering of bedding items will reduce the musky odor associated with ferrets.

Ferrets should be protected from extreme weather conditions. They do not tolerate temperatures above 90°F or below 20°F. Glass tanks are not recommended because they do not allow adequate ventilation.

Diet

Ferrets are carnivores. They require a diet based on animal-derived protein. Most commercial dog and cat foods do not have adequate protein levels and may contain plant-based protein that can lead to bladder stones. Many ferrets are fed, and will eat, a variety of fruits and dairy products, but both can contribute to digestive and urinary tract problems. Fruit produces urinary crystals that could eventually lead to the formation of kidney or bladder stones. Dairy, especially ice cream, milk, and cottage cheese, can cause diarrhea. A high-quality, fixed-formula food specifically formulated for ferrets is recommended

> **Fixed-Formula Diets Are Constant;** that is, the ingredients and percentages do not change with market availability and price. There is a standard or fixed recipe for each batch of food produced by the manufacturer.

Diets that have been specifically formulated for ferrets are also available. Many of these include blood and blood by-products, which contribute to foul-smelling stools. Whichever choice is made, abrupt or random changes in brands should be avoided to prevent diarrhea and intestinal upset. All product labels should be carefully read and evaluated not only for protein content but also for protein source.

Raw foods, including horse meat, beef, and chicken, are sometimes fed to ferrets. Whole prey items can include small rodents and chicks, but the feeding of raw meat to any non-zoo carnivore is controversial due to the potential risk of parasite and bacterial transmission. With the high standards and nutritional balance of formulated diets, feeding raw meat is neither necessary nor recommended.

Ferrets have a very short digestive tract and short transit time, and need to be fed frequently. Food should be available free choice, allowing the ferret access to food at all times. Restricting food for an extended period of time should be avoided with ferrets that have been diagnosed with insulinoma to prevent hypoglycemia (a condition of low blood sugar).

With a high-quality diet, nutritional supplements are not needed. There are vitamin and mineral products that ferrets take readily and eagerly. Many of these are liquid and may be squirted directly on the food or licked from the nozzle end of the bottle. These products can also greatly improve coat and skin condition and assist in preventing the buildup of intestinal hair and the formation of hairballs. These products can also be used as a treat, as most ferrets will abandon any activity or hiding place for the proffered bribe.

Handling and Restraint

Ferrets can be picked up easily in one hand, just behind their shoulders. With all four feet away from a surface, they usually relax and may be carried in this manner without a struggle. For more than removing from or returning a ferret to its cage, the hind end should be supported and the ferret held close to the body. They should not be expected to ride around on shoulders or in the back of a hooded sweatshirt or handled in any way that would subject them to a fall.

Restraint for an examination can be difficult because of their flexibility and quickness. The scruff should be the first approach as this is effective and safe in all but those ferrets that are very frightened, injured, or in pain. As with any species, the least amount of restraint required is the best restraint to use. Many ferrets relax so completely with a maintained scruff that they can be laid down and positioned for radiographs without moving and the use of a general anesthesia. This technique is also used for a quick examination of the oral cavity. There is a yawn reflex associated with being scruffed, and the ferret will open its mouth wide as it yawns.

Scruffing also provides adequate restraint for giving injections and taking rectal temperatures. In large hobs, the scruff may be a little harder to grasp as the skin is not only very tight, but tough. If the ferret starts to twist and squirm, the other hand should be placed around the rear limbs, just below the pelvis (**FIGURE 4-2**). Ferrets should not be held directly over the floor during any type of restraint. Should they break free from the scruff and fall to the floor, serious injury could result.

Medical Concerns

Respiratory Disorders

Ferrets are very susceptible to *canine* distemper virus (CDV), and there is a 100 percent mortality rate if they contract the virus. This virus can be airborne, transmitted directly from infected dogs, or carried on clothing, shoes, or other contaminated

FIGURE 4-2 Correct restraint of a ferret. A ferret should be restrained over a table in case it breaks free from the scruff and falls.

items. Ferrets can be protected against this disease through a series of vaccinations. When the kit arrives in the pet store, it may have had one vaccination to guard against canine distemper. If this is the case, it is a partial vaccine and does not provide full immunity. Kits need to be fully vaccinated through a series of vaccinations given at 6 to 8 weeks and 10 to 12 weeks, and given a yearly booster vaccine thereafter.

Currently, two vaccines are approved by the U.S. Department of Agriculture (USDA) for use in ferrets: PureVax® (CDV) and IMRAB-3® (rabies), both produced by Merial. Merial produces the CDV vaccine only at certain times of the year, and this can cause a shortage during the summer months. Whenever the vaccine is not available, Nobivac DPv, also known as The Puppy Shot, is recommended. The vaccine Galaxy D has been used extensively in Europe and contains the ferret distemper strain. Even though this is the same distemper virus that affects dogs, canine-specific "combo vaccines" should not be substituted for use in ferrets. Whenever using vaccines off-label, caution needs to be taken with reaction potential and effectiveness in preventing the disease.

Some ferrets may have a vaccination reaction. The reaction can be mild, severe, or life-threatening. If **anaphylaxis** (an extreme allergic reaction) occurs post-vaccination, it is usually within 20 to 30 minutes. Owners should remain at the clinic during this period to ensure immediate and possibly life-saving treatment if anaphylaxis occurs. Depending on the severity of anaphylactic shock, antihistamines or short-acting steroids are administered. In an extreme case, the ferret may need to be intubated and given respiratory support.

Ferrets are unusual in that in addition to CDV, they can become infected with human influenza. While there are many different strains of the influenza virus, all cause respiratory infections and the signs are similar: elevated temperature, coughing, sneezing, ocular and nasal discharge, lethargy, and dehydration. Sick ferrets may also be reluctant to eat. Treatment can include prophylactic use of antibiotics (even though influenza is viral), fluid therapy, and supplemental feedings. Most ferrets recover without further complications. If left untreated, however, severe cases of influenza may progress to pneumonia. This can be problematic when human household members are also suffering from the flu and the ferret does not receive needed care. Owners with the flu or the onset of a common cold should avoid close contact with their ferrets.

In recent years, there has been an increase in the number of ferrets diagnosed with the **Mycoplasma**, which can lead to a chronic respiratory infection. Signs include a dry, nonproductive cough; occasional conjunctivitis; and in severe cases **dyspnea** (difficulty breathing). Stress is the main factor that predisposes ferrets to contracting this disease.

Where Ferrets Are Legal, Rabies Vaccinations Are also Required by Law.

The approved vaccine for use in ferrets is IMRAB-3® produced by Merial. Ferrets should be vaccinated against rabies at three months and be given an annual booster thereafter.

In the rare instance when a ferret bites someone, despite proof of current rabies vaccination, the ferret may be confiscated and euthanized to be tested for rabies. This is contrary to dog and cat laws where an animal is quarantined and observed for a period of 10 days. If vaccination protocol is followed, inoculated ferrets are no more likely to be carriers of rabies than any other warm-blooded species. Fortunately, with greater public awareness and education, the majority of laws have been amended to provide for quarantine rather than mandatory euthanasia and rabies testing.

Gastrointestinal Disorders

Many ferrets are presented with vague signs of lethargy, anorexia, and diarrhea. One of the first rule-outs (r/o) would include a foreign body in the gastrointestinal (GI) tract. Foreign body obstructions are very common in ferrets because of their curious nature and because they like to chew on certain items. Rubber products are the most commonly ingested foreign bodies. Many ferrets also show more specific signs, including a hunched body posture, reluctance to eat, gagging, and abdominal pain, which becomes more evident with gentle palpation. If a foreign body is suspected, radiographs should be taken to determine not only if there is a foreign body but also the exact location and size of the obstruction. Two views are taken: lateral (Lat) and ventral/dorsal (V/D). Surgical removal of the foreign body is required, and often the retrieved item is very recognizable to the owner.

It is not uncommon for ferrets to develop gastric and duodenal ulcers involving the stomach and the small intestine. Ulcers have a variety of causes: stress, illness, loss of a cage mate, changes in diet, excessive time in the cage, dehydration, and surgical complications. Some medications also predispose a ferret to developing gastric ulcers. Evaluation should include a history of foreign body ingestion, prescribed steroid or nonsteroidal anti-inflammatory drugs (NSAID) use, and household items the ferret may have been exposed to, including plants, carpet fresheners, and cleaning products. Clinical signs include teeth-grinding (a sign of pain in many animals), weight loss, vomiting, pale mucous membranes, and hypersalivation (associated with nausea and pain). The stool may be black and tarry (melena), indicating blood in the feces. Fecal material may also adhere to the thermometer tip when obtaining a rectal temperature. Treatment may include antibiotics, fluid therapy, and antacids. Force feeding of bland foods may also be required.

Helicobacter spp. is a curved or spiral-shaped gram-negative bacteria that affects almost 100 percent of ferrets during the weaning stage. This bacteria grows on the gastric lining and can cause gastritis and ulcers. Not all ferrets show signs unless they are under stress due to diet change or the presence of other diseases. Clinical signs include anorexia, diarrhea, melena, vomiting, weakness, abdominal pain, excessive salivation, and teeth grinding. Treatment includes fluid therapy, antibiotics, and bismuth subsalicylate to protect the intestinal lining.

Inflammatory bowel disease (IBD) is a chronic condition of the GI tract and can develop in ferrets with a history of metabolic or infectious disease or a hypersensitivity to some foods. It may also be bacterial in origin. It can cause inflammation of the entire digestive tract, but most commonly the lower bowel is affected. Signs include vomiting and weight loss, dehydration, and loss of muscle tone. It is accompanied by a characteristic green diarrhea.

> **There Is a Disease Condition of Ferrets Known as ECE,** or epizootic catarrhal enteritis. This is commonly called green slime disease. ECE is distinctly different from IBD, and the presence of green, slimy diarrhea should *not* be reason to immediately presume a diagnosis of ECE. Most ECE outbreaks are confined to breeding facilities and ferret/pet shows or in a ferret that has been recently exposed to ECE-positive ferrets. The mortality rate for ECE is high, especially in young kits. The causative agent of ECE is thought to be a coronavirus, but there may also be secondary bacterial infections contributing to clinical signs.

Diagnosis of IBD requires surgical biopsies of the intestinal tract and lymph nodes. Lymph node biopsy results will help determine the treatment if a bacterial infection is also implicated. Nonbacterial IBD is often treated with supportive care, fluid therapy, anti-inflammatory drugs, and a change in diet. Nutritional support is usually recommended while the patient convalesces.

Food should be just wet enough to fit through the tip of a 6-ml syringe without clogging up or pouring through the tip. Recovering ferrets should be fed as much as they will consume every three to four hours. Specifically formulated diets for supportive nutrition include Oxbow Critical Care Carnivore Diet® and Emeraid® Carnivore from Lafeber.

Endocrine Disorders

Adrenal gland disease has become an increasing problem with pet ferrets. The exact cause of the high incidence of adrenal gland tumors is unknown. One theory that has been suggested several times is that these tumors are directly related to premature spaying and neutering. Breeding facilities spay and neuter ferrets at around six weeks of age, which prevents normal endocrine and hormonal development. This may be a contributing factor, but more current thought is that tumor suppressor genes have become ineffective due to inbreeding for temperament.

One of the first signs of adrenal gland disease is alopecia (hair loss, which starts on the rump, spreads across the hips, and progresses down the tail). As the disease advances, so too does the alopecia. Some ferrets with adrenal gland disease become devoid of all hair. Pruritus (itching and the resultant scratching) and erythematosis (red and inflamed skin) may also occur within the areas of hair loss (**FIGURE 4-3**).

In the jill, the vulva becomes swollen and it may appear as if she is in heat. Hobs with adrenal gland disease have trouble urinating, which may indicate an enlarged prostate and cystitis. Both sexes may exhibit edema (**FIGURE 4-4**).

Adrenal gland disease is usually suggested by clinical signs. Basic blood work, chemistry panels, and radiographs do not provide enough information to confirm a diagnosis. A hormone panel from a specialty laboratory is required to provide a

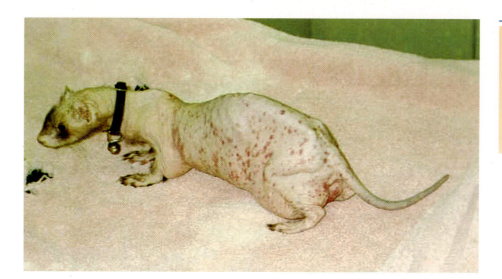

FIGURE 4-3 When adrenal gland disease progresses and is left untreated, there can be a complete loss of hair with erythematosis. *(Courtesy of Eric Klaphake, DVM)*

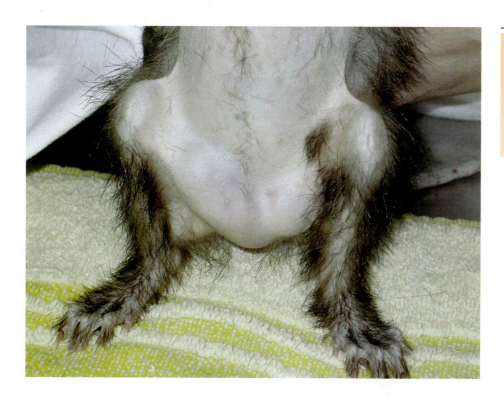

FIGURE 4-4 A typical pattern of hair loss (alopecia) and edema, which occurs in both jills and hobs with adrenal gland disease. *(Courtesy of Eric Klaphake, DVM)*

definitive diagnosis. If the ferret is otherwise young and healthy, surgical removal of the affected adrenal gland can delay the progression of the disease, but it may not be curative. The left adrenal gland is more easily removed, while the right adrenal gland is more difficult to remove because it is adjacent to the caudal vena cava. There are medical options, but the expense of the drugs used combined with the complications of timing the injections, monitoring their effectiveness, and multiple office visits needs to be considered carefully. Leuprolide acetate (Lupron Depot) is currently a treatment option and requires painful monthly intramuscular (IM) injections. The injections may control the clinical signs, but they are not curative. In July 2012, the FDA approved a new medication, Suprelorin®, as another option for the treatment of adrenal gland disease in ferrets. Produced by Virbac Labs, it is a long-acting implant that contains deslorelin.

Jills are seasonally polyestrous induced ovulators. If jills are not bred or artificially stimulated to ovulate, they will remain in constant heat, which will result in hyperestrogenism (overproduction of estrogen). Ferrets that remain in estrus for longer than a month are at risk for developing estrogen-induced bone marrow hypoplasia. Clinical signs are similar to adrenal gland disease and those of being in estrus. When the bone marrow starts to become affected, the ferret will become lethargic and weak and will have a decreased appetite, pale mucous membranes, and ecchymosis (hemorrhagic spots). The recommended treatment is to spay the jill.

Ferrets are often diagnosed with insulinomas (pancreatic tumors). The tumors can be difficult to approach surgically, and medical management is usually attempted first. Diet and drug therapy can help control the hypoglycemia, which is a consequence of insulinoma.

Mast cell tumors are relatively common in the ferret. Mast cells mediate a variety of reactions in the body and are often associated with allergies. They first appear as small, hairless areas on the surface of the skin, which may be mistaken for a slight scratch or insect bite (FIGURE 4-5). In later stages, they may be swollen, black, and crusty. In the early stages, mast cell tumors can be removed with little consequence and are usually benign (FIGURE 4-6).

Parasites

Although parasites are not common in ferrets, they can contract internal protozoan infections including coccidia and giardia. External parasites such as fleas, ticks, and ear mites may also infest ferrets, especially those housed in less-than-ideal conditions. *Otodectes cynotis* is the commonly diagnosed ear mite in ferrets.

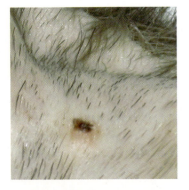

FIGURE 4-5 An early-stage mast cell tumor in the axilla of a ferret. *(Courtesy of Eric Klaphake, DVM)*

FIGURE 4-6 A young ferret with an oral tumor. *(Courtesy of Eric Klaphake, DVM)*

This ear mite is directly transmitted between animals. Ferrets may shake their heads or scratch their ears. The otic discharge appears as a thick brown, waxy exudate. This type of exudate can also appear in ferrets that are mite free. Microscopic examination is recommended to confirm the diagnosis and positive cases are treated with ivermectin injections.

Ctenocephalides spp. fleas infest dogs, cats, and ferrets. Affected ferrets may show mild-to-intense pruritus and alopecia over the back and near the neck and shoulders. Flea dirt may appear reddish-black in the fur. Treatment involves eliminating the fleas on the ferret and in the environment. Commercial topical solutions should be used with caution due to the small size of the ferret. Only the products approved for use in ferrets should be considered.

Heartworm disease is endemic in many areas. It is vector transmitted by the same mosquitoes that infect dogs and cats. Clinical signs of infection include cough, lethargy, pale mucous membranes, anorexia, and dyspnea. Treatment in ferrets can be difficult because of the risk of developing a fatal clot in the lungs. A heartworm positive ferret has a guarded prognosis. All ferret owners should discuss heartworm preventatives with a veterinarian.

Clinical Procedures

Vaccinations should be administered subcutaneously. When held by the scruff, most ferrets will not object too much. Medications should be administered orally, whenever possible as IM injections are extremely painful and will elicit a loud scream from the ferret. Also, in giving IM injections, there is a greater chance of losing the scruff and causing injury to the ferret as it attempts to escape the source of the pain.

Although blood may be drawn from various sites, the lateral saphenous and cephalic veins provide ease of access for small amounts. Larger volumes may be collected from the right jugular vein and the anterior vena cava (**FIGURE 4-7**).

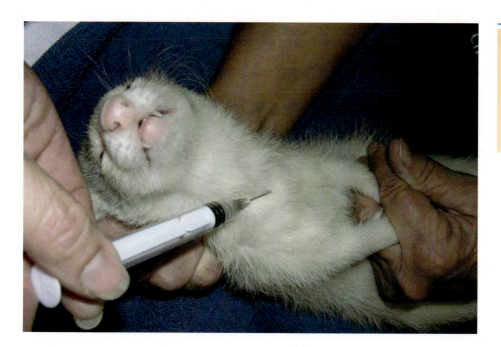

FIGURE 4-7 Restraint of a ferret and approach for blood collection from the vena cava. The vena cava is located more caudally in ferrets than in most mammals.

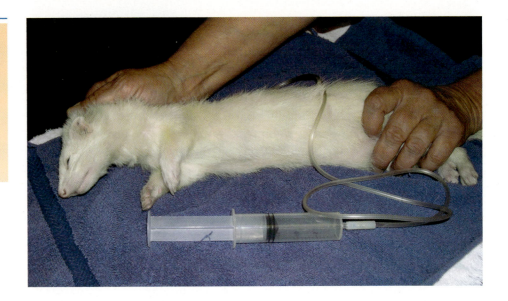

FIGURE 4-8 Restraint and setup for the administration of SQ fluids. The extension set allows the ferret some movement without dislodging the needle or having it accidentally penetrate deeper, causing more pain.

If attempting an anterior vena cava collection, general anesthesia is recommended to prevent movement and potential laceration of the vena cava.

Fluid therapy in the ferret is often required in a debilitated patient as part of supportive therapy. The daily fluid maintenance requirement is 75 to 100 ml/kg/day. Fluids can be easily given subcutaneously to a mild-to-moderate dehydrated patient. When administering any amount of subcutaneous fluids, it is simpler to attach a fluid extension set to the syringe rather than attempting direct delivery (**FIGURE 4-8**). This allows the ferret some movement without pulling the needle out of the skin and eliminates the possibility of pain caused by the needle penetrating deeper than SQ when the ferret moves. Severely dehydrated patients need fluids administered via an intravenous or intraosseous catheter. Lactated Ringer's solution and Normosol are recommended for fluid replacement in the ferret. Additional support can be provided by adding vitamin B, potassium, or dextrose to the fluids.

The use of gas inhalant anesthesia in a ferret is preferred over injectable agents and is similar to anesthesia in dogs and cats. Ferrets should be fasted before a procedure requiring anesthesia. Due to their rapid gastrointestinal time, ferrets should not be fasted for longer than four hours. Short fasting times will lower the risk of hypoglycemia secondary to insulinoma. Ferrets are initially masked down or placed in an induction chamber prior to intubation. Intubating the ferret is easier than with many other small mammals as the epiglottis is easily visualized (**FIGURE 4-9**). Use of a 2 to 3.5 mm endotracheal tube is recommended due to the small tracheal size of ferrets. Oxygen flow rate of 0.8 to 1 L/min will normally maintain a ferret during anesthesia. Maintaining a ferret under anesthesia can be accomplished at 3% isoflurane or 5% sevoflurane.

Ferrets lose body heat very quickly under anesthesia and develop hypothermia. Several commercial devices can be used to maintain the body heat of a ferret while under anesthesia.

Close monitoring of the ferret is essential for a successful recovery. Depth of anesthesia can be assessed through a toe pinch or palpebral reflex. An ultrasonic doppler or a small esophageal stethoscope is used to monitor the heart rate and rhythm.

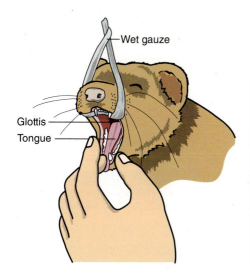

Wet gauze

Glottis

Tongue

FIGURE 4-9 When a suitable plane of anesthesia is achieved, the ferret may be easily intubated. A length of gauze placed just behind the canines aids in opening the mouth and helps avoid a *clamp bite* should the ferret respond to the stimulus of tube placement. If the gauze has been wetted, the threads are less likely to snag on the teeth.

Radiographic techniques in ferrets are similar to the techniques used in dogs and cats. Symmetry can be a problem in positioning the small ferret. Contoured pieces of foam can be used to help obtain correct positioning. Radiolucent tape and brown gauze can be used to help with limb and head positioning. Sedation is often used to reduce stress, and this can be achieved with a quick mask-down of an inhalant anesthesia

Ferrets have varying soft tissue densities in the abdomen and a small pelvis, which makes it impossible to obtain a single radiograph of the entire pelvic area, limb, and distal limb without creating some distortion. Separate views should be taken of the pelvis and of the distal limb. The head of the ferret is slightly conical and may prevent a good lateral position. The patient's head may need to be elevated slightly by lifting up on the nose.

Due to rapid respiratory rates in ferrets, shorter exposure times (less than 1/60th of a second) are required. When performing a gastrointestinal contrast study, liquid barium sulfate (10 to 15 ml) is given orally or by an orogastric tube. Strawberry-flavored or palatable liquid meat-based food may help with the administration of the barium. This technique is useful to detect a hairball.

Ultrasonography can be used to detect heart disease, neoplasia, adrenal tumors, and foreign bodies in the gastrointestinal tract. Endoscopy is commonly used to examine the lining of the upper respiratory tract and the gastrointestinal lining. This diagnostic technique is useful to detect ulcers in the ferret.

Summary

The domestic ferret has become a popular companion animal. The regulations for keeping a ferret may vary from area to area. Ferrets can be simply housed and are easily cared for with an understanding of their behavior and dietary requirements. Ferrets should be vaccinated against canine distemper and rabies. Ferrets, while similar to dogs and cats in the treatment of disease and injury, have medical concerns that are unique to the species. Procedures for restraint and handling are discussed as well as treatment protocols for injections, oral medications, and fluid therapy. Techniques for blood sampling, radiology, and anesthesia protocol are presented.

Many of the diseases seen in ferrets are common to other species, but often their management is quite different. Ferrets should have regular, yearly veterinary wellness examinations that may include a complete blood count (CBC), blood chemistries, and radiographs. Early detection of medical issues can often prevent serious consequences.

fastFACTS

FERRETS

WEIGHT
- **Hobs:** 1–2 kg (2–4.5 lbs)
- **Jills:** 0.6–0.9 kg (1–2.5 lbs)

LIFE SPAN
- 5–8 years average

REPRODUCTION
- **Sexual maturity:** 6–12 months
- **Gestation:** 41–43 days
- **Litter size:** 1–18 (8 average)
- **Weaning age:** 6–8 weeks

VITAL STATISTICS
- **Temperature:** 37.7–39.7°C (100–103.5°F)
- **Heart rate:** 180–250/min
- **Respiratory rate:** 33–36/min

DENTAL
- Total of 20 teeth
- Dental formula 2 (I 3/3, C 1/1, P 3/3, M ½)

VACCINATIONS
- **Canine distemper:** Vaccine schedule: 6–8 weeks; 9–12 weeks, 14–16 weeks; yearly thereafter.
- **Rabies:** Vaccine schedule: 4–6 months; yearly thereafter.

ZOONOTIC POTENTIAL
Bacterial
- Salmonella
- Campylobacter

Viral
- Influenza
- Rabies (rare, only if exposed)

Fungal
- Dermatophytes (ringworm)

Parasitic
- Giardia
- Cryptosporidia
- Fleas
- Scabies mite
- Toxocariasis

Review Questions

1. What is the result when ferrets are bathed too frequently?

2. Normal play for ferrets can be rough. Acceptable behavior with humans needs to be taught at an early age. Explain the correct way to discipline a ferret.

3. What are the recommended vaccinations for ferrets?

4. Describe the clinical signs exhibited by a ferret with adrenal gland disease.

5. Intestinal blockages are frequently seen in ferrets. What items are commonly found in the GI tract?

6. If food is restricted in a ferret with existing pancreatic disease, what is the medical consequence?

7. Why would a ferret still have an odor even though it has been de-scented?

8. What is meant by *ferret-proofing* the home?

9. What is a disadvantage of feeding some commercially produced ferret foods?

10. Discuss the reasons ferrets are illegal in some areas.

Case Study I

History: The owner of a four-year-old spayed jill states that the ferret has been losing hair over the past three to four weeks, but other than that, the ferret seems to be normal. The owner also mentioned that the pet store told her that the hair loss was "normal because they shed twice a year."

Physical Examination: There is alopecia of the tail and thinning of the hair over the pelvic area. The jill has a swollen vulva and an enlarged abdomen. On palpation, the spleen is larger than normal.

Laboratory Findings: The CBC is normal but the estrogen level is slightly elevated.

a. What is the likely diagnosis by the veterinarian?

b. Describe how the signs and history help confirm the diagnosis.

c. Explain the common treatment protocols for this problem.

d. What are the problems with asking pet stores for medical advice?

Case Study II

History: The owner is concerned about her five-year-old neutered male ferret. The ferret is pawing at its mouth constantly and the owner believes there is something stuck in the roof of the mouth. At times, the pawing becomes frantic. With further open-ended questions (questions that cannot be answered with yes or no), the owner also recalls episodes where the ferret seems weak and walks with a stagger "like he is drunk."

Physical Examination: There are muscle tremors and evidence of muscle wasting, but no foreign body is found in the mouth.

Laboratory Findings: Blood glucose level is 65 mg/dl.

a. What is the veterinarian's likely diagnosis?

b. Describe how the signs, history, and lab results support the diagnosis.

c. What is the likely outcome for this ferret?

For Further Reference

Antinoff, N., Williams, B.H., & Weiss, C.A. (2011). Neoplasia. In Quesenberry, K.E. and Carpenter, J.W. (Eds.), *Ferrets, Rabbits and Rodents, 3rd Edition* (pp. 103–121). St. Louis, MO: Imprint of Elsevier Science.

Barron, H.W., & Rosenthal, K.L. (2011). Respiratory Diseases. In Quesenberry, K.E. and Carpenter, J.W. (Eds.), *Ferrets, Rabbits and Rodents, 3rd Edition* (pp. 78–87). St. Louis, MO: Imprint of Elsevier Science.

Burgess, M., & Garner, M. (2002). Clinical Aspects of Inflammatory Bowel Disease in Ferrets. *Exotic DVM, 4*(2), 29–34.

Hess, L. (2005). Ferret Lymphoma: The Old and the New. *Seminars in Avian and Exotic Pet Medicine, 14*(3), 199–204.

Hess, L. (2006, January 7–11). *Clinical Techniques in Ferrets.* Conference Notes, The North America Veterinary Conference, Orlando, FL.

Hoefer, H.L. (2006, January 7–11). *Cardiac Disease in Ferrets.* Conference Notes, The North America Veterinary Conference, Orlando, FL.

Hoefer, H.L., & Bell, J.A. (2011). Gastrointestinal Diseases. In Quesenberry, K. E. and Carpenter, J. W. (Eds.), *Ferrets, Rabbits and Rodents: Clinical Medicine and Surgery, 3rd Edition* (pp. 24–45). St. Louis, MO: Imprint of Elsevier Science.

http://www.anapsid.org (accessed February 23, 2014).

http://www.fda.gov (accessed March 28, 2014).

Johnson-Delaney, C.A., & Orosez, S.E. (2011). Ferret Respiratory System: Clinical Anatomy, Physiology, and Disease. *Veterinary Clinics of North America Exotic Animal Practice, 14,* 357–367.

Mayer, J.D. (2006, January 7–11). *Update on Adrenal Gland Disease in Ferrets.* Conference Notes, The North America Veterinary Conference, Orlando, FL.

Mayer, J.D. (2006, January 7–11). *Update on Ferret Lymphoma.* Conference Notes, The North America Veterinary Conference, Orlando, FL.

Meredith, Anna and Redrobe, Sharon, eds. John Wiley and Sons, Hoboken, NJ.

Meridith, A., & Johnson Delany, C. (Eds.). (2010). *BSAVA Manual of Exotic Pets, 5th Edition.*

Mitchell, M.A., & Tully, T.N. (2009). Ferrets. In *Manual of Exotic Pet Practice* (pp. 345–373). St. Louis, MO: Imprint of Elsevier Science.

Morrisey, J.K., & Kraus, M.S. (2011). Cardiovascular and Other Diseases. In Quesenberry, K.E. and Carpenter, J.W. (Eds.), *Ferrets, Rabbits and Rodents, 3rd Edition* (pp. 62–77). St. Louis, MO: Imprint of Elsevier Science.

Murray, J., DVM. (2010). Summary of Topics from the Exotic Animal Conference, San Diego, CA. http://www.smallanimalchannel.com (accessed July 25, 2013).

Pollock, C.G. (2011). Disorders of the Urinary and Reproductive Systems. In Quesenbury K.E and Carpenter, J.W. (Eds.), *Ferrets, Rabbits and Rodents: Clinical Medicine and Surgery, 3rd Edition* (pp. 46–41). St. Louis, MO: Imprint of Elsevier Science.

Powers, L.V., & Brown, S.A. (2011). Basic Anatomy, Physiology and Husbandry. In Quesenbury, K.E. and Carpenter, J. W (Eds.), *Ferrets, Rabbits and Rodents: Clinical Medicine and Surgery, 3rd Edition* (pp. 1–12). St. Louis, MO: Imprint of Elsevier Science.

Quesenberry, K.E., & Orcutt, C. (2011). Basic Approach to Veterinary Care. In Quesenberry, K.E. and Carpenter, J.W. (Eds.), *Ferrets, Rabbits, and Rodents: Clinical Medicine and Surgery, 3rd Edition* (Chapter 2, pp. 13–26). St. Louis, MO: Imprint of Elsevier Science.

Resau, J.H., Garner, M.M., & Bolin, S.R. (2012). Mycoplasmosis in Ferrets. *Emerging Infectious Diseases, 18*(11), 1763–1770.

Rosenthal, K.L., & Wyre, N.R. (2011). Endocrine Diseases. In Quesenberry, K.E. and Carpenter, J.W. (Eds.), *Ferrets, Rabbits, and Rodents: Clinical Medicine and Surgery, 3rd Edition* (Chapter 7, pp. 86–102). St. Louis, MO: Imprint of Elsevier Science.

Schoemaker, N.J. (2002). Ferrets. In Meredith, Anna and Redrobe, Sharon, (Eds.) *BSAVA Manual of Exotic Pets, 4th Edition* (pp. 93–101). Hoboken, NJ: John Wiley and Sons.

RABBITS

OBJECTIVES

After completing the chapter, the student should be able to

- Correctly describe the housing and husbandry for a rabbit.
- Provide client education to companion rabbit owners.
- Provide basic nursing care for a rabbit.
- Assist in the anesthesia of a rabbit.
- Demonstrate appropriate restraint techniques for rabbits.
- Correctly identify the sex of a rabbit.
- Obtain basic knowledge of common medical problems in rabbits.

KEY TERMS

lagomorph
kit
warren
bolt hole
ferreting
feral
hutch
doe
buck
inguinal
crepuscular
cyanosis
pododermatitis
gut motility
urolith
timothy
alfalfa
coprophagic
cecotrophs
dewclaw
obligate nasal breather
dewlap
induced ovulator
parturition
kindle
dystocia
altricial
rhinitis
snuffles
torticollis
slobbers
enterotoxemia
gut stasis
hepatic lipidosis
trichobezoar
lactation
ataxia

Introduction

The domestication of rabbits can be traced back to the Phoenicians over 1000 years ago. The name of the Iberian Peninsula means the *land of the rabbit*. Rabbits were kept by the Romans in walled areas called *leporia*. (Iberia, once Hispania, is now the country of Spain.) Although previously included with rodents, rabbits were reclassified nearly a century ago and belong to the order lagomorph, which also includes cottontails (wild rabbits) and hares. Domestic rabbits are the same species as the European wild rabbit, *Oryctolagus cuniculus*.

Rabbits are grouped by breed and type, depending on the purpose for which they are bred: meat, fur, wool, or, in the case of the smaller dwarfs, as companion animals. The American Rabbit Breeders Association (ARBA) recognizes 45 distinct breeds. ARBA determines the standard for each breed. Standards include body type and acceptable coat colors, patterns, texture, and eye color. There are also many individual specialty clubs for rabbits. 4-H and FFA both offer opportunities for youth. The House Rabbit Society is an organization that promotes education and owner responsibility for rabbits housed indoors.

Many rabbits of mixed and uncertain breeding are just as appealing as rabbits with a known pedigree. It is useful to know which breed to choose as there are generalizations regarding temperament and breed disposition. There are dwarf breeds and giant breeds; *lops* with low, drooping ears and rabbits with ears erect; woolies; angoras; and the new Lionhead (**FIGURE 5-1**), each with its own personality type and physical characteristics.

mastitis
conjunctivitis
cannulation
metastasis
uroliths
cuterebra
fly strike
maggots
exudate
turbid
hematuria
apnea

Rabbits vary in size and weight from two to eight pounds, a consideration when buying an immature rabbit **kit**. Becoming familiar with different breeds and their traits is beneficial not only in choosing a pet rabbit but also in providing information to veterinary and rescue personnel.

Behavior

Wild rabbits live in underground **warrens**. A single warren may be home to as many as 40 or 50 rabbits. The size of the warren increases with the population, and an *old* warren may extend to nearly one-quarter of an acre. In appearance, a warren is a mound of earth with many entrances and exits called **bolt holes**.

FIGURE 5-1 A relatively new breed of rabbit, the Lionhead.

> ## On the English Downs, Many Separate Warrens Are Found in Close Proximity.
>
> **Ferreting** is a popular sport that involves the use of a ferret. With the exception of one or two, the bolt holes are covered with purse nets. A jill is sent down into the open holes to *ferret out* the rabbits. In their panic to escape, the rabbits run from their bolt holes straight into the purse nets where they are captured and dispatched. Hunting with a ferret is illegal in the United States.

Domestic rabbits allowed access to a large enclosure also dig warrens and could potentially escape, quickly establishing a **feral** population. Feral animals are those that were previously domesticated but have been able to adapt to life in the wild. Domestic rabbits exhibit digging behavior in cage bedding and on the floor of the **hutch** (an outdoor pen) or, in the case of a house rabbit, by digging at the carpet and flooring.

Rabbits are a prey species and can be easily frightened or injured in an attempt to escape capture or restraint. The eyes of a rabbit are positioned laterally, giving a nearly 360-degree field of vision. They are constantly on the lookout for predators. At the first hint of danger, rabbits thump loudly with a hind foot to alert other rabbits, even those underground. Caged rabbits also *thump* in alarm when startled by the presence of other species or unknown scents.

In general, rabbits are quiet animals and vocalize only when provoked, when in pain, or when extremely frightened. Growling is unmistakable, aggressive, and not just an empty threat. Rabbits will charge, growling and clawing with the front feet. The scream of a rabbit in pain is loud and piercing, a sound unforgettable to those who have heard it. A frightened, restrained rabbit may scream and kick wildly with powerful hind legs, raking with the rear claws. Many rabbits have been severely injured during attempts to escape.

Although they can be affectionate and playful with early socialization, rabbits are also very territorial and will not hesitate to confront an intruder, whether it is a child or another household pet. Territorial aggression is common in rabbits of both sexes, especially those that have not been castrated or spayed. Typically, **does** (female rabbits) that have not been spayed are more aggressive than unneutered **bucks** (male rabbits).

Rabbits use scent glands to mark territory. They have submandibular, or *chin glands*, and both males and females have glands in the folds of skin on either side of the genitals. These **inguinal** glands normally produce an unpleasant-smelling, thick, and somewhat waxy substance.

Rabbits are **crepuscular**, meaning they are more active at dawn and dusk. They take many short naps during the day, and for this reason, owners of house rabbits say that they feel more comfortable leaving their pet alone than they would if they had a dog. Rabbits are highly intelligent and enjoy interacting and playing games with people, such as chasing balls and tossing objects into the air in play. They should be given suitable play items, including blocks of wood of various shapes and sizes, cardboard tubes, and boxes.

House rabbits should never be left unsupervised. Many household items can pose dangers, including electrical cords, painted wood, and furniture. Injuries and fractures can occur easily by being accidentally stepped on or rolled over by someone seated in a desk chair. Rocking chairs also present a danger (**FIGURE 5-2**).

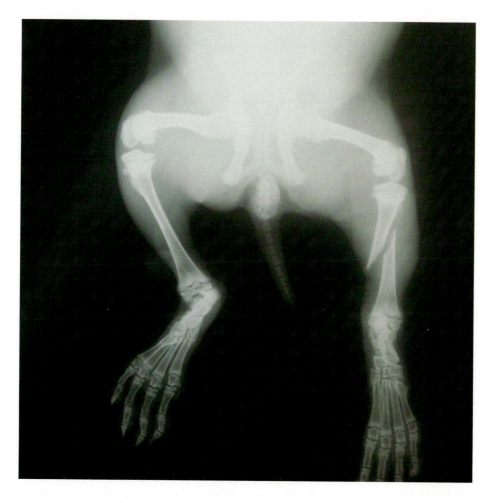

FIGURE 5-2 This house rabbit suffered a fracture to the left tibia/fibula as the owner sat on the chair and the rabbit was accidentally caught under the rocker.

Rabbits allowed outdoors also need supervision. It is not just dogs that may present a danger, but the larger owls and hawks have also been known to take rabbits from enclosures without a secure covering, such as the portable pet pens.

As strange as it may seem, many rabbits enjoy water and they are quite capable of swimming. The toes of the hind feet are webbed. Some rabbits willingly hop into a shallow tub of water, apparently to cool off. Providing a rabbit with a pool is not recommended, but there are countless reports of rabbits intentionally entering water.

Former President Jimmy Carter Had to Fend Off a Rabbit *Attack* When a Rabbit Swam Out to Him While He Was Fishing in a Boat.
The intent of the rabbit was unknown (April 20, 1979).

House rabbits will use a litter box. Owners can take advantage of the animal's natural behavior, similar to that of cats, and provide what is needed rather than actually *training* the animal. Rabbits normally urinate and defecate in one area, whether in a cage, hutch, or tray. Plain newspaper should be used instead of cat litter or any other scented substrate. Using a scented product may create tray avoidance or, depending on the type of substrate, may contribute to respiratory problems. Soiled bedding or fecal material initially placed in the tray will encourage the rabbit to use it.

The Most Heavily Used Area of a Metal Tray Can Be Sprayed with a Vegetable Oil.
It will help to prevent metal corrosion from urine and make cleaning the tray considerably easier. Alternatively, the entire tray may be enclosed in a large trash bag. When cleaning the tray, the bag can simply be pulled off (inside out) so that all the contents are contained within the bag. This should be done only if there is no way the rabbit would be able to reach any part of the bag and chew on it, for example, a dropped collection tray beneath the wire floor.

Housing

Rabbits housed outdoors need shelter from heat and cold. Temperatures below 40°F make them susceptible to hypothermia. Winter temperatures lower than that will cause frostbite, especially to the tips of the ears. Rabbits should not be left exposed to heat greater than 80°F. At temperatures higher than this, rabbits quickly succumb to heat stroke. Rabbits have a limited ability to cool themselves as they are unable to sweat and do not pant. A small amount of heat is dissipated through the ears and nares, but it is ineffective in conditions of high heat and humidity. A rabbit with heat stroke will be found collapsed with increased respiration and cyanosis, a blue or gray coloring of the gums due to low oxygen. Rabbits with these clinical signs and a core body temperature approaching 105°F are unlikely to survive and are usually humanely euthanized.

When temperatures are elevated, hutches need to be situated in an area that is shaded and well ventilated. Frozen containers of water placed in the hutch so that the rabbit is able to stretch out or sit next to it may help. Some owners use a timed misting system, which gently sprays the hutch with water for short *cool down* periods. Ideally, rabbits should never be exposed to temperatures that place them at risk.

Rabbits are easily environmentally stressed, not just by temperature but also by visual and olfactory stimuli. When rabbits are hospitalized for treatment or placed in a boarding facility, there are many considerations. Rabbits need to be housed in a quiet area and away from the odor and sight of predators including dogs, cats, snakes, ferrets, and large birds. They should not be placed in cages facing predatory species or where there may be visual contact, including high traffic treatment areas.

When selecting a cage or a hutch, the size and weight of the rabbit as an adult needs to be considered. Floor space should be at least three times the length of the rabbit. Rabbits often lay on their sternum, with legs extended laterally. Another way of determining the minimum size of the cage is the *three hop rule*, meaning it would take the rabbit three full hops to go from one end of the cage to the other.

Many hutches have wire-bottom floors so urine and fecal material fall into a collection tray underneath. This makes the hutch easier to clean and keeps the rabbit away from soiled material. Regardless, there needs to be an area with solid flooring so the rabbit is not constantly on the wire. With no relief from the pressure of the rabbit's weight on wire, sores develop on the bottoms of the feet, a condition called **pododermatitis**, or *sore hocks*. Instead of foot pads, rabbits have a thick covering of fur on their feet where the pads would be located. When the fur is worn away, there is no protection and the skin breaks down. Fur loss exposes the thin skin of the rabbit to urine, feces, and contaminated wire and eventually the feet become infected with bacteria. Rabbits suffering from pododermatitis exhibit abnormal weight bearing and lameness, trying to compensate and avoid further pain.

All cages or hutches should be provided with a *hide box*, or bolt hole. For the doe, it also serves as a nesting site (**FIGURE 5-3**).

FIGURE 5-3 A Lionhead doe with her kits in the nest box.

Rabbits use sipper tube water bottles, which should be washed and disinfected daily. Most rabbits consume 50–150 ml/kg of water daily, so the water supply bottle must be of adequate size for this volume of water consumption. Metal food hoppers attached to the side of the cage provide for a measured supply of pellets. There are also small hay racks to keep the hay clean and away from bedding material.

Diet

A pet rabbit's diet consists of pellets and grass hay with supplemental fresh leafy greens and vegetables. Rabbit pellets contain analyzed amounts of crude protein and fiber. The amounts, or percentages, are based on the specific dietary needs of the rabbit. Pellets that are high in fiber help reduce hairball formation and stimulate **gut motility** (the intestinal movement of food) but are not recommended for extended periods of time. This type of diet may disrupt the balance of normal gut bacteria.

A high-protein/low-fiber diet increases the potential for diarrhea and **urolith** (bladder stone) formation. Pellets made from **timothy** hay (a grass) rather than **alfalfa** (a legume) have a better fiber/protein ratio and are more suitable as a maintenance diet. One should always read the food label to determine which diet is compatible with the needs of the rabbit. Always follow veterinary advice, as a recommended diet can be critical in managing some health problems.

The color and consistency of rabbit urine is affected by the mineral and protein content of the diet. A diet high in protein causes an orange or red color in the urine. Owners often mistake this color change for blood in the urine. Rabbit urine is normally clear to creamy in appearance, but excessive calcium can create *bladder sludge* (hypercalciuria) and also give the urine a creamy appearance.

Many rabbits are fed pellets *free choice*, which greatly increases the risk of obesity. A daily, measured ration of one half cup of pellets per five pounds body weight is sufficient when given with one cup of fresh vegetables that are high in fiber and an unlimited amount of grass hay. Vegetable choices may vary, but there should never be an abrupt change in the diet. New foods should be introduced gradually. Whole carrots; carrot, turnip, and beet tops; parsley; and dandelions are readily consumed.

Many owners routinely give their rabbits a small amount of pineapple juice to aid in the removal of hairballs. Rabbits are unable to vomit and so cannot remove the hairball from the stomach as a cat would. Pineapples (and papayas) contain the enzyme papain that has been shown to be beneficial in preventing or reducing the size of hairballs so that they can be passed with the feces.

If a balanced, nutritionally complete diet is provided, there is no need for vitamin and mineral supplements. Many breeders recommend adding salt or mineral wheels. Not all rabbits use them but many will lick them, gradually reducing them in size. Small pieces of the *lick* should be removed from the cage and the wire fastening checked regularly to prevent injury.

Rabbits are **coprophagic**; that is, they normally consume their own feces. Rabbits produce two types of fecal pellets: hard pellets are passed during the day and consist of undigested waste; *night feces*, **cecotrophs**, are soft and encased in a mucous membrane. They are consumed by the rabbit, directly from the anus. Consumption of cecotrophs is essential to the rabbit's health as they contain quantities of vitamins B and K, beneficial bacteria, and protein.

Handling and Restraint

The most important thing to remember when handling or restraining a rabbit is how easily the rabbit can be injured.

The hind limbs have a powerful kick. Coupled with a long, but relatively inflexible, vertebral column, the force of a rabbit's kick can be enough to fracture its back. The hind limbs should always be supported and restrained securely. When removing a rabbit from its cage, one hand should be placed underneath, grasping the back legs. The other hand is used to support and hold the front legs while the rabbit is gently lifted out of the cage. Rabbits should never be lifted by the ears or picked up by the scruff alone. When the rabbit is out of the cage, it should be held close to the body with the head tucked under the arm of the restrainer (**FIGURE 5-4**).

FIGURE 5-4 When carrying a rabbit, the rabbit should be held close to the restrainer's body with the rabbit's head tucked under one arm of the restrainer and the rear legs held securely to prevent it from kicking out.

It Is Possible to Induce a Trance-Like State in a Rabbit by Covering Its Eyes and Slowly Turning It Onto Its Back.

It is important to restrain the hind legs and maintain a scruff while keeping the rabbit's eyes covered. With the rabbit on its back, keeping the eyes covered, gently stroke the abdomen. The rabbit will relax, making an examination of the ventral area very easy. This technique should not be used for any procedure likely to cause pain.

When the rabbit is returned to the cage, it should be placed in backward with the hind limbs in first to reduce the danger of the rabbit kicking out should it decide to bolt to the back of the cage. Most spinal fractures occur between the lumbar vertebrae, L6/L7. The result is paralysis and euthanasia.

Rabbits have five clawed toes on each foot. On each foreleg, is a **dewclaw**, a short claw on the medial side of the foreleg. Because the dewclaws do not make contact with any surface, they can become ingrown if not trimmed. Nails need to be trimmed regularly, especially for rabbits housed indoors. The simplest method is to place the rabbit on a solid surface and secure it in a towel (the *bunny burrito*). Each foot can be gently pulled out of the towel and the nails clipped with any of the standard variety of nail trimmers.

Tapping or touching the nose as a distraction can trigger aggressiveness or avoidance in rabbits. They are **obligate nasal breathers**, which means they can breathe only through their nasal cavity. Tapping or touching their nose can block their airway and cause distress in the rabbit.

Sexing rabbits can be a little more difficult, especially with young kits. The buck's testicles descend around 12 weeks of age (**FIGURE 5-5**).

FIGURE 5-5 The external genitalia of a buck with the testes descended.

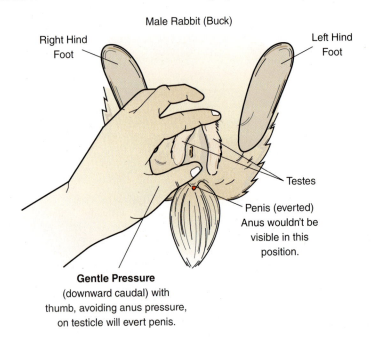

Male Rabbit (Buck)

Right Hind Foot

Left Hind Foot

Testes

Penis (everted)
Anus wouldn't be visible in this position.

Gentle Pressure
(downward caudal) with thumb, avoiding anus pressure, on testicle will evert penis.

This is complicated further because the inguinal canal remains open, allowing the testicles to retract into the abdomen. Bucks have a rounded urethral opening and a sheath that covers a relatively small penis. The vulva of a doe is elongated and the opening is vertical (**FIGURE 5-6**). It helps to wet the hair around the genitals to more easily visualize the area when determining sex. Both sexes have nipples; those of the male are more rudimentary than those of the female, but a young unbred doe may initially appear no different. Accurate sex determination is important to avoid unwanted litters.

Does are sexually mature around four to eight months, approximately one to two months earlier than bucks. The age of sexual maturity is influenced by breed. As a general rule, small breeds mature earlier than large breeds. With maturity, the doe in many breeds may have a pronounced **dewlap** (a roll of loose skin under the chin and neck). The dewlap is rarely seen in bucks (**FIGURE 5-7**).

Does are **induced ovulators**, which means that mating must occur prior to the ova being released from the ovaries for fertilization. They are capable of breeding at any time once they reach sexual maturity. Bucks have tiny penile spines that

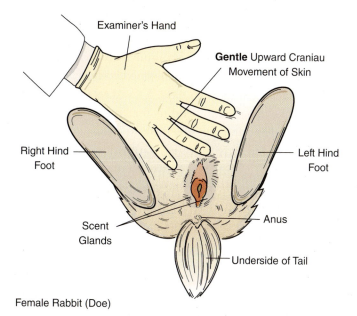

Examiner's Hand

Gentle Upward Craniau
Movement of Skin

Right Hind
Foot

Left Hind
Foot

Anus

Scent
Glands

Underside of Tail

Female Rabbit (Doe)

FIGURE 5-6 The external
genitalia of a doe. It helps
to wet the hair around
the genital area when
determining the sex of
young rabbits.

FIGURE 5-7 A doe and
buck New Zealand White.
The doe, on the right, has a
pronounced dewlap.

irritate cervical tissue, inducing the ovaries to release the eggs. Fertilization occurs in the oviduct. Does do not have a uterine body, but two elongated uterine horns for the development of multiple kits. Ovulation occurs within 10 hours after breeding. Copulation is very short. One sign that a successful mating has occurred is that the buck squeals and dismounts with a little backward flip. Due to their territorial nature, does are always taken to the buck and their interaction carefully monitored. The doe is returned to her own cage after mating. The presence of the male during gestation may be so stressful for the doe that the embryos can be reabsorbed. The buck should never be kept with the doe and her kits as bucks are known to kill the young. Cannibalism of the kits by the doe can also occur. It is usually associated with stress, poor nutrition, or environmental concerns. Once a doe starts to cannibalize, she will likely continue with each litter-consuming normal neonates as well as the stillborn or malformed young.

Gestation for the rabbit is 29 to 35 days. A few days prior to **parturition** (giving birth), the doe makes a nest with bedding material and hay and begins to pull fur from her dewlap and abdomen to line the nest. Most does **kindle** (the time of delivering the young) in the early morning. Parturition usually lasts, on average, around 30 minutes. **Dystocia**, difficulty delivering the young, is rare in rabbits. An average-sized litter is five to eight kits. The young are born **altricial**; that is, they are hairless, blind, and deaf (**FIGURE 5-8**). Around 10 days old, they are lightly furred and their eyes and ears begin to open.

FIGURE 5-8 Rabbit kits the doe has placed in a basket. Kits are hairless when they are born. Their eyes and ears are unopened.

Many rabbit owners are concerned when they do not see the doe nursing her kits or in the nest with them. They become alarmed and assume that the doe has abandoned her litter and remove the kits from the nest in an attempt to bottle feed them. Does nurse their young only for an average of three to five minutes once daily. The kits nurse on their backs and consume approximately 20 percent of their body weight per feeding. Rabbit milk is very high in fat and difficult to replicate. Attempting to feed kits from a bottle, on their backs, often results in aspiration when milk is inhaled into the lungs. Orphaned kits are very difficult to rear to weaning age and should not be removed from the nest in an attempt to bottle feed.

Medical Concerns

Rabbits have become very popular household pets, yet they are still considered exotic pets because they are different in anatomy, physiology, and behavior from more traditional companion animals. They have a variety of medical and husbandry concerns unique to them.

A rabbit in pain will demonstrate some or all of the following: an exaggerated hunched posture, face toward the back of the cage, lethargy, apprehension, dragging hind legs, and sometimes aggression.

Respiratory Disorders

Pasteurellosis is one of the most common upper respiratory infections found in rabbits. It is caused by *P. multicodia*, and this same bacteria can invade any organ of the body. The bacteria is spread through the air, by direct contact, and *fomite* transmission. Neonates can be exposed to *P. multicodia* at the time of birth or by nursing. Asymptomatic carriers can harbor the bacteria in their reproductive organs, ears, nares, lungs, and conjunctiva.

Signs of *pasteurellosis* are variable, and include nasal and ocular discharge, genital swelling, dermal ulcerations, weight loss, and sudden death. What is common to most affected rabbits is the environment. Rabbits that are older, stressed, exposed to extreme temperature changes, and housed in poor conditions are predisposed to *pasteurellosis*.

Rhinitis, or *snuffles*, is also frequently caused by *P. multicodia*, although other bacteria have been found in conjunction with this condition. Snuffles is characterized by a severe nasal discharge that is thick and yellow or white. It adheres to the nares, and as the rabbit attempts to groom, it will stick to the fore paws and may be spread to other body parts. Rabbits with snuffles sneeze frequently, spreading the infection, and will ultimately show signs of dyspnea and chest congestion. Snuffles may be chronic or the condition may progress to internal abscesses, or torticollis, a condition also known as *wry neck* that affects the nerves of the neck or inner ear. The rabbit becomes unable to eat or drink (**FIGURE 5-9**).

FIGURE 5-9 A rabbit with torticollis, a condition commonly called *wry neck*. (*Courtesy of Eric Klaphake, DVM*)

Gastrointestinal Disorders

Gastrointestinal disorders can begin with dental problems, malocclusion, overgrowth of incisors, and tooth abscesses. The most common dental problem seen in rabbits is overgrowth of the incisors, but molar malocclusion can also be a problem requiring veterinary intervention and treatment. Rabbit teeth are open rooted and grow continuously. Smaller breeds may be more genetically predisposed to malocclusion. The lower incisors tend to grow out and up while more often the upper incisors curl back into the oral cavity. Rabbits with this chronic condition may need to have routine teeth trims every six to eight weeks. In extreme cases, the veterinarian and owner may discuss the surgical removal of upper and lower incisors (**FIGURE 5-10**).

Overgrown teeth may be clipped back to the appropriate length with small wire nippers, but rotary powered dental cutting wheels are preferred. Cutting

FIGURE 5-10 Extremely overgrown lower incisors, a result of malocclusion. The left incisor is fractured and a large hairball has formed around both incisors. (*Courtesy of Eric Klaphake, DVM*)

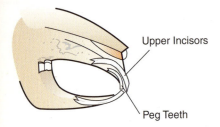

Upper Incisors

Peg Teeth

FIGURE 5-11 Rabbits have two sets of upper incisors. The smaller set (peg teeth) is located directly behind the primary incisors.

wheels operate smoothly and quickly and help prevent fractures of the teeth during the procedure. When using the clippers, great care must be taken to avoid fracture of the tooth. Each tooth should be clipped individually (do not attempt to clip across both incisors at once). The nipper should be sharp and clean and the clip made in one decisive move. Both the restrainer and the person performing the procedure should wear protective eye glasses as the nipped-off tooth can easily become a projectile. The use of a dental burr can help smooth down rough edges. Incisor trims can be performed without the use of anesthesia with appropriate restraint. Molar tooth malocclusions can be more complex. Due to the small oral cavity of the rabbit, general anesthesia is usually administered. Anesthesia not only facilitates placing the mouth gag and cheek spreaders but also avoids the stress for the patient.

Rabbits have two sets of maxillary incisors, one behind the other. The second set of incisors are called peg teeth. Peg teeth are sometimes found to be loose but do not usually require clipping (**FIGURE 5-11**).

A complete dental exam requires the use of anesthesia, specialized mouth gags, and dental instruments because the oral cavity of a rabbit is very small and hard to visualize. Dental packs are available and include molar nippers, rasps, probes, elevators, and gags. Cheek spreaders are usually of the spring type. The mouth gag is placed on the upper and lower incisors. A thumb screw is used to adjust the gag and maintain the opened mouth (**FIGURE 5-12**).

FIGURE 5-12 A dental pack designed for rabbits: left to right, molar nippers, double-ended dental rasp, dental probe, and mouth speculum. The two other instruments are different-sized cheek spreaders.

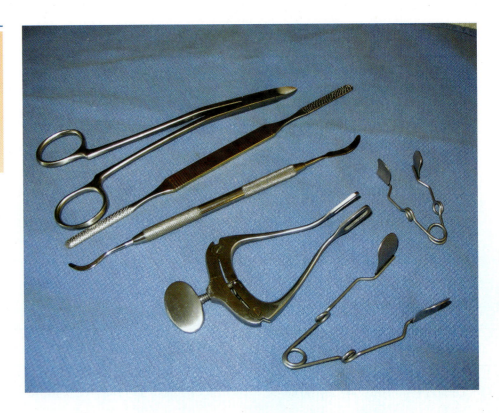

Dental problems may contribute to a condition commonly called slobbers, a chronically wet dewlap. Slobbers can create moist dermatitis that is not always easy to treat. Large does with highly developed dewlaps are more prone to slobbers. The dewlap may be continually wet from being dragged through a water bowl. Rabbits should be checked for dental problems and water bowls should be replaced by sipper bottles.

Rabbits are monogastric, hindgut fermenters. This means that they are single-stomached with microbes in the cecum that ferment (convert) organic material into usable carbohydrates. In this way, they are more similar to a horse than a dog or a cat. The cecum is a part of the large intestine and one of the largest organs in the rabbit's body (FIGURE 5-13). The microbes are normal gut flora (essential bacteria) for digestion and assimilation of food. Indigestible waste is separated out by the colon and eliminated as hard, dry, fecal pellets. Digestible foodstuffs continue metabolic breakdown in the cecum through the action of bacteria. Long-term use of antibiotics reduces normal gut flora, preventing food from being digested. When the balance of healthy bacteria is disrupted, other conditions can develop and lead to enterotoxemia (a condition of having intestinal toxins in the blood).

Enterotoxemia is caused by the overgrowth of detrimental or *bad* bacteria. Signs include watery, greenish-brown diarrhea, bloat, depression, or sudden death. *Clostridium spiroforme* is most often implicated. Supportive care with the selective use of antibiotics, fluid therapy, and nutritional support has limited effect, and mortality rates are high, especially in newly weaned kits.

Rabbits are unable to vomit due to the anatomical placement of the stomach and lack of esophageal muscle strength. Relative to other species, they also have a very short small intestine. Because of this, rabbits are prone to blockages and gut stasis, a slowing down or complete absence of gut motility (FIGURE 5-14).

Gut stasis is associated with diets low in fiber, pain, environmental stress, lack of exercise, and hairballs. Signs include anorexia, an enlarged stomach, and decreased fecal production. Fluids are given to rehydrate the rabbit, and either

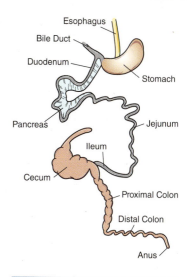

FIGURE 5-13 Digestive tract of a rabbit. The cecum is located at the proximal end of the colon.

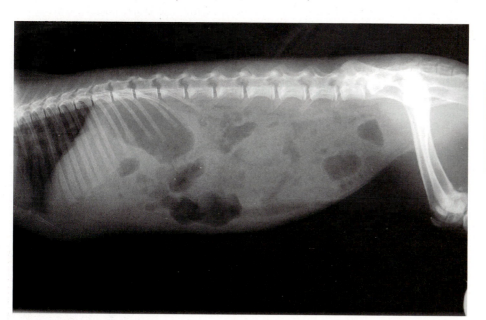

FIGURE 5-14 Rabbits have a very sensitive GI tract. This rabbit has multiple air pockets throughout the intestinal tract. In addition to gut stasis, air pockets can compromise respiration and cardiac function.

metoclopramide or cisapride (GI stimulants) may be prescribed by the veterinarian for short-term use to help the rabbit become more comfortable until normal gut motility is restored. Force feeding a high-fiber diet may also be required along with an appetite stimulant such as vitamin B.

Obesity has become a problem in pet rabbits and has been associated with commercial diets that are high in protein and fat. Reduced exercise and free choice food are contributors to the obesity problem. Obesity puts rabbits at a higher risk of developing **hepatic lipidosis**, fatty liver disease. Anorexia is the main clinical sign associated with fatty liver disease. Dental disease, GI disease, and stress-related illnesses should be ruled out prior to diagnosing hepatic lipidosis. Treatment involves force feeding a high-quality critical care diet (e.g., Oxbow®) through a feeding tube or syringe.

Tyzzer's disease is a bacterial infection that primarily affects the cecum. It is a common disease among rabbits and rodents but difficult to positively diagnose and treat. Most diagnoses are based on history and clinical signs. *Bacillus piliformis* causes inflammation of the cecum, which, in turn, causes diarrhea and fecal soiling of the rectal area. Tyzzer's disease can occur in rabbits as young as six weeks. A compromised immune system, overcrowding, poor diet, and poor husbandry all contribute to this disease. Oral antibiotics are used to treat the condition.

Hairballs, or **trichobezoars**, form in the digestive tract. They are most commonly found in the stomach. They are caused by ingestion of hair due to excessive grooming, boredom, or barbering of cage mates. Hairballs can cause impaction, anorexia and weight loss, generalized discomfort, and depression. Trichobezoars, as with any foreign body, require surgical removal.

Rabbit digestive problems can be complex. A veterinarian who is experienced with rabbit medicine should always be consulted at the first sign of medical concern.

Nondigestive Disorders

While dystocia is rare in rabbits, other problems with pregnancy and **lactation** (milk production) are more common. Pregnancy toxemia can occur, usually during the last week of gestation. Signs are acute and include weakness with **ataxia**, a lack of muscle coordination. The doe shows no interest in food or nesting and is generally depressed. If medical intervention is not immediate, toxic shock can cause abortion and seizures. With the onset of seizures, the prognosis is grave. Treatment options are limited, but warmed IV fluids may help deter the progression of toxic shock.

Lactating does, especially those living in dirty environments, can develop **mastitis**, a condition where the mammary glands become inflamed and hard. The doe spikes a fever, often as high as 104°F. *Streptococcus*, *Staphylococcus aureus*, and *Pasteurella* bacteria have all been implicated in mastitis. Due to the bacterial infection, abscesses are likely to develop in the mammary glands and the infection can become systemic. Oral antibiotics and nutritional support for the doe are important, as is a cleaner environment.

Rabbits are susceptible to a variety of bacterial infections, many of which can be directly related to unsanitary housing, confinement to outdoor hutches with no environmental protection, and overcrowded living conditions. Bacterial infections can create abscesses that affect the internal organs. External abscesses are surgically removed. Because rabbit pus is very thick, abscesses cannot simply be opened and allowed to drain as with other species. Postsurgical antibiotics and analgesics may be prescribed. Internal abscesses are far more difficult to diagnose and treat. There may be no clinical signs before the rabbit dies. Most internal abscesses are diagnosed on necropsy. Bacteria commonly implicated in rabbit abscesses are *Pasteurella multocida*, *Pseudomonas aeruginosa*, *S. aureus*, and *Proteus*.

Conjunctivitis, an inflammation in the soft tissues around the eye, can also be caused by bacteria, especially when rabbits are housed outdoors in unsanitary conditions. Bacteria, and the resulting pus, can completely occlude the lacrimal duct and prevent the eye from being kept moist. Lacrimal duct cannulation is required to unblock the duct. A very small tube (cannula) is inserted directly into the lacrimal duct in order to flush out the bacterial and accumulated pus.

Rabbit syphilis or *vent disease* is caused by a spirochete bacteria. It is transmitted from rabbit to rabbit via direct genital contact. Rabbits with this disease can be asymptomatic carriers until stressed. Lesions appear on the genitalia, perineum, eyelids, and mouth. There may be edema, ulcerations, and scab formation. Diagnosis can be made with skin scrapings or biopsy of the affected tissue and a plasma region card (PRC) test. This condition can be treated with injectable penicillin. Rabbit syphilis is *not* zoonotic.

Reproductive adenocarcinomas are the most common neoplasias in female rabbits. The risk of carcinomas can be reduced when an ovariohysterectomy is performed before they are sexually mature. Uterine and mammary neoplasia are the most common sites. The range of clinical signs include depression, anorexia, soft tissue masses, hematuria, and vaginal discharge. If metastasis has not occurred in the lungs, the prognosis is guarded. An ovariohysterectomy is recommended along with chemotherapy to treat the rabbit.

Urine color and consistency is a good indicator of rabbit health. An average rabbit produces 30 to 35 ml of urine daily. Normally, the urine can be clear or creamy. Urine that is clear may indicate that the body has a higher demand for calcium. Rapid growth, pregnancy, and lactation all create higher calcium requirements. Some foods may affect the color of urine, for example, plant and vegetable pigments. Orange or rust-colored urine alarms many owners, creating concern that there is blood in the urine. In most instances, the color indicates a diet high in protein. Excess protein and calcium in the diet can lead to the formation of bladder stones, or urolithiasis.

Uroliths are more common in rabbits with little exercise and high-protein diets. They may also be genetically predisposed to stone formation. Rabbits with urolithiasis often have a history of being fed vitamin and mineral supplements or unlimited amounts of alfalfa, both pellets and hay. Stones are more common in bucks because of their long urethra, preventing smaller stones and grit material from being eliminated.

A rabbit that sits with an exaggerated hunched-over posture, strains to urinate, and grinds its teeth when attempting to urinate may be exhibiting signs of stone formation. Other indicators include anorexia, weight loss, and depression. There may be blood in the urine due to the irritation of the bladder wall and urethra.

Radiographs (**FIGURE 5-15**) will be able to confirm the presence of stones or a gritty sludge in the bladder, which can precipitate stone formation

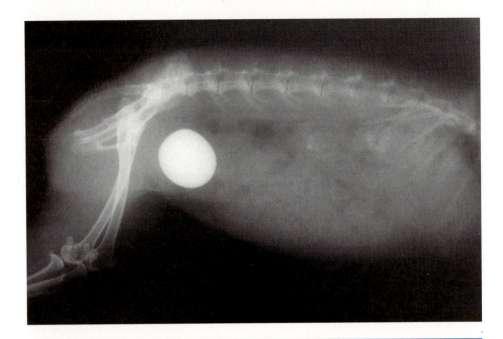

FIGURE 5-15 This large urolith was discovered during a routine wellness examination. The rabbit had exhibited no signs suggestive of having a bladder stone. Six-monthly wellness examinations can potentially avert disaster.

FIGURE 5-16 Surgery was successfully performed to remove the stone, shown here, and the rabbit recovered without incident.

(**FIGURE 5-16**). The majority of uroliths are composed of calcium salts, calcium oxalates, and calcium phosphates, all of which are radiopaque and easily visualized on a radiograph. Surgery is required for the removal of the stones and grit, followed by a change in diet. Rabbits should be fed grass hay, and all alfalfa and vitamin/mineral supplements should be eliminated to prevent further stone formation.

Parasites

Many parasites affect rabbits. They are susceptible to ear and fur mites, lice, cat fleas, internal protozoans, and intestinal worms. Definitive diagnoses are made by fecal floatation, skin scraping, and microscopic examination.

Rabbits housed outdoors can become hosts to cuterebra, the larva of many large fly species. The eggs are deposited on the fur. When the larva hatches, it burrows under the skin of the rabbit, where it matures. The larva forms a cavity and may be seen protruding from a sinar hole (a small hole) in the skin. Larva can also migrate into the nasal sinuses and ear canals. Cuterebra larvae produce toxins that cause muscle weakness in the rabbit. The larva must be carefully removed. It is very important not to crush the larva in the process as it will cause anaphylactic shock potentially killing the rabbit. The entire body should be checked for cuterebra. Usually there are only one or two, but as many as seven cuterebra larvae were removed from the abdomen of a young doe.

Fly strike occurs when the house fly lays eggs on the fur. It is most common in the rectal area, especially in rabbits with fecal soiling and dirty enclosures. The eggs hatch into maggots, which feed on the traumatized skin and exudates, the leaking fluid from damaged blood vessels. Maggots must be removed, one by one, with forceps. This can be facilitated by flushing through maggot holes with hydrogen peroxide. The maggots will crawl out of the skin, and are more easily collected and destroyed. The area should then be thoroughly flushed with sterile saline.

Clinical Procedures

Rabbits will accept oral medications given as a liquid or semi-liquid much better than they will accept a capsule or tablet. Medication can be mixed with fruit juice, mashed fruit, or pureed human infant vegetable food. The mixture can be drawn into a syringe and administered orally if the rabbit refuses it. The rabbit should be held close against the restrainer's body, with the rabbit's rear end toward the restrainer. Lean over the rabbit, put the tip of the syringe in the side of the mouth, and administer the amount slowly and steadily. Do not place the rabbit on its back or cradle it, as there is a risk of aspiration (**FIGURE 5-17**).

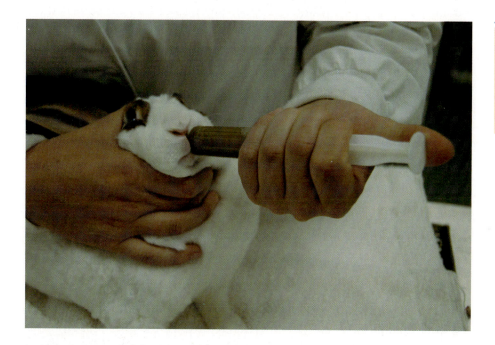

FIGURE 5-17 Correct restraint and positioning for administering supplemental feeding and administration of oral medications to a rabbit.

Administering intramuscular injections causes pain and may elicit a scream from the rabbit. It can also cause prolonged muscle inflammation and potential necrosis of muscle tissue. If there is no alternative route for the specific drug, the quadriceps muscles of a hind leg are used.

Rabbits vary a great deal in size and weight, as does their blood volume. It is important to obtain an accurate weight at the time of the blood draw to calculate the amount that may be safely drawn from that particular rabbit at that particular time. As a *general* guideline, no more than 6 to 10 percent of the blood volume (1 ml/100 g body weight) should be collected. Microtainer tubes are adequate as the amount of the sample will be small. Clots can form quickly and the sample needs to be added to the appropriate collection tube and rocked with the anticoagulant to avoid rendering the sample unusable. For a complete blood count (CBC), an EDTA tube is used, but for a biochemistry panel, a lithium heparin tube provides more accurate results. All supplies should be immediately at hand and ready for use.

Rabbit skin is thin and can be easily torn when inserting a needle for blood collection. The veins are thin walled and fragile, and hematomas form easily at the puncture site. Collection sites in rabbits are similar to those in other small animals. The lateral saphenous or cephalic veins usually provide the easiest access and least amount of trauma and stress to the rabbit. The marginal ear vein or central ear artery are often cited for possible blood collection, but in reality, it is very difficult to obtain an adequate sample from either of these sites without causing trauma to the sensitive tissue of the ear, collapsing the vein, and creating a potentially large and painful hematoma. For the pet rabbit, ear draws should be avoided. The jugular vein can be accessed, but because of the restraint technique required, this procedure will cause a great deal of stress to the rabbit and potentially compromise an already critical patient. Sedation is necessary when collecting from the jugular vein.

Fluid therapy is administered for dehydration and in treating chronic diseases or trauma. Rabbits generally drink 50 to 150 ml/kg daily. The cephalic vein is the best choice for intravenous catheter (IVC) placement using a 24- to 26-gauge catheter. Intraosseous (IO) catheter placement is recommended in severely dehydrated patients. The 22-gauge catheter is placed in the greater tubercle of the humerus. When preparing the site for IO catheter placement, sterile technique must be used to prevent possible contamination of the bone canal. Careful observation of the patient is necessary as a rabbit can quickly chew through IV lines and pull out the catheter. An E-collar may be necessary to prevent access.

A urinalysis will provide information that supports findings from the physical examination, clinical signs, and laboratory tests. A free catch or cystocentesis can be used to obtain a urine sample. Cystocentesis is the preferred method of collection and will provide a sterile sample. The rabbit is placed on its back and the bladder is palpated, just cranial to the pelvis. A 22- to 25-gauge needle connected to a 6 ml syringe is used to collect the sample while the bladder is stabilized with the opposite hand.

Rabbit urine is straw colored and clear to turbid. If the urine is red in color, hematuria or porphyrin pigmentation will need to be ruled out. A urine dipstick or a woods lamp can be used to differentiate between the two. Porphyrin will fluoresce under a woods lamp. Turbid or cloudy urine is usually due to crystals and excessive calcium.

Many Practices Routinely Administer Presurgical Atropine to Their Small Animal Patients.

Atropine is an anticholinergic drug that is used for muscle relaxation and to dry up or slow down body secretions including the flow of saliva. Most rabbits produce atropinase, an enzyme that blocks the effects of atropine. Because of atropinase, most veterinarians prefer to use glycopyrrolate rather than atropine in presurgical patients. Glycopyrrolate is also an effective anticholinergic drug with predictable effects in the rabbit patient.

Rabbit anesthesia is usually a little more of a challenge than with other mammals. Being aware of some of the idiosyncrasies can make induction and recovery more successful and less stressful for the rabbit. Inhalant anesthetic agents used for rabbits are isoflurane or sevoflurane. Both are well tolerated by rabbits, but an advantage of sevoflurane is that it does not have an odor and the patient is less resistant when being masked down. Rabbits often hold their breath when being induced and resist breathing in the anesthetic agent because of the strange odor. This is especially evident when rapid induction is practiced. For the rabbit, a slow and calculated induction is preferable. Even before applying the face mask, allow a few minutes of pure oxygen flow, which will help flush any residual odor in the mask. With the oxygen still at 100 percent, gently position the face mask around the nose and mouth to secure a correct fit with no ocular pressure points. Allow the rabbit to breathe 100 percent oxygen for another few minutes before introducing the anesthetic agent. When the anesthetic gas is slowly introduced in 0.5 percent increments, the occurrence of apnea (breath holding) will decrease, making for a smoother and less stressful induction.

> **A Word of Warning: *Never* Assume That a Rabbit Has Reached a Suitable Plane of Anesthesia by Visual Parameters Alone.**
>
> Many a rabbit has been positioned in dorsal recumbency, shaved, and scrubbed for the surgery table, apparently *anesthetized*, only to bolt off the table in an attempt to escape.

Intubation of a rabbit can be difficult because the oral cavity is so small. Additionally, the tongue, while small, is thick and muscular and further blocks visualization of the epiglottis, which is more caudal than in many species. These anatomical differences all contribute to the challenge of intubation. The technique needs to be mastered, but repeated attempts to intubate the same patient are not recommended. Stimulation during the attempts disrupts the plane of anesthesia already achieved and will further increase the frustration for all involved and potentially cause trauma to the rabbit. The rabbit needs to be positioned by the restrainer in a manner that optimizes access to the epiglottis. The use of a laryngoscope will assist in the placement of an ET tube (**FIGURE 5-18**). Once intubated, the rabbit can be positioned for surgery and connected to a nonrebreathing anesthesia delivery system (**FIGURE 5-19**).

Rabbits are obligate breathers and cannot breathe through their mouths. Another method of inducing and maintaining anesthesia is nasal intubation. A small red rubber feeding tube; a flexible, noncuffed endotracheal tube (1.0 to 1.5 mm); or a nasogastric tube (3 Fr) can all be used successfully. Slowly pass the lightly lubricated tube through one nostril. Lubricating the tube with either plain warm water or a water-soluble gel will help in the placement and decrease discomfort for the patient. When placed with care, into the trachea should enter the nasal septum, which leads to the back of the throat and to the trachea. Confirm that placement is correct and that it has not entered the esophagus.

FIGURE 5-18 Restraint position required in a lightly anesthetized rabbit for placement of the ET tube. A laryngoscope is used to help visualize the epiglottis.

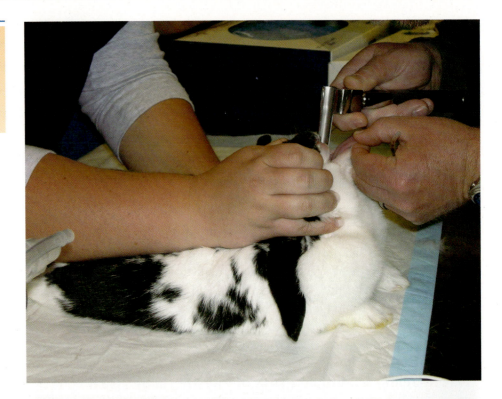

FIGURE 5-19 A rabbit positioned for surgery and connected to a non-rebreathing anesthesia unit.

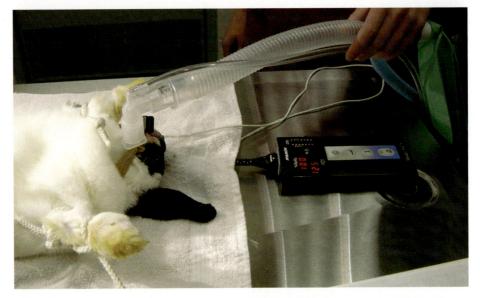

A light breath into the end of the tube will cause the thorax to rise, confirming that the tube has entered the trachea. Under no circumstances should tube advancement be forced.

Radiographic studies of the rabbit usually begin with two views, lateral and ventro-dorsal of the abdomen and thorax, and can often be taken without sedation. Skull radiographs are necessary to completely evaluate dental disease in the rabbit. Lateral, dorso-ventral, and rostro-caudal views are taken to provide information on possible abscess or malocclusions.

Abdominal radiographs are taken to evaluate the gastrointestinal tract and reproductive organs. Contrast studies are frequently performed to rule out gastric foreign bodies such as a trichobezoars or masses. Thoracic radiographs are performed to evaluate the cardiac and pulmonary systems. Ultrasonography provides diagnostic images of the abdomen, urinary system, and cardiac system. These images aid in diagnosing pathologies associated with the various organ systems in the rabbit, including foreign bodies and soft tissue masses. Caution needs to be taken when shaving the area for the ultrasound due to the thin skin layer and prevent tearing of the skin.

Summary

Rabbits have become increasingly more popular as companion animal house pets. There is a great variety of rabbit breeds and each has its own unique physical characteristics and behavioral traits. While generally quiet and docile, rabbits can become territorial and aggressive. Because they are prey species, they can be easily frightened by strange scents and sounds. Rabbits need to be handled and restrained correctly to avoid injury and potential spinal fractures. Rabbits are induced ovulators and can breed at any time once they reach sexual maturity. Rabbits have many medical concerns unique to them, many of which are directly related to incorrect diet and husbandry practices. Care must be taken when performing a blood collection: their skin is very thin and the veins are small and fragile. Anesthesia can sometimes be a little more challenging than with other species and rabbits can be difficult to intubate.

*fast*FACTS

RABBITS

WEIGHT
- **Adult buck:** 2–5 kg (4.4–11 lbs) average
- **Adult doe:** 2–6 kg (4.4–13.2 lbs) average

LIFE SPAN
- 7–9 years

REPRODUCTION
- **Sexual maturity:** bucks: 22–25 weeks; does: 22–25 weeks
- **Gestation:** 30–32 days
- **Litter size:** 4–8 kits
- **Weaning age:** 4–6 weeks

VITAL STATISTICS
- **Temperature:** 38.3–40°C (101–104°F)
- **Heart rate:** 130–325 beats/min
- **Respiratory rate:** 35–60 breaths/min

DENTAL
- Total 28 teeth
- Dental formula 2 (I 2/1, C 0/0, P 3/2, M 3/3)

VACCINATIONS
- None

ZOONOTIC POTENTIAL
Fungal
- Dermatophytes

Bacterial
- Salmonella
- Tularemia
- Leptospirosis
- Streptobacillus
- Pasteurella

Review Questions

1. Under what circumstances would a rabbit vocalize?
2. What is the best way to encourage a rabbit to use a litter box?
3. What are cecotrophs and why are they important to rabbit health?
4. What are the causes of pododermatitis?
5. Why is it important to control the hind feet of a rabbit?
6. Explain the term *induced ovulator*.
7. What are the main parasites of rabbits?
8. What is the average life span of a companion rabbit?

Case Study I

History: A two-year-old buck has been sneezing and there is a thick mucous discharge around the nares. The owner reports that the rabbit has not been eating. It is housed in an outdoor hutch.

Physical Examination: There is dried mucous on the medial side of the forepaws. The nasal passages are partially clogged with mucous and the breathing is labored. The footpads are heavily soiled with fecal material. The rabbit's temperature is 104.5 °F.

a. What is the most immediate problem that comes to mind?
b. Is this disease transmitted to other rabbits, and if so, what is the mode of transmission?
c. What can predispose a rabbit to this condition?

Case Study II

History: A four-year-old doe has become lethargic and weak. The owner has noticed a decreased appetite over the past four or five days and that very few fecal pellets are being produced. The small amount of fecal material produced is soft and watery. The doe sits with an exaggerated hunched posture and is very tense and fearful.

Physical Examination: The veterinarian palpated the doe and noted a distended abdomen. The doe's temperature was below the normal range.

a. What could be the possible causes of these signs?
b. Describe how the signs and history would help confirm a diagnosis?
c. What is the recommended treatment/prevention for this condition?

For Further Reference

Campbell-Ward, M.L. (2011). Gastrointestinal Physiology and Nutrition. In Quesenberry, K.E and Carpenter, J.W. (Eds.), *Ferrets, Rabbits and Rodents: Clinical Medicine and Surgery, 3rd Edition* (pp. 183–193). St. Louis, MO: Imprint of Elsevier Science.

Divers, S.J. (2011). Exotic Mammal Diagnostic and Surgical Endoscopy. In Quesenberry, K.E. and Carpenter, J.W. (Eds.), *Ferrets, Rabbits, and Rodents: Clinical Medicine and Surgery, 3rd Edition* (pp. 485–501). St. Louis, MO: Imprint of Elsevier Science.

Fisher, P.G., & Carpenter, J.W. (2011). Neurologic and Musculoskeletal Diseases. In Quesenberry, K.E and Carpenter, J.W. (Eds.), *Ferrets, Rabbits and Rodents: Clinical Medicine and Surgery, 3rd Edition* (pp. 245–256). St. Louis, MO: Imprint of Elsevier Science.

Graham, J., & Mader, D.R. (2011). Basic Approach to Veterinary Care. In Quesenberry, K.E. and Carpenter, J.W. (Eds.), *Ferrets, Rabbits, and Rodents: Clinical Medicine and Surgery, 3rd Edition* (pp. 174–182). St. Louis, MO: Imprint of Elsevier Science.

Harcourt-Brown, F. (2004). *Textbook of Rabbit Medicine.* Great Britain: Butterworth/Heinemann.

Hawkins, M.G., & Pascoe, P.J. (2011). Anesthesia, Analgesia, and Sedation of Small Mammals. In Quesenberry, K.E. and Carpenter, J.W. (Eds.), *Ferrets, Rabbits and Rodents: Clinical Medicine and Surgery, 3rd Edition* (pp. 429–451). St. Louis, MO: Imprint of Elsevier Science.

Hess, L. (2006, January 7–11). *Clinical Techniques in Rabbits.* Conference Notes, The North America Veterinary Conference, Orlando, FL.

Hess, L., & Tater, K. (2011). Dermatologic Diseases. In Quesenberry, K.E. and Carpenter, J.W. (Eds.), *Ferrets, Rabbits and Rodents: Clinical Medicine and Surgery, 3rd Edition* (pp. 232–244). St. Louis, MO: Imprint of Elsevier Science.

Hoefer, H.L. (2006, January 7–11). *Urolithiasis in Rabbits and Guinea Pigs.* Conference Notes, The North America Veterinary Conference, Orlando, FL, 2006.

http://www.accu.vt.edu (accessed December 18, 2013).

http://www.petdoc.ws (accessed November 22, 2013).

http://www.rabbit.org (accessed May 28, 2014).

Klaphake, E., & Paul-Murphy, J. (2011). Disorders of the Reproductive and Urinary Systems. In Quesenberry, K.E. and Carpenter, J.W. (Eds.), *Ferrets, Rabbits and Rodents: Clinical Medicine and Surgery, 3rd Edition* (pp. 217–231). St. Louis, MO: Imprint of Elsevier Science.

Lennox, A.M. (2011). Respiratory Disease and Pasteurellosis. In Quesenberry, K.E. and Carpenter, J.W. (Eds.), *Ferrets, Rabbits and Rodents: Clinical Medicine and Surgery, 3rd Edition* (pp. 205–216). St. Louis, MO: Imprint of Elsevier Science.

Mayer, J. (2006, January 7–11). *Analgesia and Anesthesia in Rabbits and Rodents.* Presented at the North America Veterinary Conference, Orlando, FL.

Mayer, J.D. (2006, January 7–11). *Analgesia and Anesthesia in Rabbits and Rodents.* Conference Notes. Presented at The North America Veterinary Conference, Orlando, FL.

Murray, M.J. (2006). *Practice Tips: Rabbits.* Conference Notes. Presented at the North America Veterinary Conference, Orlando, FL.

Murray, M.J. (2006, January 7–11). *Rabbits: It's Not Always Pasteurella.* Conference Notes. Presented at the North America Veterinary Conference, Orlando, FL.

Vella, D., & Donnelly, T.M. (2011). Basic Anatomy, Physiology, and Husbandry. In Quesenberry, K.E. and Carpenter, J.W. (Eds.), *Ferrets, Rabbits and Rodents: Clinical Medicine and Surgery, 3rd Edition* (pp. 157–173). St. Louis, MO: Imprint of Elsevier Science.

GUINEA PIGS

OBJECTIVES

After completing the chapter, the student should be able to

- Correctly house a guinea pig.
- Provide appropriate client education to new guinea pig owners.
- Understand the importance of providing the correct diet for guinea pigs.
- Provide basic nursing care to a guinea pig.
- Assist in the anesthesia of a guinea pig.
- Demonstrate appropriate restraint techniques for guinea pigs.
- Understand common medical disorders in guinea pigs.

Introduction

There are probably few children who do not know what a guinea pig is. They will tell you that guinea pigs are cute, soft, and gentle and make good classroom pets. Yet ask anyone where guinea pigs come from and the response is usually *New Guinea*. Their species name, *porcellus*, means *little pig*, but it is uncertain how this South American species came to be called a guinea pig, when New Guinea is so distant from their natural habitat. Another thought is that these little mammals were purchased for a guinea, an old coin. *Pig* may be because of their pig-like squeals and chunky body type, but the real derivation for the prefix *guinea* remains unknown.

The guinea pig is more accurately called a cavy, *Cavia porcellus*, and belongs to the suborder *Hystricomorphic*, or *porcupine-like* rodents. They are found throughout the western, mountainous regions of South America. They live in family groups with one dominant male, using rocky outcrops and ledges for shelter with no permanent dens or nesting sites. Strictly herbivorous, they forage from site to site, grazing on grasses and leafy plants.

Long used in research, the phrase *does anyone volunteer to be a guinea pig* is not without reason and indicates apprehension about what is to come. Guinea pigs have contributed greatly to the understanding of many human diseases, in the development of vaccines, and in many other areas of medical research.

Today, there are guinea pig clubs and associations, guinea pig shows, and exhibitions with nationally recognized judges and breed standards. This little pig has become a household favorite and makes an ideal small pet, 4-H club or FFA project. They are docile and rarely bite in any situation. They are easy to handle and require little in the way of housing or specialized needs.

Because of their timid nature, they should be kept away from other household pets, including dogs, cats, and ferrets. However friendly the family dog may seem, to the guinea pig it is a predator and can cause a great deal of stress. Stress is a major contributor to health problems commonly seen in guinea pigs and can be one cause of anorexia. Anorexia is more than simply *going off feed*, but for the cavy, it can lead to major medical concerns and mortality.

The American Cavy Breeders Association recognizes 13 different breeds of guinea pigs, with many varieties in each category. Not officially recognized, two hairless mutations have been developed, the Skinnies and the Baldwins. Skinnies have no hair except on their heads and lower legs, while the Baldwins are born with a full coat that is completely lost by the time they are weaned. These two hairless mutations have been a center of controversy, and while they were originally bred for research, there are concerns about their hardiness and the strength of their immune systems.

There is great variety in the breeds recognized: the Silkie with its very long silky hair; the Abyssinian, a shorter-haired guinea pig with whorls all over its body; the Teddy, with short fuzzy hair; and the Textel, with a very kinky coat. There are numerous coat colors within each group and an endless assortment of patterns, patches, and markings, all contributing to their appeal (**FIGURE 6-1**).

FIGURE 6-1 Different breeds and varieties of cavies are available in pet stores. As illustrated here, they should not be housed with rabbits.

Behavior

Guinea pigs are not only social with each other but actively respond to human companionship and recognize different people. While their eyesight is poor, their hearing is highly developed, and familiar footfalls often elicit a variety of chirps, whistles, and *chuckles*.

Guinea pigs produce a large range of vocal sounds. The whistle can be a call of greeting or of alarm, the exact meaning determined by the length and pitch of the call. Owners closely involved with their guinea pigs are able to not only interpret situations but also know which guinea pig in a group is whistling. Chirps are lower pitched and equally varied in tone and modulation. They are the sounds of contentment. Very distinct from other sounds is **drilling**, a low, almost guttural and rapid *d-r-r-r-r* produced by their molars. Drilling is a warning, an alert to other guinea pigs of possible danger. **Boars** (males) in pursuit of a **sow** (female) also drill, challenging other males. Drilling is also heard when a guinea pig is in pain. Drilling should never be interpreted as a contented purr.

While guinea pigs rarely fight, two adult males placed together are likely to engage in minor, mostly bluff squabbles to establish dominance. A single boar with one or more sows is compatible, as are groups of two or more females.

Guinea pigs have large **perianal** (near the anus) sebaceous glands used for scent marking. These are especially pronounced in unneutered boars. They drag their rumps, pressing down on the floor surface and leaving a scent trail. When excessive marking occurs in males, a build-up of oil coats the hair around the glands, giving the hair a greasy appearance. The perianal sac around the scent glands can easily become impacted with fecal and bedding material, resulting in a foul smell. The rump area should be checked regularly. Mineral oil can be applied with cotton-tipped swabs to help soften the impaction and the area gently cleaned with warm water.

Guinea pigs have little in the way of self-defense. Their oral cavity is small, which makes biting difficult. They have four clawed toes on the front feet and only three on the hind feet. Scratches may be likely in an attempt to escape. The normal group (or herd) response to fear is to flee, to stampede. Often a single guinea pig will freeze in place, not moving at all until the perceived danger has passed. Attempts to capture a loose guinea pig can become very stressful to the animal. It is best to tempt them with food. In most instances, the guinea pig will approach, and if it has been handled regularly, it can be scooped up in two hands, one around the shoulders and the other supporting the rump. The guinea pig should be held close to the body where it feels more secure and is less likely to struggle (**FIGURE 6-2**).

Determining the sex of a guinea pig is fairly easy compared to some small mammals. The anogenital area of the sow is more rounded in appearance, and has a *Y* shape. The *Y* configuration extends from the anus to the vaginal opening (**FIGURE 6-3**).

Boars have prominent, slender testicles on either side of the anal area. However, due to an open inguinal canal, the testes are not always apparent and young guinea pigs may be sexed incorrectly. Applying gentle pressure with the thumb and a **caudally** (downward) directed stroke on the lower abdomen will cause the penis to protrude and confirm the sex. The anogenital distance is longer in the boar (**FIGURE 6-4**).

Early determination of sex is important to avoid unwanted litters. In the male, sexual maturity is reached at 9 to 10 weeks, while in the female it is unexpectedly

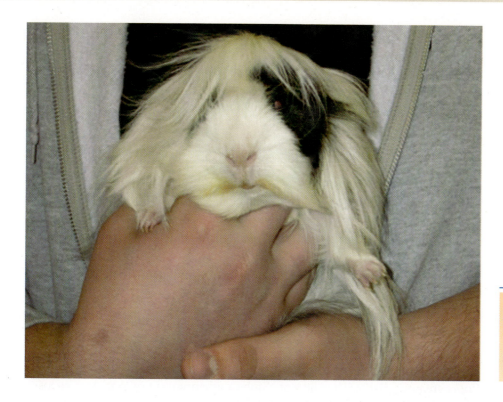

FIGURE 6-2 Guinea pigs should be held with two hands, one around the shoulders and the other supporting the rump.

FIGURE 6-3 The external genitalia of a female guinea pig with a distinctive Y shape.

FIGURE 6-4 The external genitalia of a male guinea pig. The penis may be everted with gentle thumb pressure on the lower abdomen.

earlier, at four to six weeks. If the boar is left with the group, father–daughter matings will occur.

Sows are polyestrous in that they have many breeding cycles throughout the year. Estrus cycles last an average of 16 days. A vaginal plug appears after copulation but falls away within a few hours after mating. The vaginal plug is also common to many rodents, but the exact purpose is unknown. It is made up of ejaculate and thought to prevent sperm from leaking from the vagina, ensuring greater fertility, or to prevent another male from mating with the same sow.

Both sexes have a single pair of inguinal nipples. Despite having only two mammary glands, sows are able to take care of an average-sized litter of three or four young. Unweaned guinea pigs patiently await their turn with no aggression toward one another. It is not necessary to remove the boar from the presence of the neonates except to prevent rebreeding. Boars are very tolerant of their antics and will often groom the young guinea pigs.

Guinea pigs are born **precocial** (fully developed at birth) after a gestation period of 63 days. Their eyes and ears are open, they have a full coat of hair, and their teeth are erupted. They are very active from birth and will start to nibble at hay and the fecal pellets of adults within a few days of birth. Guinea pigs are coprophagic. Fecal ingestion by the pups is necessary to help establish normal gut flora (adults eat soft feces directly from the anus as do rabbits). Even though guinea pig pups are precocial, for the first 7 to 10 days, the sow licks the **perineum** (the area between the anus and the genitals) to stimulate urination and defecation. Pups should be weaned at six weeks.

Housing

Guinea pigs can be simply housed. Elaborate cages with platforms and wire floors are not suitable. They do not climb or attempt to jump out. Any enclosed area with sides that are at least 8 to 10 inches high will keep them safely inside. The floor space should be at least 36 inches long, providing ample room for exercise and food bowls. Guinea pigs urinate and defecate without a location preference and bowls are frequently contaminated. Using a food hopper and sipper tube bottle will prevent this. Guinea pigs drink a lot of water and need to be provided with sipper bottles large enough to meet their daily needs. Water bottles should be cleaned out on a daily basis and checked frequently. Guinea pigs drink when they have mouthfuls of food, often pushing chewed material up into the sipper tube where it swells and totally blocks the water flow. Owners should be alerted to monitor water intake, as guinea pigs can quickly become dehydrated.

All small animals should be provided with a hide box, a place to escape to and feel safe, and guinea pigs are no different. A variety of commercial small mammal hide boxes, half round logs, and plastic houses are available, but guinea pigs are just as comfortable with an upside-down cardboard box with an entrance cut into it. The box can be easily replaced when soiled, and there is no danger of chewed pieces of plastic becoming imbedded in the oral cavity.

A variety of cages are available. A good choice would be one that has an easily removable wire frame cover that clamps down over a plastic tub. This allows for plenty of air circulation, is easy to clean, and provides some protection from other household pets (**FIGURE 6-5**).

Guinea pigs should not be housed in aquariums. In such enclosed environments, ammonia and nitrogen, by-products from their urine, can quickly accumulate and cause respiratory problems. Cedar bedding should not be used for guinea pigs or any small animal. When cedar shavings become wet, they release toxic fumes (cyanic acid) that cause severe respiratory problems and can lead to death. Newspaper or aspen shavings are safe.

Diet

Guinea pigs are herbivores and can be fed a variety of fresh greens and vegetables. Feeding excessive amounts of cabbage or kale can cause bloating. Beet greens are not recommended because they are high in oxalates, which contribute to the formation of bladder stones. Excessive amounts of fruits with natural sugar can lead to bladder problems and gastrointestinal bacterial imbalances. Daily offerings include guinea pig pellets supplemented with timothy or other varieties of grass hay. While appearing to be the same as rabbit pellets, guinea pig pellets have added vitamin C. Ascorbic acid (vitamin C) is an essential part of the guinea pigs' daily diet as they are unable to synthesize it from other foodstuffs. Guinea pigs lack a specific liver enzyme, *L-gulonolactone*, an essential component of the D-glucose L-ascorbic acid pathway. Ascorbic acid is necessary for the metabolism of cholesterol, amino acids, and carbohydrates. Without adequate daily amounts of the vitamin in the diet, the signs of vitamin C deficiency can appear in as little as 10 days to 2 weeks. Guinea pigs with vitamin C deficiency become weak and refuse to eat; their joints become swollen with severe lameness; they lose weight and rapidly deteriorate. If immediate veterinary care is not received and the diet corrected, the guinea pig will die.

The disease caused by vitamin C deficiency is scurvy, and it can be easily prevented with a correct diet. Always check the expiration date on the bag when buying pellets. Although vitamin C is added, it has a relatively short shelf-life. The level of ascorbic acid can be reduced dramatically by exposure to light, heat, long-term storage, and dampness. Pellets should never be purchased from bulk bins or in quantities larger than can be used within three months or by the expiration date stamped on the bag.

A variety of vitamin C supplements are available, including drops and tablets, and some of them are formulated to be added to water. This can create problems, as many guinea pigs refuse to drink the treated water. Also, ascorbic acid is a water-soluble vitamin and so quickly degrades in water.

The best solution is to always ensure that the guinea pig receives an adequate amount of vitamin C in the foods offered. Foods high in vitamin C and readily consumed by guinea pigs are kale, parsley, spinach, and broccoli. Once introduced to them, guinea pigs will also eat red and yellow peppers, tomatoes, kiwi fruit, and orange segments. Dandelions are always well received, but it is important to be sure that the dandelions have not been sprayed with weed killer or they are not collected from an area of lawn that has been fertilized. Care must be taken not to overwhelm guinea pigs with new foods or abrupt changes in diet. Both will cause diarrhea and lead to further problems with the digestive tract. Iceberg lettuce should never be fed as it has very little nutritional value and will cause intestinal upset and bloating.

Handling and Restraint

Guinea pigs are easy to handle and are often vocal when picked up. Unless fully supported, they may struggle in attempts to escape. Holding them securely with two hands and close to the body makes them feel more secure. There is very little scruff, and attempting to use this method will only cause alarm and pain as it results in more of a pinch than a scruff. Care must be taken that they do not fall or walk off the edge of an examination table as their eyesight and depth perception is poor. Placing a towel on the table reduces the slippery surface for them, but still, one hand should always be on the body. When carrying a guinea pig, the rear legs can be cupped in one hand. This not only supports them securely but also reduces a possible scratch from the hind claws should they attempt to struggle. The other hand should be placed under the thorax and around the shoulders.

Medical Concerns

While guinea pigs are generally hardy if fed the correct diet and housed appropriately, some serious medical issues can develop. Guinea pigs are adept at hiding illness and should have routine examinations by a veterinarian to detect problems early. Many complications arise from the gastrointestinal tract and are further complicated by sensitivity to antibiotics. Guinea pigs in pain or discomfort may exhibit a hunched-over posture, fearful attitude, urgent repetitive squeals, drilling, and dragging the hind quarters.

Respiratory Disorders

Bordetella bronchiseptica is the bacterial agent responsible for kennel cough in dogs and cats, but in guinea pigs it causes severe upper respiratory tract infections. It is also responsible for abortions and stillbirths in pregnant sows, and mortality is high. It spreads via direct contact, by fomite transmission, and through the air. Rabbits can be asymptomatic carriers of *Bordetella* and should never be housed with guinea pigs. Clinical signs include anorexia, dyspnea, and ocular and nasal discharge. As a further complication, *Streptococcus pneumoniae* often develops. The prognosis is grave. Sulfa drugs and supportive isotonic fluids (0.9% sodium chloride) can be used in an attempt to treat *Bordetella* infections.

Streptococcus pneumoniae is commonly associated with upper respiratory tract infections. Transmission is through direct contact or aerosol distribution of the bacteria from infected guinea pigs. Guinea pigs may be asymptomatic carriers or show multiple clinical signs including sneezing, nasal and ocular discharge, coughing, anorexia, and depression. The disease is progressive and mortality is high. *Streptococcus* pneumonia is not always associated with *Bordetella*, and can occur as an independent infection.

Gastrointestinal Disorders

Dental examinations of guinea pigs can be difficult because of their small oral cavity. Guinea pigs do not have cheek pouches but have cheek folds that accumulate food. Care must be taken when examining the oral cavity to prevent food in the mouth from being pushed into the trachea.

All teeth are **hypsodontic** (open rooted) and grow continually. Correct alignment of the incisors can easily be checked and may indicate potential problems if they do not appear straight and evenly worn (**FIGURE 6-6**). There are a total of 20 teeth, including the incisors. Mandibular molars grow toward the tongue, while maxillary molars grow toward the cheek. Both are convex. This arrangement can predispose guinea pigs to malocclusion, which results in possible tongue

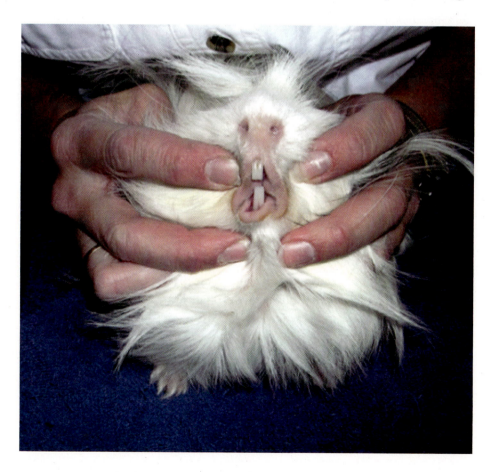

FIGURE 6-6 A cursory oral examination shows correct alignment of guinea pig teeth.

entrapment, making eating and drinking difficult. An otoscope and buccal pad separator (cheek dilator) can be used to assist in visualizing the molars. The rodent/rabbit dental packs can also be used with guinea pigs.

Submandibular abscesses occur in so many guinea pigs that the condition is commonly known as *lumpy jaw*. The correct term for this condition is *cervical lymphadenitis*. Lymph nodes under the jaw become abscessed, and the ensuing swellings create the lumps. The abscesses occur when *Streptococcus* bacteria invade oral wounds and abrasions. A frequent cause is consumption of coarse feed items, hay stalks, straw, and the hulls of seeds and peanuts that penetrate or abrade the oral mucosa. (No seeds, peanuts, or nuts of any kind should ever be offered, but are frequently found in guinea pig *mixed diets* or have been recommended as a treat.)

Another cause of *lumpy jaw* is malocclusion, caused when teeth penetrate the **buccal** (cheek) surface when the guinea pig unintentionally bites into the cheek while trying to chew. Bacteria enter through these wounds, creating an abscess that drains directly into the cervical lymph nodes. Veterinary treatment involves a cutaneous incision to open the abscess, allowing it to drain and be thoroughly flushed. Follow-up care is often required, especially in cases caused by malocclusion.

Guinea pigs are monogastric. Their stomach is proportionally larger than that of a rabbit and the cecum is well developed. Because of the position and size of the stomach relative to the diaphragm, guinea pigs are unable to vomit. Normal gastrointestinal flora, necessary for digestion, is gram positive. Any disruption of normal flora and gut motility can quickly lead to gut stasis and an overgrowth of toxin-producing gram-negative bacteria.

Clostridium is commonly implicated in cases of bacterial enterotoxemia. Most incidences of gut stasis involve the cecum rather than the stomach and can be attributed to a low-fiber diet. Reduced fiber leads to reduced gut motility, which in turn causes gut stasis and may also account for pockets of air and gasses trapped within the intestinal tract (**FIGURE 6-7**). The combination of gut stasis and the overgrowth of gram-negative bacteria is a common cause of death. Gastrointestinal disorders occur rapidly and there may be few clinical signs. Diarrhea, anorexia, and decreased water consumption are always a cause for alarm. Fluid therapy is required to treat the dehydration, and short-term antibiotics may be used in an attempt to control the overgrowth of gram-negative bacteria. **Probiotics** that promote the growth of healthy bacteria, such as lactobacillus and live yogurt, can be fed in an attempt to restore normal microflora.

Many bacterial infections are seen in guinea pigs, and treatment is difficult because they do not tolerate many of the commonly used antibiotics. Adverse reactions include endotoxemia and death due to the disruption of normal gut flora. Broad-spectrum antibiotics and those that primarily target gram-positive bacteria will destroy beneficial bacteria, while allowing the more pathogenic gram-negative bacteria to flourish. One example of an antibiotic that destroys gram-positive bacteria in the guinea pig is penicillin. Death may be sudden or occur within a week or less. Many topical antibiotics should also be avoided as they may be licked off and ingested. One example of an antibiotic that may be prescribed is enrofloxacin. It is well tolerated by guinea pigs and can be given orally; however, when administered as an injection, it can cause tissue necrosis at the injection site. Antibiotic choices and therapy protocol should be made only by the veterinarian. Owners need to be

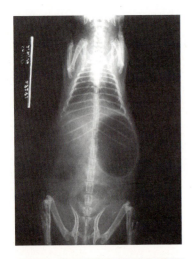

FIGURE 6-7 A radiograph showing a large air pocket on the right side. The guinea pig was presented with abdominal distension and discomfort, decreased amount of fecal droppings, and difficulty breathing. (*Courtesy of Brandi Orr, Bird and Exotic All-Pets Hospital*)

cautioned to never administer any antibiotic, for whatever reason, whether it is something recommended from a pet store or found in their own medicine cupboards. Other choices that may be considered by the veterinarian include chloramphenicol, trimethoprim-sulfa, and aminoglycosides that target gram-negative bacteria. Careful observation of the guinea pig should be done to detect early signs of toxicity, which include diarrhea, dehydration, decreased appetite, or sudden death.

> **Fecal Pellets from a Healthy Guinea Pig Are Sometimes Mixed with a Supportive Diet and Fed by Syringe.**
> *Coprophagic therapy* has been used incidentally and with some success. However, the risk of introducing Salmonella to an already compromised guinea pig needs to be carefully considered.

Guinea pigs are highly susceptible to Salmonella bacteria. Salmonella infections are commonly associated with bacterial enteritis, which is transmitted, as are most Salmonella infections, via the fecal/oral route. The most common causes of infection include eating infected feces and access to contaminated water and food or bedding. In guinea pigs, Salmonella bacteria can also enter through the conjunctiva. The incidence may be higher than in other small mammals due to behavior. Being enclosed, frequently urinating and defecating in food and water bowls, and normal consumption of fecal material are all contributing factors. Clinical signs include diarrhea, anorexia and associated weight loss, loss of coat condition, ocular discharge, and abortion in pregnant sows.

Nondigestive Disorders

A common reproductive problem is dystocia. If the sow does not have her first litter before she is seven to nine months old, the cartilaginous pelvic midline, which normally relaxes prior to parturition, begins to ossify. Fat pads can also block the pelvic opening, preventing delivery. Caesarian delivery is necessary to prevent the death of the sow and her litter. The young develop in both uterine horns, rather than in the small body of the uterus.

Pregnancy toxemia is another reproductive disorder and is most often seen during the last two weeks of gestation or in the first week postpartum. It is more common with obese sows. Signs are acute and include weakness and depression with accompanying ataxia, dyspnea, anorexia, and a dramatic drop in water consumption. The normally thick and slightly creamy consistency of urine becomes acidic and clear. If left untreated, this condition in the pregnant sow quickly progresses to abortion and seizures. With the onset of seizures in the pregnant or postpartum sow, the prognosis is grave.

Medical intervention includes warmed IV fluids to help stabilize the sow and prevent the progression of toxic shock. Blood chemistry values change dramatically as the sow becomes **hypoglycemic** (low blood sugar), **hyperkalemic** (increased levels of potassium), and **ketonic** (ketones in the blood). Sows at this stage may also develop **hepatic lipidosis** (fatty liver disease).

It is always better to guard against and possibly prevent this condition by feeding a correct diet and not allowing the sow to become obese. For a pregnant sow, vitamin C should be provided in greater amounts.

Ovarian cysts may also occur in females from two to four years old and are a source of discomfort and pain. Cysts develop spontaneously and may become evident only when the sow shows signs of pain warranting further investigation with an ultrasound examination. An ovariohysterectomy may be required, but as with any surgical procedure for the guinea pig, the potential anesthesia complications need to be carefully considered.

Poor husbandry contributes to poor health. Pododermatitis, commonly called *bumble foot*, is associated with animals, especially obese guinea pigs, housed on wire floors and kept in unclean environments. The constant weight of the guinea pig's hairless, unprotected feet on the wire creates pressure sores that soon become inflamed and infected. The **plantar** surfaces of the feet (the soles of the feet) become painful and swollen with crusty areas of dried blood and exudates. If left untreated, pododermatitis may progress to necrosis of the foot.

The guinea pig with pododermatitis should be provided with a solid floor; soft, nonabrasive bedding; and meticulous husbandry practices. Affected areas should be carefully cleaned and treated as open wounds. Staphylococcus is the most commonly found bacteria in cases of pododermatitis. This condition can be prevented by providing the correct housing and with careful husbandry practices including frequent cage cleaning and providing ample fresh, dry bedding.

Parasites

Guinea pigs are often presented with alopecia and need to be examined carefully for mites and lice. Some hair losses can be attributed to barbering by cage mates, due to either boredom or, more likely, hierarchal reinforcement. Clinical signs that warrant further investigation with skin scrapings and microscopic examination include intense scratching, secondary skin infections, and lesions.

Trixacarus caviae is a zoonotic burrowing mite that causes scabies in guinea pigs and should be a primary rule-out in any dermatological problem. There are two species of host-specific guinea pig lice: *Gliricola porcelli* and *Gyropus ovalis*.

Transmission is through direct contact with infested guinea pigs or bedding. Guinea pigs with lice have usually been housed in poor conditions and suffer from other illnesses concurrently. Oral or injectable ivermectin can be given upon the recommendation of a veterinarian. The environment must be thoroughly cleaned and disinfected prior to the return of the guinea pig to eliminate the chance of reinfestation. Any wood items should be destroyed and the remaining bedding supply should be bagged, tightly sealed, and thrown out.

Dermatophytosis (ringworm) occurs frequently in guinea pigs, especially those housed in crowded conditions. While it is a fungal infection rather than a parasitic infection, transmission easily occurs through direct contact, contaminated bedding, and fomites. Lesions that are inflamed and hairless should always be carefully examined as ringworm has zoonotic potential. A tentative diagnosis should not be made with the use of a Woods lamp (black light) as not all types of dermatophytes fluoresce under this type of UVB light. A definitive diagnosis can be confirmed only with a DTM culture of the lesion and surrounding hairs. The environment needs to be completely decontaminated. Affected guinea pigs should not be returned to the group until lesions are healed. Disposable gloves should be worn to prevent human contact while applying a prescribed anti-fungal cream. Gloves should be removed so that they are turned inside-out and disposed of in a biohazard container.

Clinical Procedures

Guinea pigs have small, fragile blood vessels that easily collapse and allow only small volumes to be collected. The lateral saphenous and cephalic veins are the most commonly used for blood sampling. The cranial vena cava and jugular vein can be used as a collection site for larger blood volumes if the patient is sedated prior to collection. Using a tuberculin or insulin syringe with a 25-gauge needle, no more than 0.5–0.7 ml/100 gm bodyweight should be collected. Examination of the blood film should include recognition and recording of **Kurloff bodies**, which are cytoplasmic inclusions in some monocytes. They are normal in the guinea pig and may increase in number and become more obvious in a pregnant sow. The function of Kurloff bodies is unknown, but researchers speculate that they may give guinea pigs certain properties to resist cancer. Typically, guinea pigs have a relatively greater number of lymphocytes, compared to dogs or cats.

Peripheral catheters for fluid administration can be difficult because of the small, fragile vessels. With severely dehydrated patients, intraosseous catheters are often placed within the greater trochanter of the femur. Subcutaneous fluids can be given between the scapulas. The skin is tough and tight, with no scruff and very little subcutaneous space. An intrascapular fat pad makes the procedure more painful than in other small mammals and the patient will vocalize loudly. Care must be taken not to overhydrate with fluids and the patient must be carefully monitored. The recommended total volume is 50–100 ml/kg of bodyweight per day. The choice of fluids is decided by the veterinarian depending on blood chemistry values, but most frequently isotonic fluids (e.g., LRS, Normosol-R, or 0.9% NaCl) are administered.

Guinea pigs are extremely difficult to intubate because of their small, long, and narrow oral cavity. They can be masked down with either isoflurane or sevoflurane. Guinea pigs should not be fasted for more than two to four hours prior to general anesthesia. Though they lack the ability to vomit, regurgitation under anesthesia is common and may be frequent. As soon as a light plane of anesthesia is achieved, the oral cavity must be swabbed out, removing food debris from their cheeks and molars. This may be necessary several times during the procedure to prevent aspiration.

Recovery should take place in a quiet, undisturbed area with unobtrusive observation. Guinea pigs sometimes fail to recover due to dramatic metabolic changes and the stress of induction. This is especially true when attempting a cesarean delivery, where the primary focus is often to save the litter. Rapid induction decreases stress time. Guinea pigs with any medical concerns must be handled minimally because of the potentially catastrophic results of stress.

The urine of guinea pigs is typically yellow to amber in color. It may be darker in color due to the patient's diet. Hyperpigmentation (orange) may be mistaken for hematuria. A urinalysis will determine the pH of guinea pig urine, which should normally be alkaline (8.0–9.0), which is typical with herbivores. If the microscopic examination reveals **crystalluria** (crystals in urine), the guinea pig should be further examined for the presence of urinary calculi.

Lateral and ventral/dorsal whole-body radiographs are diagnostic views preferred by veterinarians treating guinea pigs. Carefully extending the limbs will help prevent the patient from rotating and creating an oblique view. The use of sedation in a guinea pig can reduce patient stress during positioning and increase the ability to obtain diagnostic radiographs.

Dental malocclusions and possible bone involvement can be diagnosed with the use of skull radiographs. Lateral oblique views in addition to a dorsoventral view can help localize lesions associated with malocclusions.

Ultrasonography is useful in diagnosing guinea pig diseases such as urinary calculi, ovarian cysts, and gastrointestinal disorders. Sedation may be required to obtain good positioning without stressing the patient.

Summary

Guinea pigs (cavies) make excellent small animal companions. They are social with each other and people, and when correctly housed, they are easy to care for. One of the most important things to remember about their diet is that they need a daily intake of vitamin C to prevent scurvy. Sows must have their first litter before reaching seven to nine months of age, when the pelvic midline begins to ossify preventing normal delivery.

While generally hardy, serious medical concerns can develop. Many bacterial infections are seen in guinea pigs, and treatment can be difficult because they do not tolerate most of the commonly used antibiotics. Guinea pigs should not be housed with rabbits because they are highly susceptible to *Bordetella bronchiseptica* and rabbits may be asymptomatic carriers. Bordetella causes severe respiratory infections in the guinea pig and the bacteria can cause stillbirths and abortions.

Submandibular abscesses are frequently seen in guinea pigs, a condition commonly called *lumpy jaw*.

Blood collection can be difficult due to the tiny and very fragile veins. Examination of blood films should include recognition of Kurloff bodies within the monocytes. Guinea pigs are masked down for anesthesia because they are extremely difficult to intubate. Though they lack the ability to vomit, regurgitation is common during general anesthesia, and the oral cavity should be swabbed out frequently during the procedure to prevent aspiration. Guinea pigs sometimes fail to recover from anesthesia due to the dramatic metabolic changes that occur. Recovery should take place in a quiet undisturbed area with unobtrusive observation.

fastFACTS

GUINEA PIGS

WEIGHT
- Boars: 900–1200 g
- Sows: 700–900 g

LIFE SPAN
- 5–7 years

REPRODUCTION
- **Sexual maturity:** boars: 9–10 weeks; sows: 4–6 weeks
- **Gestation:** 63 days (average)
- **Litter size:** 2–5 pups
- **Weaning age:** 3 weeks (21 days)

VITAL STATISTICS
- **Temperature:** 37.2–39.6°C (99–103.1°F)
- **Heart rate:** 230–280 beats/min
- **Respiratory rate:** 42–100 breaths/min

DENTAL
- Total 20 teeth
- Dental formula 2 (I 1/1, C 0/0, P 1/1, M 3/3)

ZOONOTIC POTENTIAL
- Bacterial
- Salmonella
- Pasteurella spp.
- Streptobacillus
- Staphylococcus
- Viral
- LCM—lymphocytic choriomeningitis (rare)
- Rabies (potential; highly unlikely due to cavy exposure)
- Parasitic
- Giardia spp. (protozoan)
- Coccidian spp. (protozoan)
- Fungal
- Dermatophytosis (ringworm)

Review Questions

1. What is the importance of vitamin C in a guinea pig diet?
2. Describe the signs of pregnancy toxemia.
3. Why should antibiotics be prescribed only by a veterinarian?
4. Rabbits should not be housed with guinea pigs. What is the reason?
5. What are Kurloff bodies?
6. Describe *lumpy jaw*. What is the correct name and what is the cause?
7. Why should a sow have her first litter before she is seven to nine months old?
8. What is *drilling*?
9. Which is better for guinea pigs, alfalfa or grass hay? Why?
10. Why is it necessary to check guinea pigs' water bottles frequently?

Case Study I

History: A two-year-old sow is brought in because she seems lethargic and the client has not seen her eating for several days. When the cavy attempts to move, she collapses and seems reluctant to get up again. The owner reports that she was rescued and lives compatibly with his pet rabbit. They are both fed rabbit pellets and some hay.

Physical Examination: The joints above both hocks are swollen. The patient has lost weight and has very little energy. An examination of her oral cavity reveals that there is evidence of bleeding around all four of her incisors.

a. With a review of the history and the diet offered to both the guinea pig and the rabbit, what would be the most obvious reason for this condition?

b. How could this problem have been prevented?

Case Study II

History: A six-month-old intact boar, recently purchased at a pet fair, is presented with a yellow discharge from his eyes and nares. The owner says it is refusing food and hardly ever sees him drink. The owner reports that he would return the guinea pig, but the people selling bunnies and guinea pigs are gone.

Physical Examination: The temperature is below the normal range and there is audible congestion when auscultating the lungs. The yellow discharge is thick and crusty around the eyes and nares.

a. Considering the history, what would be the most likely cause of this condition?

b. What clue was given in the history?

c. How is this disease spread?

For Further Reference

Capello, V. (2006). *Clinical Approach to the Anorectic Guinea Pig* (pp. 1697–1699). Conference Notes, Presented at the North America Veterinary Conference, Orlando, FL.

Hoefer, H.L. (2006). *Urolithiasis in Rabbits and Guinea Pigs.* Conference Notes, Presented at the North America Veterinary Conference, Orlando, FL.

http://www.vcahospitals.com (accessed November 15, 2013).

http://www.netvet.wustl.edu (accessed November 15, 2013).

http://www.ehs.uc.edu

http://www.guinealynx.info

Mayer, J. (2006). *Analgesia and Anesthesia in Rabbits and Rodents* (pp. 1740–1742). Conference Notes, The North America Veterinary Conference, Orlando, FL.

O'Malley, B. (2005). *Clinical Anatomy and Physiology of Exotic Species.* New York: Elsevier Saunders.

Quesenberry, K.E., Donnelly, T.M., & Mans, C. (1997). Biology, Husbandry and Clinical Techniques of Guinea Pigs and Chinchillas. In Hillyer, E.V. and Quesenberry, K.E. (Eds.), *Ferrets, Rabbits and Rodents: Clinical Medicine and Surgery, 3rd Edition* (pp. 279–294). Philadelphia, PA: W.B. Saunders Co.

Standard of Perfection. *Standard Bred Rabbits & Guinea Pigs.* American Rabbit Breeders Association, Inc. http://www.arba.net (accessed November 30, 2013).

CHINCHILLAS

KEY TERMS

ecosystems
mosaic
hair follicle
fur slip
urethral cone
pinna
pelters
choke
dry heaves

Introduction

South America is home to many diverse species in **ecosystems** as varied as the lowland scrubs and tropical rainforests to the great heights and freezing climate of the Andes Mountains. The Andes extend for more than 5000 miles and have many peaks higher than 22,000 feet. This rugged and stark region is the natural habitat of the chinchilla.

As far back as the sixteenth century, chinchilla pelts were used to decorate the ceremonial dress of the Chincha Indians. The word *chinchilla* literally means *little chincha*. In the 1800s, with the arrival of the Europeans, chinchillas were hunted to near extinction for their pelts. Wild populations did not recover from this mass exploitation. Today, their range is greatly diminished and they are rarely seen in the wild. Wild chinchillas are listed by CITES as an endangered species. Their current status in the wild is unknown, but there are reported colonies of *Chinchilla laniger* in the mountains of northern Chile. Attempts to reintroduce chinchilla into the wild have failed, but the Chinchilla National Preserve in Chile provides protection for the remaining known colonies.

All domestically bred chinchillas are descendants of 11 animals that were captured and brought to the United States in the 1930s. An American, Mathias F. Chapman, lived in the Andes of Peru and Chile, where he became interested in the financial reward of chinchilla pelts. He reported that he lived at an altitude of 11,300 feet and that the trappers went a mile higher to capture the chinchillas. Because their numbers were already so depleted, locating this elusive little animal was not easy. They were so few in number that, in three years of searching, 23 men brought Chapman less than a dozen chinchillas. For two years, he housed his small collection at an altitude of 11,000 feet before trying to acclimatize them to lower altitudes and eventually to the extreme heat and humidity of sea level. He stopped again at an altitude of 8000 feet, where he remained for a year before completing the descent down the mountains. Sometime during captivity, one chinchilla gave birth. Despite being under the cooperative protection of four South American countries, Chapman somehow managed to obtain an export permit. These 12 chinchillas became known as the *Chapman chinchillas* and were the first captive breeding *herd* of chinchillas in the world.

While there are still farms where they are bred for their pelts and for medical research, chinchillas have firmly established their place, and possibly their survival as a species, by becoming popular as companion animals.

The natural color of a chinchilla is bluish-gray with lighter underbelly fur, usually white or cream (**FIGURE 7-1**). Over time, many color mutations have appeared and chinchillas have been selectively bred to produce many varieties, even some with patterned coats described as **mosaic**. Some of the colors now available are beige, violet, white, silver, and the much sought after black velvet. Other than cost, there is no difference; this appealing and gentle rodent is still *C. laniger*.

FIGURE 7-1 The natural coat color of a chinchilla is bluish-gray with a lighter underbelly. Many color mutations are now available.

Chinchillas are small, compact animals, but their dense fur makes them appear bigger and rounder. The fur is very soft with as many as 60 hairs growing from one hair follicle. Chinchillas have large, rounded ears; long whiskers; and big, dark eyes. They have a brush tail, covered with hair that is only slightly coarser than that on the body. Chinchillas have four toes on each foot with nail pads that never need trimming. They weigh from 400 to 500 g (approximately 1 lb). Females are slightly larger than males. The average life span is 14 to 15 years, but some have lived as long as 20 years.

Behavior

Chinchillas are very curious animals, a trait that may have led to their ease of capture. At the first sign of danger, they scamper away, yet within a few moments they return to the area or situation that caused them to leave. They have very long hind feet and are capable of jumping up and over items many times their own length. Chinchillas are very quick and agile. Their long tail acts as a rudder as they leap from place to place. Though not strictly nocturnal, chinchillas are more active at dusk and during the night.

Chinchillas have little means of defending themselves. One mechanism that helps them to escape capture is known as fur slip. If grabbed or frightened, they are able to release patches of fur. It may take six weeks or more for the fur to regrow and completely cover the area of fur loss.

Females can become cage territorial, especially when they have young kits. They stand upright on their hind legs, spitting and barking. If this fails to deter an intruder, they squirt urine (with great accuracy) at the perceived danger, be it another household pet or a person. Even if provoked to this point, they rarely bite.

Chinchillas are seasonally polyestrous. The female breeding cycle is every 28 to 35 days. Pairs may be kept together for months or even years before they breed. Many owners have been surprised when a pair that has been together for some time produces a completely unexpected litter. Chinchillas are not nest builders, and there may be no indication that the female is pregnant. Gestation is an average of 111 days. Parturition is very short and usually occurs in the morning. It may last only 20 to 30 minutes for each kit.

Chinchilla kits are precocial. At birth, they are fully furred with their eyes open and teeth erupted. They are active within minutes and nibble on solid food within a few hours of birth (**FIGURE 7-2**). Both parents tend to the young and there is no need to remove the male; however, the female may be rebred within three days postpartum. The average litter size is two, but a litter of three is not that unusual, with one kit being much smaller than the other two. Many of these smaller kits do not survive. Kits are fully weaned between 8 and 10 weeks.

Chinchillas are coprophagic and produce two types of fecal pellets. They are not encased in a mucous membrane, like those of the rabbit, but they are produced and eaten during the night. Young chinchillas eat the fecal droppings of adults, contributing to the healthy gut flora of the kits.

Sexual maturity is variable and can develop anywhere from 4 to 12 months. Both sexes have a urethral cone, which may be mistaken, in the female, for a penis (**FIGURE 7-3**). Male and female may at first appear to be similar, but the anogenital

FIGURE 7-2 Chinchilla kits are born precocial. This kit is one day old.

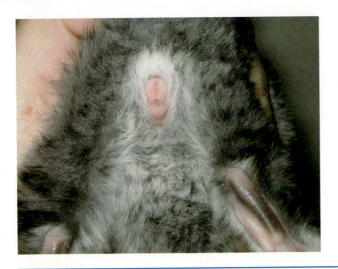

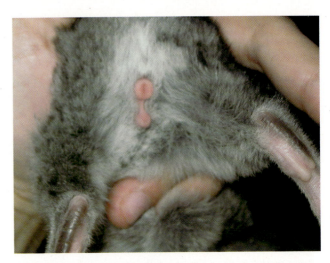

FIGURE 7-3 Chinchillas of both sexes have a urethral cone. Sex determination is confirmed by the anogenital distance, which is greater in the male. *Courtesy of Tabitha Lindsey*

distance is longer in the male. Males do not have a true scrotum. The testes are contained within the inguinal canal in paired sacs next to the anus. Females have three pairs of mammary glands, but these are difficult to visualize unless she is nursing kits.

Housing

Chinchillas are active and need a cage large enough for them to be able to exercise. It should have a solid floor and a hide box and different levels with platforms for jumping. Any connecting ladders to the different levels need to be removed to prevent the chinchilla from becoming entangled and possibly fracturing a limb. Some items used for exercise in other small mammals can be a hazard for chinchillas. They will use large wheels but can become trapped in the side braces. Fracture and potentially the amputation of a hind leg are the most common injuries caused by these items (**FIGURE 7-4**). Because of the normally hunched sitting posture of the chinchilla, fractures and injuries to the rear legs often go unrecognized by the owner until the leg has become infected and odorous. Giant exercise balls are not recommended. Chinchillas hop and *bounce* from place to place. Being enclosed in a rolling ball, able only to run, can cause the chinchilla to quickly become stressed and overheated.

Supervised out-of-cage play time can be very rewarding for both the chinchilla and the owner. A safe place with multiple levels is the bathroom. There are usually no exposed electrical cords and the only hazard to a chinchilla on rebound is the toilet. If using this room for exercise, *always* ensure that the lid of the toilet seat is down. Chinchillas do not swim!

The cage should be placed in an area that is cool and dry and receives no direct sunlight. Chinchillas are most comfortable in temperatures between 65°F and 75°F. The cage should not be exposed on all sides but kept with one side against a wall, giving the chinchilla a place to retreat and feel protected when not in the hide box.

Because of their dense fur coat, chinchillas need to be given access to a dust bath to keep the fur clean. The dust should be about 1 inch deep and be provided

FIGURE 7-4 The fractured leg of a chinchilla that was caught in an exercise wheel. *Courtesy of Eric Klaphake, DVM*

in a shallow pan or bowl. Chinchilla dust is very fine powdery pumice. A good way to provide chinchillas with a dust bath is to place a one-gallon, flat-sided fish bowl in the cage flat side down. It provides ample room for their rolling antics, running in and out, and keeps the dust somewhat contained. It is not necessary to leave the dust in the cage at all times; offering it two or three times a week is sufficient. Dust can be purchased from most pet stores and the standard gray color is readily available. With the color varieties now seen in chinchillas, color-specific dust can be ordered (**FIGURE 7-5**).

FIGURE 7-5 A chinchilla enjoying a dust bath. *©H Schmidbauer/Blickwinkel/ age fotostock*

The bottom of the cage can be layered with plain newspaper or aspen shavings. Crock pottery dishes are suitable to use for pellets and treats as they are easily cleaned and resist being chewed. Chinchillas need a water bottle that is cleaned, disinfected, and refreshed on a daily basis.

Diet

Chinchillas are herbivorous and should be fed a quality, high-fiber diet consisting of formulated pellets and free choice timothy hay. They eat mainly at night, ingesting 70 percent of their daily diet in the dark. Pellets made for rabbits and guinea pigs are usually not as high in protein as those specifically made for chinchillas. Another important factor is the length of chinchilla pellets. Chinchillas do not eat directly from a food bowl, but pick up and hold individual food items. Chinchilla pellets are longer so that they are easier for the chinchilla to hold. A small amount of unsalted *trail mix* with unshelled nuts and dried fruit can be offered as treat food, but the amount needs to be carefully controlled to prevent obesity and selective feeding. The mix should not contain any chocolate. Raisins are a known favorite, and these should be limited to two per day. Unsweetened dry cereal can also be given for variety. Sunflower seeds should not be fed as they are very high in fat and the shells can cause oral trauma and intestinal blockage. New foods and food changes should not be abrupt as dramatic changes in diet can cause gastrointestinal upset, diarrhea, or, more commonly, constipation.

Handling and Restraint

Chinchillas need to be handled gently. When removing a chinchilla from its cage, place one hand under the abdomen and let the body rest in the palm of your hand. Wrap the index finger of your other hand around the base of the tail. In this way, if the chinchilla tries to jump away, it is safely restrained until it relaxes and can be safely released for play time (**FIGURE 7-6**).

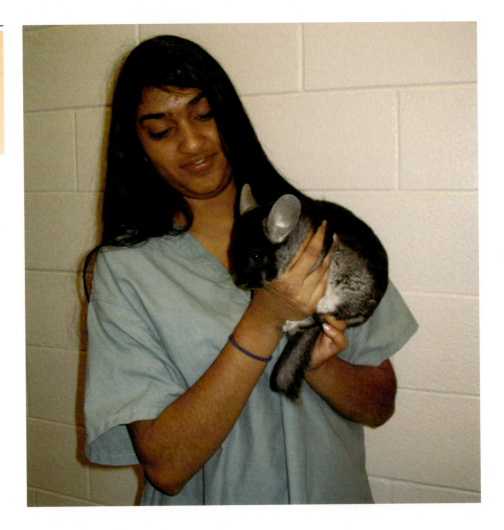

FIGURE 7-6 The correct way to hold a chinchilla; one hand supports the body around the thorax and the other hand supports the rump, with one finger wrapped around the tail.

In some circumstances, it is also appropriate to lift the chinchilla by the tail and place the body in the palm of the other hand. This technique can be used with chinchillas that are not as socialized or may be frightened by the strange surroundings of a clinic (**FIGURE 7-7**). While it may appear to be a bit harsh, it causes no harm to the chinchilla and avoids the potential of fur slip and returning the patient to its owner with bald patches. No attempt should ever be made to lift or restrain a chinchilla by the ears. This inhumane practice causes damage to the fragile vessels of the **pinna** (outer part of the ear), but it is a common practice with **pelters** (fur farmers) to avoid damaging the product, the fur pelt.

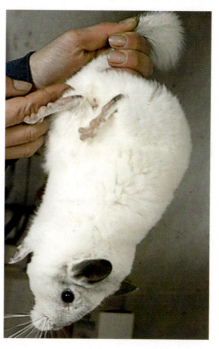

FIGURE 7-7 Chinchillas that are not used to being handled or may be frightened may be held by the tail for a short period until placed on the table for restraint.

Never scruff or *make a grab* for a chinchilla; it will result in nothing but a handful of hair and a chinchilla with a large bald spot (fur slip). A secure method of holding a chinchilla for an examination is to place one hand around the thorax from underneath and maintain the *finger wrap* around the tail. The examination table should have a towel placed on the top to make it less slippery (**FIGURE 7-8**).

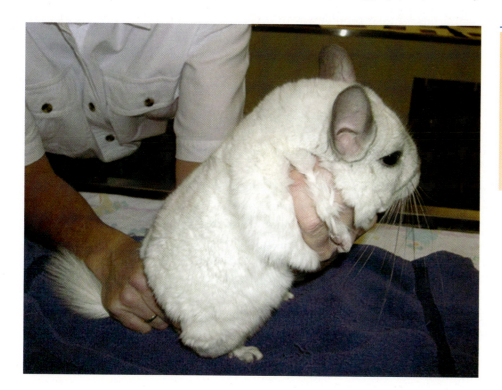

FIGURE 7-8 When restraining a chinchilla for an examination, the table should be covered with a towel, making the surface less slippery. One hand supports the thorax from underneath. The other hand securely grasps the tail.

Medical Concerns

Chinchillas that are housed and fed appropriately are usually hardy, but as with all species, they can develop medical problems unique to them. Many concerns are difficult to detect immediately due to the density of the coat, for example, weight loss and pregnancy. Chinchillas are adept at hiding illness and should routinely have an examination by a veterinarian to detect problems early. Careful observation of the chinchilla's normal behavior and activity level will help to determine when something is abnormal.

Gastrointestinal Disorders

Chinchillas have a relatively long digestive tract and, like many rodents, they are unable to vomit. They are, however, susceptible to **choke.** It may appear that the chinchilla is trying to vomit, but choke is an esophageal blockage. Signs of choke include retching, **dry heaves** (nonproductive attempts to vomit), excessive salivation, pawing at the mouth, and difficulty breathing. Choke can be caused by pieces of food (e.g., raisins, nuts) or a hairball. A chinchilla with choke, or suspected choke, should be seen by a veterinarian immediately, as this condition can obstruct the airway.

Chinchilla teeth, like those of all rodents, are hypsodontic in that they grow continually. If the diet and chew items are inadequate to wear the teeth down normally, dental problems will develop. The incisor teeth can easily be examined for overgrowth and trimming by gently pulling back the lips; however, examination of molar teeth is very difficult without anesthesia due to the very small oral cavity and the concerns of restraint. Diagnoses of molar teeth health or problems are confirmed with radiographs. Under anesthesia, the veterinarian can assess and correct many molar problems with the use of the rodent dental pack. Signs of dental problems include loss of appetite, dropping food from the mouth, excess salivation (drooling), swelling, and abscesses. There may also be a discharge from the eyes and nares, especially with molar teeth overgrowth and tooth infection. The chinchilla loses weight and condition because of its inability to eat. Chinchillas exhibiting any of these signs should have a full dental examination by the veterinarian.

Normal chinchilla droppings look like grains of rice and may be either black or dark brown, depending on diet and water consumption. An incorrect diet will cause either diarrhea or constipation. Constipation is seen more frequently and is usually the result of not enough fresh food or a combination of lack of exercise and obesity. Adding a small amount of fresh fruits, gradually adjusting the diet, and providing for ample exercise can help with constipation. Diarrhea, too, is usually caused by dietary changes either by feeding too much fresh food or by a sudden change in diet. Stress and excessive environmental temperatures can also cause diarrhea.

Gastric ulcers have occurred in young chinchillas due to feeding course, fibrous roughage, or moldy feeds. Kits become anorexic and lethargic, and the result is usually premature death. Correct dietary management is the best prevention and treatment.

Enterotoxemia is seen in chinchillas that are in distress. The patient may present with lethargy, recumbency, diarrhea, fever, and respiratory distress. With the acute form they may die without showing clinical signs. *Clostridium perfringens* is commonly associated with this disease when the intestinal microflora is altered. The treatment is supportive therapy including fluids and caloric support.

Enteritis is a common disorder in chinchillas and can occur with diet changes, improper antibiotic use, stress, protozoans, and opportunistic bacteria. Clinical signs include diarrhea, abdominal pain, rectal prolapse, decreased feces, and possible intussusception. Treatment includes identifying the cause and treating the condition.

Nondigestive Disorders

Because of the dust bath, chinchillas can develop conjunctivitis from irritation caused by dust particles. The chinchilla may keep one or both eyes closed, and there may be a sticky ocular discharge that prevents the eye from opening. The eye can be flushed with sterile saline. Dust baths should be limited until the inflammation is gone. A veterinarian should examine the eye for any surface abrasions.

Generally, bacterial diseases in pet chinchillas are not common. *Fur-ranched* chinchillas are likely to suffer more from bacterial infections due to stress, overcrowded conditions, and less-than-desirable hygiene practices. Any suspect bacterial infection should be cultured. Bacterial infections include *Yersinia* spp., *Listeria*, and *Pseudomonas*.

Male chinchillas can accumulate a tightly woven ring around the urethral cone and penis. Eventually, as the hair builds up and the ring becomes tighter, the chinchilla will be unable to retract the penis. It can become swollen due to a tourniquet effect of the hair ring, which cuts off blood supply and causes the penis to become necrotic, and the area should be examined regularly. The fur ring can be removed with the use of a sterile lubricant and then gently rolled off the penis. Never use scissors in an attempt to cut away the hair ring. Should the chinchilla move, it could result in inadvertently nicking or even possibly amputating the penis.

Female chinchillas, especially those that are obese, may be prone to vaginal prolapse. It may occur during delivery, postpartum, or in a female that is not pregnant. It is usually attributed to stress and a lack of exercise (**FIGURE 7-9**). This condition requires immediate veterinary care and probable surgical treatment to repair the prolapse.

Chinchillas are native to cooler regions of South America, and do not tolerate high heat and humidity. Animals exposed to excess heat, higher than 80°F, can develop life-threatening heatstroke. Clinical signs include panting, lethargy, and excessive salivation. Treating this condition involves lowering the core body temperature in the patient. This can be done by administering cool water baths, applying alcohol to the feet and ears, and providing fluid therapy (subcutaneous or intravenous).

Parasites

Giardia is common in many chinchilla colonies, and even healthy chinchillas are likely to have giardia in low numbers. Stressful conditions can cause a dramatic increase in the numbers of giardia cysts and become problematic. It causes diarrhea and intestinal upset and is transmitted by the fecal/oral route

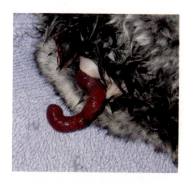

FIGURE 7-9 A female chinchilla with a vaginal prolapse. This condition requires immediate veterinary care. *Courtesy of Eric Klaphake, DVM*

and through contaminated water. Veterinary treatment includes a fecal examination and treatment if high numbers of giardia cysts are seen under microscopic examination. Metronidazole has usually been the drug of choice for giardia; however, there are recent reports of this drug causing liver failure in chinchillas, and the veterinarian may decide to use an alternative drug. Giardia has zoonotic potential; veterinary staff and handlers should wear disposable gloves when cage cleaning. Water and food bowls should be washed separately and disinfected.

External parasites are not a concern with chinchillas because the fur is so dense it creates a natural barrier to blood-feeding parasites.

Clinical Procedures

Medications given orally are preferable due to the restraint required for injections, the pain involved, and the potential of fur slip. If there is no alternative, the quadriceps of a rear leg can be used. A 25-gauge needle, as small as it is, will still cause pain and potential necrosis at the injection site.

One way to help minimize fur slip is to wet the site with warm water. A soaked cotton ball, squeezed over the site, will *wet down* the fur and separate the hairs and provide easier visualization of the injection site.

Veins are difficult to access due to their small size and the restraint required. The lateral saphenous vein is usually a good choice for blood sampling. A 25-gauge needle and a tuberculin syringe are recommended. The volume should not exceed 0.5 ml/100 g bodyweight. If larger volumes are needed, sedating the patient will be required. The jugular, cranial vena cava, and the femoral vein are sites used for larger blood volumes.

Peripheral catheters are equally difficult to place. A small bolus (3–4 ml) of fluids may be given SQ between the shoulders, but long-term fluid therapy requires placement of an intraosseous catheter into the proximal end of the femur. For accurate placement, and pain avoidance, the chinchilla will need to be anesthetized.

The intraperitoneal cavity can be used as an alternative site for fluid administration in a moderately to severely dehydrated chinchilla. Large boluses of fluids can be administered by this route, but care needs to be taken to avoid ascites from forming in the body cavity.

Chinchillas can be masked down with either isoflurane or sevoflurane. They are not generally given a preanesthetic due to the potential of muscle necrosis at the injection site. Prior to anesthesia, the cheeks should be emptied of food to prevent aspiration. Food should not be withheld for more than two to four hours prior to anesthesia as hypoglycemia could result. Chinchillas, like other small exotic mammals, are difficult to intubate because of their long, narrow oral cavity. A 2.7 mm endoscope or a small laryngoscope can assist with the visualization of the airway and placement of the endotracheal tube. A 2.0–4.0 mm endotracheal tube is recommended in chinchillas. Anesthesia can be maintained in chinchillas with isoflurane between 1.5 percent and

3 percent or sevoflurane between 3.0 percent and 5.0 percent at an oxygen flow rate between 0.8 L and 1 L/min.

Careful monitoring during anesthesia is important. Like most small mammals, chinchillas are prone to hypothermia. Supplemental thermal heat in the form of water-circulating heating pads should be provided. During recovery, the chinchilla should be wrapped in a towel and placed in a small animal incubator to prevent it from trying to bolt before fully recovering.

A urinalysis can be a helpful diagnostic tool for urinary tract disease in chinchillas. Urine samples can be collected through cystocentesis, free catch, or floor catch. The pH of the urine is alkaline (8.0–9.0) due to the herbivore diet. The color is yellow to amber, although it can be darker and more orange depending on the diet. Pigments such as porphyrin can be mistaken as blood in the urine.

Whole-body radiographs are generally taken to assist in the evaluation of a chinchilla. Care needs to be taken when extending the limbs as to not to rotate the patient or cause injury to the small limbs. Sedation may be required to obtain correct positioning. Skull radiographs are commonly taken on the chinchilla due to the incidence of dental malocclusions. In addition to the lateral and ventrodorsal views, right and left lateral obliques may be required to visualize lesions.

Ultrasound can be used in chinchillas to help diagnose various diseases or assist with obtaining laboratory samples (e.g., cystocentesis, biopsies). Computed tomography (CT) and magnetic resonance imaging (MRI) are useful in diagnosing dental problems and other medical issues. Cost and availability of the equipment limits the use of these as a diagnostic tool.

Summary

Chinchillas are native to South America but are now endangered in the wild. They were exploited to near extinction because of their luxurious pelts. Today they have become popular companion animals. Chinchillas are very curious and gentle and have the ability to jump many times higher than their own length. They are very quick and agile and an attempt to capture an escaped animal will result in fur slip, the willful release of hair. Chinchillas have little in the way of defense; they rarely bite and have no claws, only nail pads. One method of defense is to stand on their rear legs and urinate on the perceived threat. Chinchilla kits are born precocial. Both sexes have a urethral cone. Chinchillas need a large cage with room to jump and exercise and be provided with a hide box and a dust bath to clean the dense fur. They should be provided with a high-fiber diet consisting of timothy hay and formulated pellets. While generally hardy, they are susceptible to choke and conjunctivitis from particles of dust. Male chinchillas can accumulate a tightly woven ring of hair around the urethral cone and penis. Giardia is a common intestinal parasite and it has zoonotic potential.

*fast*FACTS

CHINCHILLAS

WEIGHT
- 400–500 g (approximately 1 lb)

LIFE SPAN
- 15 years average (180 months)

REPRODUCTION
- Sexual maturity: 4–12 months
- Gestation: 111 days
- Litter size: 1–3 (2 average)
- Weaning age: 6 weeks

VITAL STATISTICS
- Temperature: 37–38°C (98.6–100.4°F)
- Heart rate: 120–160 beats/min
- Respiratory rate: 50–60 breaths/min

DENTAL
- Total 20 teeth
- Dental formula 2 (I 1/1, C 0/0, P 1/1, M 3/3)

ZOONOTIC POTENTIAL
Bacterial
- Salmonella
Parasitic
- Giardia

Review Questions

1. Explain what is meant by *fur slip*.
2. When chinchilla kits are born, they are precocial. Describe the appearance of a newborn chinchilla.
3. Why is it important to provide a chinchilla with a dust bath?
4. What factors contribute to constipation in a chinchilla?
5. What problem could develop for a male chinchilla?
6. What is the difference between rabbit pellets and chinchilla pellets? Why are they different?
7. How long is the gestation period for chinchillas?
8. Describe the correct setup of a chinchilla's habitat.
9. What methods are used for chinchilla restraint?
10. How do chinchillas defend themselves?

Case Study I

History: A nine-year-old intact male chinchilla is presented because the owner has noticed it appears to have difficulty urinating and the area around the lower abdomen is always wet.

Physical Examination: On palpation, the bladder is full. The penis is visible and swollen and the chinchilla is unable to retract its penis. The wet fur on the abdomen is most likely caused by the patient's excessive grooming of the area.

a. What else is likely to be evident?

b. What is the direct cause of this problem?

c. How will this case be resolved?

Case Study II

History: A two-year-old female chinchilla has been brought in because the owner has noticed that there are very few droppings in the cage. What few fecal pellets there are seem normal and he reports that the chinchilla's appetite is good, or perhaps slightly less than normal, but the chinchilla just is not very active. When questioned further, the owner reports that the diet consists of iceberg lettuce, free choice pellets, and a raw carrot daily.

Physical Examination: The chinchilla is bloated and reluctant to move. Upon palpation, the veterinarian suspects a fecal impaction and this is confirmed with a radiograph.

a. What are the likely causes of the chinchilla's problem?

b. What are the ways to prevent this condition from occurring?

For Further Reference

C.I.T.E.S. Appendices I, II, and III, valid from June 14, 2006. http://www.cites.org (accessed August 10, 2006).

http://www.animaldiversity.ummz.edu (accessed November 30, 2013).

http://www.vetdent.edu (accessed May 24, 2014).

Mans, C., & Donnelly, T.M. (2012). Disease Problems of Chinchillas. In Quesenberry, K.E. and Carpenter, J.W. (Eds.), *Ferrets, Rabbits and Rodents Clinical Medicine and Surgery, 3rd Edition.* (pp. 311–325). St. Louis, MO: Elsevier Saunders.

Quesenberry, K.E., Donnelly, T.M., & Hillyer, E.V. (2004). Biology, Husbandry and Clinical Techniques of Guinea Pigs and Chinchillas. In Quesenberry, K.E. and Carpenter, J.W. (Eds.), *Ferrets, Rabbits and Rodents Clinical Medicine and Surgery, 3rd Edition* (pp. 232–244). St. Louis, MO: Imprint of Elsevier Science.

HEDGEHOGS

OBJECTIVES

After completing the chapter, the student should be able to

- Provide correct housing for a hedgehog.
- Advise new hedgehog owners regarding hedgehog behavior, habitat, and diet.
- Describe the correct diet for a hedgehog.
- Provide basic nursing care to a hedgehog.
- Assist in the anesthesia of a hedgehog.
- Demonstrate appropriate restraint techniques for hedgehogs.
- Understand common medical disorders in hedgehogs.

Introduction

From the time Beatrix Potter first introduced Mrs. Tiggy-Winkle in *The Tales of Peter Rabbit*, hedgehogs have held an endearing and special affection in the hearts of many people. Later, the cartoon strip *Sonic* captured a whole new generation and spawned a multitude of toys, ornaments, video games, and, unfortunately, a new *fad*, the desire to own a pet hedgehog.

Many people who acquired hedgehogs were soon confronted with reality; the spiny, hissing, nocturnal, antisocial little animals with some very odd behaviors were nothing like Mrs. Tiggy-Winkle or Sonic, and as a consequence, many pet hedgehogs were abandoned.

Laws were enacted by many cities and states prohibiting the breeding, sale, and keeping of hedgehogs. There was concern over the possibility of introducing anthrax (the disease caused by the bacteria *Bacillus anthracis*) with imported African hedgehogs and that hedgehogs, should they escape or be released, would establish a feral population in many areas. The importation of African species has been banned. It is an owner's responsibility to determine the legality of any animal, to comply with the law, obtain required permits, and thoroughly understand the specific needs and behaviors of the species.

The hedgehog has changed little in the past 15 million years. It belongs to the class of insectivores, *Erinaceidae*. Paleontology (fossil evidence) indicates that insectivores were the first placental mammals. Two species of hedgehog are generally recognized: the European hedgehog and the African hedgehog. These are further divided into subspecies. Due to their smaller size, the African hedgehog has become known as the African pygmy hedgehog, the *pet* hedgehog. The taxonomic name is *Atelerix albiventris*, and they are also known as the white-bellied or four-toed hedgehog. They are native to the savannahs of central and east Africa. There are no hedgehogs native to North America. Contrary to some popular notions, hedgehogs are not related to porcupines (rodents) or echidnas (monotremes).

Behavior

Hedgehogs are solitary and nocturnal. During the day, they can be found under logs, bushes, and piles of wood. They do not make burrows but dig out a shallow spot in which to sleep or give birth. They are not strictly territorial but use preferred pathways or routes familiar to them and travel far while foraging. They are intolerant of each other, accepting the presence of another hedgehog only during brief mating encounters.

When threatened, hedgehogs curl up into a tight ball to protect themselves. They are covered in very sharp spines, and when curled, their heads and soft, haired underbelly are completely protected. The spines are made of keratin, the same protein that forms nails and horny tissue. Hedgehog spines are continually shed and replaced. Predators, confronted with a very prickly object, soon wander off, leaving the hedgehog alone. To further dissuade would-be predators, a curled hedgehog can make a variety of alarming noises, startling the unsuspecting. Hedgehogs hiss and make a popping or clicking sound when annoyed or threatened. The whole body, while still curled, seems to vibrate and snap. Spines do not come out when touched, nor do hedgehogs have the ability to throw their spines. Spines do not contain toxins, but many people are sensitive to them and develop a rash when spines come in contact with bare skin. Screaming from a hedgehog is an extreme distress call or indicative that the animal is in pain. Clucking noises are high-pitched sounds from the dam when calling her young or during courtship. Hedgehogs can make and hear sounds above the range of humans, 40–90 kHz.

A behavior unique to the hedgehog is anointing or *anting*. When confronted with an item that has a new smell or taste, hedgehogs lick the item and begin to hypersalivate, producing a white, frothy spit. They use their tongues to paint, or anoint, their spines with the foam produced. The reason for this behavior is uncertain, but it may have significance in attracting a mate or in disguising the hedgehog's own scent. Any new smell can produce this behavior, including hand lotion and food items. When handled from a very young age, they become accepting, rather than social, and may relax enough to uncurl and explore their surroundings. Their eyesight is poor with little depth perception. Hedgehogs are frequently injured when they fall from the edge of a table or counter top (**FIGURE 8-1**).

Hedgehogs can breed throughout the year. They are sexually mature at approximately eight weeks. Sexing hedgehogs is not difficult once the genitalia is visualized. The male's penis is mid-abdomen, near the umbilicus (the navel) (**FIGURE 8-2**).

The vulva (external genitalia of a female) is close to the anus (**FIGURE 8-3**). The boar should be taken to the sow and promptly removed after breeding.

FIGURE 8-1 An albino hedgehog that accepts being held by his owner.

Penis

FIGURE 8-2 The external genitalia of a male (boar) hedgehog. The penis is located mid-abdomen, near the umbilicus.

Vulva

FIGURE 8-3 The external genitalia of a female (sow) hedgehog. The vulva is located close to the anus.

Mating is brief, but it may be repeated several times over the course of 15 or 20 minutes. If mating does not occur or if the animals are aggressive with one another, they should not be left together but be paired again on another night.

Gestation is 34–37 days, producing an average of three to four young. **Hoglets** (young ones) are born hairless, and the spines are barely visible beneath a membrane that protects the sow during delivery. Their eyes and ears are closed. Within six hours after birth, the spines begin to protrude as the membrane shrinks away. Newborn hoglets are approximately the size of a nickel and grow very quickly (**FIGURE 8-4**).

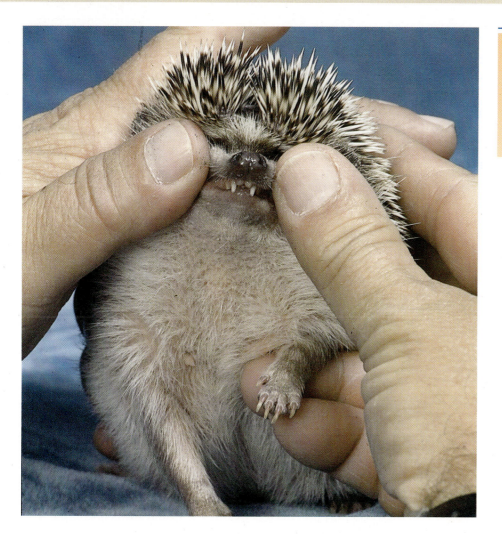

FIGURE 8-9 A cursory oral examination showing the placement of the top incisors. The space between the teeth is normal for this insectivore/omnivore. *Photo by Isabelle Francis*

The feet and limbs need to be checked carefully for threads and human hair that easily become wound around the limbs and toes, cutting off circulation. Many owners give their pets blankets or cloth to sleep under and human hair is easily collected from the carpet or directly from a handler. This is also a good time to trim the nails.

During the examination, other abnormalities may also be noted: potential tumors, areas of discoloration, or conditions that require further investigation. This is also the opportunity to obtain blood samples and radiographs, if necessary. Radiographs also provide information on fat density (**FIGURE 8-10**).

There is a high incidence of neoplasia (abnormal growth) in pet hedgehogs. All body organs are affected, and more than 75 percent of the tumors are diagnosed as malignant (cancerous) and may spread to other organs and result in death. Tumor growth may be related to age, as most are seen in hedgehogs three years old or older. Aside from a prevalence of mammary gland tumors in sows, the development of tumors seems to be unrelated to gender.

Tumors may affect the lymphatic system and all parts of the digestive tract, including the oral cavity (**FIGURE 8-11**). General signs related to tumor growth include weight loss, lethargy, and ascites. Any abnormality or change in behavior warrants early medical investigation.

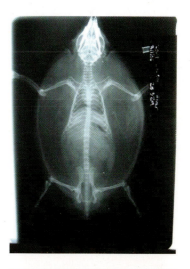

FIGURE 8-10 A radiograph reveals the fat density in an obese hedgehog. *Courtesy of Brandi Orr, Bird and Exotic All-Pets Hospital*

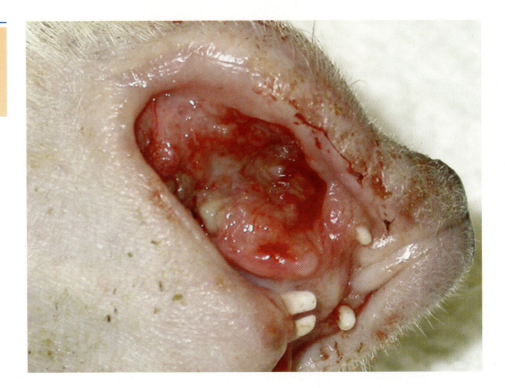

FIGURE 8-11 A large, rapidly growing oral mass prevented this hedgehog from eating. *Courtesy of Eric Klaphake, DVM*

Demyelinating paralysis, commonly referred to as *wobbly hedgehog syndrome*, is a degenerative, neurological disease and, as the name suggests, causes ataxia, weakness, and, in the later stages, seizures. The hind limbs are affected first, becoming weaker with ensuing paralysis, as the disease progresses to the front limbs. The exact cause is not known, but it is thought that the condition may be inherited. There is no effective treatment, and death is inevitable. The quality of life deteriorates as the hedgehog becomes less and less able to ambulate, unable to reach food and water. In the interest of the hedgehog, regardless of the nursing care an owner is prepared to provide, humane euthanasia should be considered.

Corneal ulcers and various ocular injuries are common in hedgehogs. Their appearance and signs are similar to those in other species. Treatment for these problems are also the same, but medicating the eye can be difficult in the hedgehog because of the hedgehog's reluctance to uncurl to expose the eyes or curling up when treatment is attempted.

Contact dermatitis is a common problem with hedgehogs kept in unsanitary bedding. Pruritus occurs with the growth of new and developing spines. Loss of spines, flaking skin, and moist erythema are other signs that can occur with contact dermatitis. Steroid injections along with addressing husbandry concerns have shown to be effective in treating the condition.

Dilated cardiomyopathy occurs in hedgehogs older than three years of age, although it has been seen as early as one year of age. Clinical signs include heart murmur, dyspnea, dehydration, anorexia, lethargy, and pulmonary edema. Radiographic evaluation will confirm the diagnosis. If the disease has progressed to congestive heart failure or renal disease, the prognosis is guarded.

Parasites

Mites are a frequently seen external parasite of hedgehogs. Signs include lethargy, loose spines and spine loss, decreased appetite, and hyperkeratosis. There is also a crusty appearance at the base of the spines and around the eyes due to mite droppings. Mites are blood-feeders and a severe infestation can cause anemia. Mites can be diagnosed with a skin scraping and microscopic examination of an oil prep slide. Treatment includes injectable invermectin and discarding the bedding and all cage furnishing as mites can come from bedding material and natural wood hide boxes. A complete change of habitat is necessary to prevent the mites from reinfesting the hedgehog.

Dermatophytosis (ringworm) in hedgehogs causes crusting around the face and ears. Diagnosis is confirmed by placing a skin scraping in DTM culture media. Treatment with topical antifungal products are effective but can be problematic due to the anointing behavior of the hedgehog.

Clinical Procedures

Hedgehogs can be anesthetized by delivering the gas agent either with the use of an induction chamber or by placing a large dog face mask over the patient's entire body. Using a mask is usually more effective as the area is smaller, but it offers less room so hedgehogs will not be unable to unroll completely. When a plane of relaxation is reached and the mouth and nose are accessible, transfer the elbow of the nonrebreathing unit directly over the mouth and nose. This is a quick and simple way of providing an appropriately sized mask. A 2 mm or less endotracheal tube or a 14 gauge over the needle catheter with the stylet removed can be used to intubate the hedgehog. A rigid endoscope can assist with the visualization of the glottis. Isoflurane and sevoflurane are both used successfully with hedgehogs. To prevent hypothermia, common with hedgehogs under anesthesia, they should be placed on a covered heating pad. Recovery should take place in an incubator, prewarmed to 75–80°F.

Subcutaneous injections should be given on the right side below the spine layer. Injections given on the right side decrease the chance of inadvertently penetrating the spleen or kidney in a tightly rolled hedgehog. The right kidney is more cranial than the left. If the injection is given in the spine layer, drug absorption is delayed because of a thick fat layer that has little blood supply. Intramuscular injections are given in the quadriceps of a hind leg. Both injections can be problematic with a rolled-up patient. Oral medications can be equally difficult. Medications can be mixed with a favorite food or juice (grape juice) or injected directly into a prey item (a mealworm or a wax worm) immediately prior to feeding it to the hedgehog. Use caution when choosing flavors to mix with medication. Some of the flavors may cause hypersalivation, resulting in the anointing behavior with the medication wiped on the spines.

The most common site for blood collection is the jugular vein. Hedgehogs have very short legs, making venipuncture difficult from peripheral sites.

Visualizing the jugular is difficult, but anatomically it is located in the same region as in any other small mammal. A tuberculin syringe with a 25-gauge needle will help prevent laceration or collapse of the vein. Preheparinizing (filling the 25-gauge needle and part of the syringe with heparin and then squirting it back out of the syringe) will prevent clotting of the sample. For larger blood volumes, the vena cava is accessible in a sedated patient. The hedgehog heart is located more cranially in the thorax than in other smaller mammals. Radiographic images for hedgehogs generally require anesthesia to be able to correctly position the patient. The spines may hinder detail and positioning, especially on the ventrodorsal view.

Ultrasound examination is performed on anesthetized patients to evaluate cardiac function and abdominal masses. Ultrasonography can assist with obtaining laboratory samples such as tissues samples or fluid samples from masses.

Summary

Hedgehogs became popular as pets but owners soon realized that because of their natural behaviors, having a pet hedgehog was nothing like expected or depicted in stories, cartoons, and video games, and so many were abandoned. Laws in many areas prohibit the breeding, sale, and possession of hedgehogs because of the possibility of introducing anthrax from imported animals and the concern that if they were intentionally released or escaped, they could establish feral populations in some areas. Hedgehogs are insectivores and have changed little in the past 15 million years. They are solitary and nocturnal and may object to being handled. When threatened, hedgehogs curl up into a tight ball to protect themselves. They are covered in very sharp spines, and while they do not contain any toxin, many people have developed a skin rash from contact with the spines. A behavior unique to the hedgehog is anointing; when the hedgehog is confronted with a new smell or taste, it begins to lick the item and produce a foamy saliva and use their tongues to paint or anoint their spines. Hedgehogs are normally very active at night, and confining them to a small enclosure causes behavioral problems such as weaving and compulsive pacing. A major medical concern with pet hedgehogs is obesity, and they can become so overweight that they are unable to curl into a ball. The correct diet and ability to exercise must be provided. Because of their spines, hedgehogs are difficult to handle and even simple procedures like nail trims may require general anesthesia. Salmonella is a problem with hedgehogs, and they may be asymptomatic carriers, which is a public health concern. There is a high incidence of neoplasia in hedgehogs with all body organs being affected. A disease unique to them is a degenerative neurologic disease, commonly called *wobbly hedgehog syndrome*. There is no treatment and the disease is progressive, resulting in humane euthanasia.

fastFACTS

HEDGEHOGS

WEIGHT
- Dams: 300–600 g
- Sows: 400–600 g

LIFE SPAN
- 4–6 years (48–72 months)

REPRODUCTION
- Sexual maturity: 8–10 weeks
- Gestation: 34–37 days
- Litter size: 3–4
- Weaning age: 4–6 weeks

VITAL STATISTICS
- Temperature: 35.5–37.2°C (96–99°F)
- Heart rate: 180–280 beats/min
- Respiratory rate: 25–50 breaths/min

DENTAL
- Dental formula 2 (I 3/2, C 1/1, P 3/2, M 3/3)

ZOONOTIC POTENTIAL
Bacterial
- Salmonella
- Anthrax (rare, reportable)

Fungal
- Dermatophytes

Review Questions

1. What is anointing, and when does it occur?
2. Which bacteria can be a problem in hedgehogs and a potential public health concern?
3. Describe hedgehog behaviors regarding sleep, social nature, and handling.
4. Hedgehogs do not throw their spines. What is a common cause of spine loss?
5. Hedgehogs belong to which classification of animals?
6. Why is it important to provide an exercise wheel for a hedgehog?
7. What are two common behavioral disorders seen in captive hedgehogs?
8. What is the protocol for handling a suspected case of anthrax?
9. Describe two methods of getting a hedgehog to unroll.
10. Neoplasia is common in hedgehogs. Define neoplasia.

Case Study I

History: A four-month-old hedgehog is brought in by the young owner and his mother. The mother's chief concern is that the hedgehog has rabies because whenever the hedgehog is allowed to wander around the family room, it starts "foaming" at the mouth and wiping the foam all over itself.

Physical Examination: Weight and vital signs are normal. The hedgehog is very active and during the examination it is more interested in exploring its surroundings than curling into a ball. It also demonstrated the very same behavior that has concerned the boy's mother.

   a. What will the veterinarian explain to the mother?

   b. This visit to the veterinarian's office is justified for at least three reasons. Explain how all concerned benefited.

Case Study II

History: A one-year-old female hedgehog was presented by the owner who states that "she seems to be dragging her hind legs." The sow has difficulty getting up and is reluctant or unable to curl into a ball.

Physical Examination: There is a definite weakness in the hindquarters with a slow reflex response. The front limbs also exhibit a diminished reflex response. The hedgehog is alert but unable to stand up on all four legs.

   a. What is a likely cause for this hedgehog's weakness?

   b. Explain the prognosis for this patient

   c. What is the recommended treatment for this condition?

For Further Reference

Carpenter, J. W. (2010). *Diseases and Treatment of Pet Hedgehogs*, CVC in Baltimore Proceedings. April, 2010.

Heatley, J. J. (2005). A Review of Neoplasia in the Captive African Hedgehog. *Seminars in Avian & Exotic Pet Medicine* 14:182–192.

Ivey, E., & Carpenter, J.W. (2012). African Hedgehogs. In Quesenberry, K.E. and Carpenter, J. W. (Eds.), *Ferrets, Rabbits, and Rodents: Clinical Medicine and Surgery, 3rd Edition* (pp. 411–427). St. Louis, MO: Elsevier Saunders.

Johnson-Delany, C.A. (2002). Hedgehogs. In Meredith, Anna and Redrobe, Sharon (Eds.), *BSV Manual of Exotic Pets, 4th Edition* (pp. 108–112). British Small Animal Veterinary Association, Hoboken, NJ: John Wiley and Sons.

Mitchell, M. A., & Tully, T. N. (2009). Manual of Exotic Pet Practice, *Hedgehogs* (pp. 433–455). St. Louis, MO: Saunders Elsevier.

DEGUS

KEY TERMS

diurnal
weeping
anthropomorphic
atrophies
elodontoma
red-nose

OBJECTIVES

After completing the chapter, the student should be able to

- Describe the correct habit for a colony of degus.
- Provide client education to new degu owners.
- Provide the correct diet for degus.
- Provide basic nursing care to a degu.
- Assist in the anesthesia of a degu.
- Demonstrate appropriate restraint techniques for degus.
- Understand common medical problems in degus.

Introduction

Degus have been kept as pets in many European countries and Canada and are now gaining in popularity in the United States. As with many exotic species, they were originally imported as research animals and later introduced to the pet market. Research studies have been particularly focused on diabetes, embryonic development, and sleep disorders.

Degus are very active, **diurnal** (active during the day) rodents from South America. Their behavior and social nature make them interesting, medium-sized pets that are quite different from other rodents because they are most active during the day. Larger than many other rodents that are kept as pets, the degu has a body length of six to eight inches and a five-inch, haired tail with a darker tuft on the tip. They have dense brown fur that is lighter on the underside (**FIGURE 9-1**). Their skin color may vary from slightly pink to very dark brown. They have five toes on each foot. The thumbs of the forefeet are nonopposable and not much more than nubs. Degu pups are born precocial and covered with black fur, which lightens with maturity. The teeth of neonates are white but gradually turn yellow as they mature. The teeth of healthy, mature degus can vary from yellow to orange in color.

The degu's taxonomic family, *Octodontidae*, refers to the *figure 8* shape seen on the molar surface (**FIGURE 9-2**). Classification of the degu within the family *Octodontidae* is debated, especially among hobby breeders, who believe that they are more closely related to guinea pigs (*caviomorphs*) and should not be grouped

FIGURE 9-1 A young degu. Degus are also called the Chilean squirrel or a brush-tailed rat.

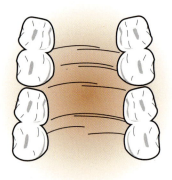

FIGURE 9-2 Degus' taxonomic family name, *Octodontidae*, refers to the *figure of 8* shape of the molars.

with other rodents. Taxonomic classification is determined by several factors, some of which include the shape and structure of the skull, jaw musculature, and teeth.

There are nine subspecies of degu, and they all belong to the same suborder, *Hystricognathi*, which includes the *porcupine-like* rodents. Degus are ground dwelling, burrowing foragers of the lowland areas of Chile. They are also commonly referred to as the *Chilean squirrel*, or *the Brush-tailed rat*.

In many areas degus are considered an agricultural pest. Once established, a colony of degus is capable of causing a great deal of damage, especially in planted fields where they dig out seeds and young shoots, devastating the crop. They will chew on anything, stripping bark from young shrubs and small trees. While ground dwelling, they are adept climbers and can reach the branches of more established plants. In addition to consuming these food items while out foraging, they carry bark, plant twigs, and other items back to their burrows. Because of the agricultural concern, their importation into the United States is restricted. All degus domestically bred or offered for sale or trade (even those given away) require a USDA permit for possession.

Behavior

Degus live in groups of 5 to 10. The group cooperatively digs a complex, communal burrow that serves as shelter, nursery, and food storage. Females share the nursery, combining several litters and nursing each other's young. Degus are very vocal and produce a variety of sounds from a high-pitched whistle to a low *chuckle*. One of their more distinctive vocalizations, described as **weeping,** may continue off and on for several hours. The significance of *weeping* is not clear. Both the solitary animal and those in colonies *weep*. It may simply be a territorial song, but assigning an anthropomorphic (human-like) interpretation, that of extreme sadness, should be avoided as the true significance is not understood.

The hierarchal structure of degus is maintained with much squabbling and fighting to establish and maintain dominance. They are territorial and socially dependent on this structure, which gives the group unity in fending off other degu colonies. Dominant males haul all manner of debris to the burrow, building large mounds around the entrance. The larger and more elaborate the mound, the higher the status of the builder. Both sexes scent-mark established territory with feces and urine. Degus dig shallow areas in the soil around the burrows, creating dusty wallows. They urinate copiously in the depression before rolling in it, covering themselves and each other with a group scent. They also urinate on the heads of each other, reinforcing the scent of the hierarchy and group as a whole. Because of their socially dependent nature, they should never be kept alone (**FIGURE 9-3**). Solitary animals of either sex often become aggressive and self-mutilate.

FIGURE 9-3 Degus are socially dependent on one another and should be housed collectively.

Squabbling is common within the colony and should not be taken as a sign of incompatibility. Serious injuries rarely occur, but there is a constant dynamic to establish which degu is the *boss*. Fighting is worse in mixed groups than in single-sex colonies. One way to deter some of this behavior in a captive colony is to set aside a small amount of soiled bedding and mix it in with the new bedding. A simple axiom might be *the larger the cage, the larger the perceived territory, the more peaceful the colony*.

Degus are incredibly curious and very interested in the goings-on around them. They willingly interact with people, becoming playful and scampering up, over, and around objects in the enclosure almost as an enticement to engage a human in chase games. They exhibit no fear of other household pets, running to the front of the enclosure to investigate a cat, dog, or ferret (**FIGURE 9-4**).

FIGURE 9-4 Degus are very curious and bold. They will readily approach a stranger or a household pet to investigate.

Degus, both male and female, are sexually mature between 12 and 16 weeks of age. In captivity, breeding can occur year round. Degus are induced ovulators. Copulation is brief but frequent. Gestation is slightly variable, the average being approximately 90 days. There may be as many as 10 pups in a litter, but five or six are more usual. Pups are born precocial, and while active from birth, neonates may not open their eyes for up to 48 hours after birth (**FIGURE 9-5**). Determinig the sex of a degu sexing may at first seem difficult as both sexes have a urethral cone. In males, the anogenital distance is obviously greater when compared to females, with a very short anogenital distance.

Housing

Appropriate housing for degus is more difficult than with other small mammals. The enclosure should be large enough to accommodate several degus; provide areas for digging, climbing, collecting, and storage of food; and have an area for the communal nesting site. Captive colonies exhibit normal dominance behaviors, scent-marking not only the territory but also each other. Providing an area suitable for this ritual is necessary for the social structure and mental health of the group.

Too often, many exotic animals kept as pets are expected to adapt to their caged status without careful thought given to what the animal actually requires. If requirements cannot be met, serious consideration needs to be given to the ethical

FIGURE 9-5 Degu pups are born precocial. This pup is approximately 10 days old. Degus have a urethral cone, similar to that of the chinchilla (see **FIGURE 7-3**).

aspects of keeping any exotic animal species. This is particularly important not just with degus but with sugar gliders as well (see Chapter 13). In looking closely at ready-made cages offered for a variety of species and the potential for adapting them to accommodate degus, there is very little presently available. Condo-type cages, those designed for ferrets or chinchillas, have been suggested and may at first appear suitable. They provide ledges and ladders for climbing, places for water bottles and food dishes, and usually an area for a nest box. However, the height is greater than the floor space and ledges are placed far higher than a degu would climb. There is the danger of a fall or entrapment in the ladders. Even the largest of these cages is far too small for a degu colony, and most, if not all, have large amounts of plastic, which should never be used for degus. The gauge of the wire used in small-mammal cages would be easily chewed through by degus. A heavier gauge, nongalvanized wire would be required to keep them contained safely.

Degus need *ground room*. The open wire of these cages would not be able to contain a substrate for digging, nor is it able to confine pups. Rabbit hutches have also been used, but with little success. Although they are longer than they are high, the frame is usually made of wood. Considering degus' ability to shred wood, the hutch is soon destroyed and the animals lost. Suitable caging must be provided with the species welfare of foremost importance, and for the captive degu colony, this means cages that are custom-built.

> **Degus Will Chew Plastic,** often swallowing sharp pieces. This can cause lacerations to the oral cavity, intestinal blockage, tears in the intestinal tract, prolapse, and, potentially, death.

Dedicated keepers of degu colonies have become very inventive. Some have adapted ideas from zoos, often combining many aspects of small-mammal collections. The habitats are not only functional for the degus but can also create an attractive focal point (**FIGURE 9-6**).

Custom-built cages should be a minimum length of five feet and an approximate height of four feet. The framework can be made of tempered aluminum or one-inch-square steel pipe in conjunction with acrylic or glass panels. One advantage to acrylic is that it can be drilled for screw attachment to the frame. When using glass, a nontoxic aquarium sealant is required. Many cages incorporate large aquarium *basements*, making degu activities visible, with the constructed enclosure mounted on the top (**FIGURE 9-7**). Confining degus to just an aquarium, no matter how large, is neither suitable nor humane.

Regardless of how degus are housed, they need an area for a sand bath. A metal-bottom tray from a small-mammal cage provides an area large enough for several degus to roll in. It is easily cleaned and can help to contain urine and dust. Sand is readily available from garden centers and builders' supply stores. It is inexpensive, an important factor as it will need to be changed often due to frequent urination within the tray. This is not to say that degus can be litter trained. Urination in the sand bath is a scent marking behavior and is not associated with normal urination and defecation.

FIGURE 9-6 An example of a large flight cage that has been adapted for housing a degu colony. The habitat is approximately 6 1/2 feet long. Double glass panels prevent the degus from chewing on the wood and keep them safely enclosed.

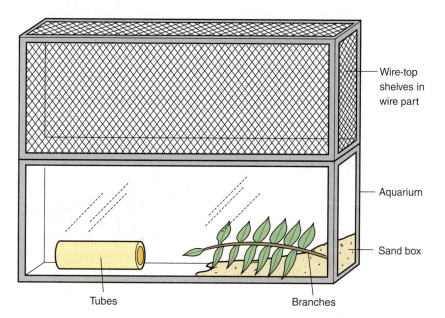

Wire-top shelves in wire part

Aquarium

Sand box

Tubes

Branches

FIGURE 9-7 An example of a cage set-up that works well for degu colonies. The bottom is a large glass aquarium with a custom-built top made of heavy gauge wire.

The floor of the enclosure should contain at least four inches of bedding. Aspen shavings, grass hay, and recycled paper products are all suitable. The degus will sort through it, arranging and rearranging little piles, carrying pieces back to their chosen nest site (**FIGURE 9-8**). Cedar shavings or cedar wood should never be used.

Enrichment items need to be included. Branches from unsprayed fruit and willow trees provide many areas of interest and varying levels to explore. Other items may include natural hollow logs, wood blocks of assorted shapes and sizes, and heavy cardboard tubes. Degus will use a large rodent wheel. Choose one that

FIGURE 9-8 Degus enjoy sorting through bedding material and other items to carry back to the nesting den.

has a grid-like running base or a solid floor, making sure that none of the parts are plastic (**FIGURE 9-9**).

Several heavy-duty ceramic bowls should be placed in the enclosure. Degus defend food bowls as a substitute for normal hoarding behavior. If there are not enough bowls, fighting over this little patch of territory will occur. Degus use water bottles, but these too need to be either ceramic or glass with metal sipper tubes and must be fastened securely to the enclosure walls with metal brackets.

FIGURE 9-9 Degus are very active diurnal rodents that enjoy using an exercise wheel. For safety, the wheel should have a wire mesh floor and be made entirely of metal.

It can be difficult to combine degus from different groups. An effective method of introducing new animals is to first exchange the bedding, rather than the animals. Change out small quantities of bedding from each cage, mixing it thoroughly, exchanging it back and forth before introducing the degus to each other. Soiled sand trays also need to be exchanged in this manner. Mix both trays together and divide the contents before returning them to separate cages. Careful observation of the degus' behavior in response to the strange scents can help to determine when the animals could be put together for the first time. This process may take a few days, or it may require several weeks. Cage proximity may also help familiarize them with each other, but the direct exchange of scent is a very effective method.

Diet

Degus are grazing herbivores, consuming a variety of grasses, plants and grains, roots, tubers, and bark. They are coprophagic and, in the wild, also consume fecal material of grazing herbivores. Compared to other species, the nutritional needs for degus are more easily met. Fresh, clean grass hay, such as timothy, should be available at all times. Guinea pig, rabbit, or chinchilla pellets can be provided free choice. Rodent mixtures that contain nuts and sunflower seeds should not be fed as they are too high in fats and the roughage and bulk of these items may cause intestinal blockage.

Fruit, either fresh or dried, should be avoided as degus have a low tolerance for sugars. Sugars that predispose degus to diabetes are derived mainly from molasses, a by-product from refining sugarcane, sugar beets, and grapes. It is also thought that the sugars in honey are not well tolerated. Fresh root vegetables are readily consumed and provide variety in the diet, as do dandelions, complete with roots. Enrichment foods such as willow branches and unsprayed fruit wood and leaves can also be provided. For treats, yams and unsugared cereals like puffed wheat, puffed rice, and corn flakes are excellent choices. In addition, specially formulated degu pellets are available. These usually have to be ordered through a supplier specializing in zoo and lab animal diets.

Handling and Restraint

As interested as they may appear in the activities of people, degus do not like to be held. If a degu must be handled, it can be scooped up in both hands and held cupped, like a guinea pig, with one hand encircling the forequarters just behind the front legs and the other supporting the hind end. Degus, like gerbils, will slough the skin of their tails in an attempt to escape. If held by the tail, a degu will spin to escape, leaving only the skin once covering the tail. Once degloved, the tail atrophies (wastes away) as the underlying tissue dies. The tail will need to be surgically removed or the degu will chew it, potentially causing more trauma. Wild degus self-amputate, but the risk of infection in a captive colony is greater simply because they are confined to an area where greater numbers of bacteria may be present.

Medical Concerns

If provided with the correct diet and housing, degus are hardy. The major concern is diabetes, which may be acquired or genetic and is best managed with diet.

Diabetes-related cataracts are fairly common in aging degus. There are many references to degus being *unable to digest sugar*, but the reason, the metabolic process, has been unstated in degu papers and little information is readily available; however, the sugars mentioned earlier in the text should be avoided.

Although degus can fight when new groups are introduced, rarely are there injuries requiring medical treatment. The most severe trauma is seen in degus that are kept alone and self-mutilate. Depending on the degree of injury, the decision must be made to attempt a repair, an amputation, or euthanasia. This problem is easily avoided by keeping degus in groups.

As with many rodents, tooth overgrowth and malocclusion can occur. Providing adequate, hard material to chew will help prevent overgrown teeth. Malocclusion occurs most commonly in closely bred colonies, suggesting that it is likely a genetic predisposition. Adult degus normally have yellow/orange teeth. Teeth that seem to be losing color may be an indication of a calcium/phosphorous imbalance or erosion of the dental enamel.

The teeth of degus grow continually throughout their lifetimes, and in addition to malocclusion, cases of **elodontoma** have also been noted. Elodontoma is a condition that may be genetic or caused as a result of malocclusion where tumor-like growths develop due to an overgrowth of molar teeth roots. The tumors can become so large that they protrude into the sinus and nasal cavities, making it very difficult for the degu to breathe. Surgical removal of the tumor can be attempted but is frequently unsuccessful, and owners may opt for humane euthanasia.

Fur barbering is common in degus that are bored. The barbering, or fur chewing, usually begins on the forepaws. Adding a variety of new enrichment items will provide enough distraction before barbering becomes more serious and leads to the trauma of self-mutilation.

Red-nose is a term most often used to describe a condition in gerbils, but it can also occur in degus. It is caused by chewing on cage bars, resulting in trauma and hair loss to the end of the nose. Correct housing will prevent this from occurring and will decrease the possibility of a bacterial infection entering the damaged tissue.

Pseudomonas is a bacterial infection seen in degus that are forced to drink from dirty water bottles. It is imperative that the water bottles are thoroughly cleaned two to three times a week to prevent this infection.

Parasites

Degus are frequently infested with ear mites. Ear mites most likely come into the colony through infested hay. Once ear mites become established, the entire colony should be treated and all cage furnishings should be bagged up and destroyed. For this reason, it is not recommended to feed field hay but offer cleaned and packaged hay available from small-animal suppliers.

Clinical Procedures

Clinical procedures, those of routine blood collections and injections, may be approached in the same manner as they are with guinea pigs. As the legs of a degu are longer, it may also be possible to access the cephalic or lateral saphenous veins. For a jugular blood collection, anesthesia is required. The lateral tail vein, under anesthesia, may be attempted with caution but remember that the tail is easily degloved.

Anesthesia may be induced with a face mask or with the use of an induction chamber. Both isoflurane and sevoflurane are suitable choices. There is little data on the use of injectable anesthetics in degus.

Anesthesia is required in most patients when taking radiographs to achieve correct positioning and to prevent injury to the patient. Lateral and ventrodorsal views of the whole body are taken to assist with diagnosing abdominal masses, fractures, and respiratory disease. Lateral oblique views are taken of the head to diagnose dental disease in degus.

Summary

Degus are very active diurnal rodents native to South America. They were originally imported as research animals for use in studies focusing on diabetes, embryonic development, and sleep disorders. All degus in the United States require a USDA permit for possession as they are considered an agricultural pest. Degus live in colonies; they are socially dependent on one another and should never be housed alone. Providing a degu habitat that is appropriate for a colony can be very difficult. They will chew through almost everything; in addition, the habitat must be large enough to provide areas that meet their unique behavioral requirements. Degus are grazing herbivores and consume a variety of grasses, plants, and grains. In general, degus do not like to be handled, and if held or caught by the tail, the animal will spin rapidly and degloving of the tail will occur. Clinical procedures are similar to those for guinea pigs.

fastFACTS

DEGUS

LIFE SPAN
- 5–8 years (120 to 192 months)

REPRODUCTION
- **Sexual maturity:** 12–16 weeks
- Males and females have a urethral cone; sexing similar to chinchillas
- **Gestation**
- 92 days (average)
- Young born precocial
- **Litter size:** 1–10 (6 average)
- **Weaning age:** 5–6 weeks
- Both males and females have eight mammary glands

VITAL STATISTICS
- **Temperature:** 39.7°C (100.9°F)
- **Heart rate:** 274 beats/min
- **Respiratory rate:** 123 breaths/min

DENTAL
- Total 20 teeth
- Dental formula 2 (I 1/1, C 0/0, P 1/1, M 3/3)

ZOONOTIC POTENTIAL
- None reported

Review Questions

1. Why are degus referred to as octodont?

2. What are some of the difficulties in providing suitable housing for degus?

3. What is the best way to introduce a new colony member?

4. Describe a suitable diet for degus. What should be fed and what should not be fed?

5. What are the most common medical problems seen in degus?

6. What are some of the factors that determine taxonomic classification of degus?

7. What is the legal requirement for possession of degus? Why?

8. What does *anthropomorphic* mean?

9. A degu should never be kept as a single animal. What behaviors are seen in solitary degus?

10. What is the purpose of a sand pit in a degu enclosure?

Case Study I

History: A four-year-old female degu is brought in by the owner. Her chief complaint is that the degu is losing hair on both front legs. The degu's appetite and activity level are reported as normal. The owner feeds the degu guinea pig pellets and grass hay. The owner has only one degu and houses the patient in a 50-gallon aquarium. The degu did have a rodent wheel, in fact, about three, before the client stopped buying them because "all she ever did was chew them to pieces."

Physical Examination: There are areas of alopecia on both forepaws that appear random and ragged, with shorter hairs further up the leg. The veterinarian also notices a similar pattern in the hair coat of the lower abdomen.

a. What is the probable cause of the alopecia and disrupted hair coat?

b. Review what the owner has described regarding the habitat. What advice should be offered to this client regarding degu habitat and behavior?

Case Study II

History: A year-old male degu was presented by the owner. His concern was that the eyes appear "cloudy-looking." The owner states that he has both parents of this degu; there are two others in the same litter and both seem normal, though he did casually mention that he has wondered sometimes if the mother was blind. The owner feeds the degu grass hay, rabbit pellets, dried fruit, and vegetables.

Physical Examination: The veterinarian performs an ophthalmic examination and confirms that the patient has early-stage bilateral cataracts. She also requests a blood glucose level to which the owner consents, although not fully understanding the reason. As suspected by the veterinarian, the degu is hyperglycemic.

a. With the confirmation of cataracts, why did the veterinarian request the blood glucose level?

b. What is the connection between the development of cataracts and the blood glucose level?

c. Is this problem likely to be hereditary, diet related, or both?

For Further Reference

Banks, R.E., Sharp, J.M., Doss, S.D., & Vanderford, D.A. (2010). *Exotic Small Mammal Care and Husbandry* (pp. 137–141). Ames, IA: Wiley-Blackwell.

Capello, V. (2006). *Clinical Approach to the Anorectic Guinea Pig*. Conference Notes, Presented at the North America Veterinary Conference, January 7–11, Orlando, FL.

Cloyd, E. (2003). http//animaldiversity.ummz.umch.edu (accessed December 27, 2013).

Girling, S. (2003). *Veterinary Nursing of Exotic Pets*. Chichester, England: Blackwell Publishing.

Harkness, J.E., & Wagner, J. (1989). *The Biology and Medicine of Rabbits and Rodents, 3rd Edition*. Philadelphia, PA: Lea & Febiger.

http:/www.degutopia.co.uk (accessed January 17, 2014).

Klaphake, E. (2007). DVM, DABVP, DACZS, personal correspondence, January 2007.

Long, C. (2007). Degutopia unpublished data, personal correspondence, January 2007.

Mayer, J. (2006). *Analgesia and Anesthesia in Rabbits and Rodents* (pp. 1740–1742). Conference Notes, Presented at the North America Veterinary Conference, Orlando, FL.

O'Malley, B. (2005). *Clinical Anatomy and Physiology of Exotic Species*. New York: Elsevier Saunders.

Quesenberry, K.E., Donnelly, T.M., & Mans, C. (2012). Biology, Husbandry and Clinical Techniques of Guinea Pigs and Chinchillas. In Quesenberry, K.E. and Carpenter, J.W. (Eds.), *Ferrets, Rabbits, and Rodents: Clinical Medicine and Surgery, 3rd Edition* (pp. 279–294). St. Louis, MO: Elsevier Saunders.

HAMSTERS AND GERBILS

OBJECTIVES

After completing the chapter, the student should be able to

- Describe the correct housing for a hamster or gerbil.
- Provide client education for a new hamster or gerbil owner.
- Provide a correct diet for a hamster or gerbil.
- Deliver basic nursing care to a hamster or gerbil.
- Assist in the anesthesia of a hamster or gerbil.
- Demonstrate appropriate restraint techniques for a hamster or gerbil.
- Understand common medical conditions in hamsters and gerbils.

Introduction

Hamsters and gerbils often provide the first experience many people have with exotic pets. There are similarities in husbandry and diet of these two rodents, but the differences between them are more complex.

Hamsters

Two species of hamsters are commonly available: the Golden (or Syrian) hamster and the Russian hamster, commonly called a dwarf hamster. The Russian hamster, while considerably smaller than the Golden hamster, is not a miniature version of the other species. They are separate and distinct. The word *hamster* is a derivation of a German word, *hamstern*, a verb that means *to store*. It refers to the anatomy and behavior common to all hamsters, that of storing and transporting food in their cheek pouches.

Golden hamsters are native to the desert regions of Syria. In 1930, a litter was discovered by a field zoologist, Professor I. Aharoni, who collected them and attempted to transport them to the Hebrew University in Jerusalem. All but three died, leaving a male and two females. The litter mates were allowed to reproduce, and while not verifiable, these three are attributed with being the foundation stock for all captive breeding colonies. Because of their ability to reproduce rapidly, colonies were quickly established for use in medical research around the world. The status of the wild Golden hamster is unknown. They may possibly be extinct due to erosion of habitat and overcollecting.

The Term *Dwarf* Is Specific. Dwarfism can occur in any species and refers to an abnormality of development causing an animal (or plant) to be smaller than the norm for that species. It may be attributed to a developmental abnormality, hormonal imbalance, or nutritional deficiency.

They are called *golden* to reflect their natural coat color, a rich light chestnut or a golden brown. There are now many colors and long coat varieties available, but they are still the same *Mesocricetus auratus* (**FIGURE 10-1**).

FIGURE 10-1 The Golden, or Syrian, hamster now has many color varieties.

Russian hamsters are relatively new as pets and are illegal to keep in many states. Little is known about how and when they appeared in the pet trade during the 1980s. They are small, approximately three inches long, with thick fur that makes them appear somewhat larger. They are less than half the size of the Golden hamster. The natural color of the Russian hamster is soft gray with a dark dorsal stripe. Even though different colors are now available, they still maintain the stripe down the middle of the back (**FIGURE 10-2**). Unlike the Golden hamster and other rodents, the feet of the Russian hamster are furred. Both species have cheek pouches in which to store and transport food, bedding material, and other items, sometimes moving an entire litter from one place to another.

Behavior

Hamsters are often chosen as a first pet because of their small size and the relatively low cost to feed and house them. Problems arise when new owners realize

FIGURE 10-2 The Russian hamster is naturally a soft gray color. Many varieties have been bred, but all still retain the distinctive dorsal stripe.

that they are a strictly nocturnal species that does not willingly accept being disturbed during the day, the time when children are expected to interact with and care for the new pet. Hamsters sleep very deeply, and care must be taken to ensure that the animal is fully awake before attempting to pick it up to avoid being bitten. The deep sleep during the day changes at night, when the hamster becomes very active. The biggest complaint about the nocturnal behavior of hamsters is that they *make too much noise*. Usually it is not the hamster making the noise, but the exercise wheel squeaking and tapping up against the side of the cage. Hamsters have been known to run as far as 10 kilometers (approximately six miles) in a single night. There is no method of training a nocturnal animal to meet the expectations of a diurnal pet. Many hamsters have been abandoned, returned to the pet store, or, unfortunately, treated as disposable and destroyed. Hamsters are not recommended as pets for young children because of their quickness to bite and their nocturnal activities. The temperament of individual hamsters varies greatly, and *generally* the Russian hamster tends to be more aggressive. This, along with their very small size, makes them an even less desirable choice as a child's pet.

Hamsters are fastidious in maintaining and arranging their habitats. They choose specific areas for sleeping, stockpiling food, and depositing their waste. They do not rouse during the day to defecate. When they wake up, they collect the fecal material from the sleeping den and place it in their cheek pouches to move it to a specific area where they have deposited other droppings. Tiny plastic houses are available, a variety of *hamster potties and outhouses*. If placed in the location already designated by the hamster, these tiny litter boxes will be used.

Caregivers need to be aware of the hamster's arrangement and not put items back in a different place after cleaning the cage. Hamsters become quite agitated and will immediately start to rearrange their cage.

Golden hamsters should be housed alone. They are solitary by nature and will not tolerate other hamsters. Attempting to keep two sexually mature animals together will result in fighting and the death of one of them. They should be put together only for brief, supervised periods to mate. The female is more aggressive, and the male should always be introduced to the female. Alternatively, they may both be placed in a neutral environment but still should be observed carefully for any indication of aggression. Russian hamsters tend to be more tolerant of one another and are often successfully housed in groups, provided they are placed together when they are very young.

Male hamsters have large scent glands that appear as dark brown patches on either side of the back, near the pelvis. These glands are used for territory marking. The testicles of a male hamster are very evident and are often mistaken by owners as *suddenly appearing tumors.*

Females need to be left undisturbed with a new litter as they are known to cannibalize their young. Gestation is an average of 16 days, and both species form copulatory plugs after mating. Females are seasonally polyestrous, and most do not have reproductive cycles during the winter months. A normal reproductive cycle usually lasts four days. During this time, females produce a creamy-white discharge from the vulva. The discharge normally has a very strong odor. Average litter size for the Golden hamster is four to twelve and for the Russian hamster an average litter ranges from four to eight. Weaning occurs when the litter is three to four weeks old. The Russian hamster reaches sexual maturity between five and eight weeks and the Golden hamster slightly later at six to twelve weeks.

Hamsters do not live for many years. The life span of a Russian hamster, an average of 18 to 24 months, is shorter than that of a Golden hamster. Some Golden hamsters have been known to live as long as 36 months. Other than size, there is no other indicator that could approximate the age of a hamster when purchased unless acquired from a private breeder who has recorded the delivery date.

Gerbils

Gerbils are native to the desert regions of Mongolia and northeastern China. In the 1930s, several pairs were caught on the border between China and Russia. Some 20 years later, a few breeding pairs were sent to the United States. Gerbils sold as pets are descendants of these specimens.

While approximately the same size as a golden hamster with an average body length of five inches, gerbils have haired tails that are as long as their bodies. Hamster tails are very short and hairless. Gerbils do not possess cheek pouches. Their Latin name, *Meriones unguiculatus,* means *little clawed warrior.*

The normal color of a gerbil is sandy brown with a lighter underbelly which camouflages them in their native desert habitat. Several color varieties have been bred in captivity, including pure black, snow white, pied, and albino (**FIGURE 10-3**).

FIGURE 10-3 The normal color of gerbils camouflages them in their native desert habitat.

Behavior

Unlike hamsters, gerbils are social with each other and need the companionship of others. Same-sex groups may be housed together when they are introduced to each other at a young age. A bonded male and female are monogamous, mate for life, and share in the care of their offspring. Owners need to be aware of the social bonds that have formed and should maintain paired gerbils together. Gerbils are aware of their social community and are protective of one another. If danger approaches, they thump with a hind foot in warning to the others.

Gerbils are not strictly nocturnal and are often out and very active during the day. They are playful with each other, scampering and leaping around the enclosure, stopping often to engage in bouts of digging. They enjoy small toys and many items can be included for them: cat balls with a bell inside, cardboard tubes, and hide boxes with several entrances to run through, around, over, and back again.

Both sexes have an area of alopecia on their abdomens. It is a large sebaceous (oil-secreting) gland used for scent marking. Regardless of the hair coat color, this area usually has a slightly orange pigment. Neither hamsters nor gerbils have sweat glands, but gerbils are able to tolerate slightly higher temperatures than hamsters, provided that the temperature is not accompanied with an increase in humidity. If temperature exceeds 80°F, hamsters estivate and become very difficult to rouse.

Like hamsters, female gerbils are seasonally polyestrous. The reproductive cycle lasts approximately four to six days. The gestation period for gerbils is 26 days, considerably longer than that for hamsters. As the pair remains together, rebreeding occurs postpartum. During the breeding season, it would be possible to produce one litter per month, quickly leading to overpopulation and inbreeding. The average litter size is two to six and the young are weaned completely by four weeks.

The young of hamsters and gerbils are born altricial; their eyes and ears are closed and they are hairless. Gerbils are less likely to cannibalize their young than hamsters. The average life span for a gerbil is three to five years.

Housing

The habitat for hamsters or gerbils should have solid floors and sides. Glass aquariums fitted with secure screen lids make an ideal and inexpensive arrangement. Both species are capable of climbing up and jumping out. Gerbils are more likely to jump than hamsters, as jumping is more natural to them. Gerbils have very long hind legs and are able to jump a greater distance than might be thought possible for such small animals.

Habitats for hamsters can be quite elaborate with many plastic tubes, towers, and connecting tunnels. Some of these arrangements resemble small cities, with different pieces, connecting tubes, and towers invested in over a period of time. Hamster owners still need to remember: *one hamster, one habitat*. The more elaborate and complex a unit becomes, the more time is required to clean it. It should be taken apart and cleaned completely on a weekly basis, and then the whole project needs to be reassembled.

The substrate can be aspen shavings or recycled paper products. Ground corn cob bedding should be avoided because it can allow bacteria to grow if old food or urine is left in the cage. Gerbils may also be housed on fine sand. Both hamsters and gerbils use water bottles, but due to their desert heritage, gerbils tend to drink less water. However, all water bottles need to be cleaned and refreshed daily. A hide box should always be provided. Hamsters will use a hide box, provided it is placed in the designated sleeping area. If a hamster prefers to sleep curled up in a nest of its own making, some of the bedding material collected by the hamster can be placed in the hide box and returned to the spot the hamster has chosen as the sleeping den.

Rodent wheels can also be provided for gerbils, but they should have a grid or solid running surface as gerbils tend to spring forward on them, as opposed to the scurrying run of a hamster. These wheels are safer for gerbils and help to avoid getting their feet and tail entrapped. If its tail is caught in the wheel, the gerbil will completely deglove the skin, leaving exposed tissue. If the tail is injured or degloved, it should be amputated by a veterinarian to prevent further trauma from the gerbil chewing on it and to prevent bacterial infection. Both gerbils and hamsters can be exercised with a rodent ball, but they should not be left unsupervised to roll around the house for extended periods.

Because they are rodents with continually growing teeth, gerbils and hamsters need to be provided with chew blocks. These can be purchased in various shapes and colors. Chew blocks also provide enrichment. Other enrichment items can be very simply provided. Empty toilet paper rolls, plain paper towels, and tissue boxes are readily shredded by both gerbils and hamsters. Both species will also use these materials for bedding and nest making.

Diet

Hamster and gerbil diets are similar. The basic diet should be good-quality rodent blocks. These blocks were originally developed for laboratory use and are well balanced to meet nutritional needs. In addition, since they are hard, they help to maintain dental health. Rodent mixes can also be added, provided they are low in sunflower seeds, which are very high in fat. Obesity can be a problem for both hamsters and gerbils. Unsweetened shredded wheat, rice and corn cereals, trail mix (without chocolate), plain popcorn, and cracker pieces can be offered in limited amounts. Many other treat items are available in pet stores.

Hamsters will eat fresh, chopped fruits and vegetables. If fresh foods are given and not all are consumed, the remainder should be removed the following morning. Hamsters store food. Stashed, fresh food will grow mold if not removed from the storage area. Cheek pouches of hamsters can accommodate a great amount of material and large pieces of food items can impact the pouches. Unshelled peanuts should be given with caution. Hamsters have been known to place three fully shelled double peanuts in one cheek pouch, requiring veterinary assistance to remove them.

Handling and Restraint

When first awakened, hamsters are easily startled. They stand upright or turn away making hissing/chittering noises with their eyes still closed. This is not the time to reach in and attempt to pick up a hamster. Gently prodding it away from the sleeping site will fully awaken the hamster. Once they are awake and moving, most hamsters can be scooped up in the hand or herded into a cup.

Hamsters do not tolerate restraint for very long without trying to bite. Restraint for an examination requires a scruff. Always make sure that the cheek pouches are empty before scruffing. Particles of food or bedding retained in the pouches could cause injury. All of the extra skin that forms the cheek pouches needs to be included in the scruff so the hamster cannot turn around to bite. Extended periods of scruffing may cause exophthalmosis. Scruffing causes pressure behind the eyes, pushing them out of the orbits (FIGURE 10-4).

The most important thing to remember about gerbils is that they should never be caught or restrained by the tail. They are much quicker than hamsters and initially scooping them into a cup gives the handler a better chance of keeping the gerbil confined before attempting a scruff. Gerbils may also simply be held in one hand (FIGURE 10-5). They are less likely to bite and they do not have the ability to *turn around in their skin* like a hamster because the skin is tighter and without the excess of cheek pouches.

FIGURE 10-4 Care must be taken when scruffing a hamster. The cheek pouches should be empty and the scruff sufficient enough to gather in the loose skin of the pouches to prevent the hamster from turning and being able to bite. Prolonged scruffing, or a scruff that is too tight, may cause exophthalmosis.

Medical Concerns

Gerbils tend to be a little hardier than hamsters. Both are susceptible to bacterial diseases and respiratory problems that are usually associated with poor husbandry.

Gastrointestinal Disorders

Hamsters are very susceptible to a condition called wet tail. It causes chronic wetness around the tail and genitals and quickly leads to dehydration and death. Wet tail is a proliferative ileitis caused by a Campylobacter bacteria infection in the lower part of the small intestine, the ileum. The condition can be brought on by stress, overcrowding, shipping, and poor hygiene. Treatment needs to be prompt and aggressive, with fluid therapy and antibiotics. Even with the best veterinary care, the mortality rate is high. It is endemic (prevalent) in many commercial breeding colonies and is easily transmitted from one hamster to another, or by exposing another hamster to the same environment.

Retailers often offer a *pet guarantee* that provides for replacement of an animal should it die within a specified number of days. If there is an outbreak of wet tail and the replacement hamster comes from the same source and is then placed in the same cage as its predecessor, it will most likely develop the same condition.

Over-the-counter (OTC) products sold to prevent or treat wet tail should be approached with caution. Many OTC products interfere with the action of veterinary prescription drugs, delaying or preventing vital medications from being absorbed.

FIGURE 10-5 Gerbils can be held securely in one hand. They are less likely to attempt to bite than hamsters.

Nondigestive Disorders

Bacterial pneumonia is another common disease in hamsters. Signs include rhinitis, ocular discharge, and a purulent discharge consisting of pus from the nasal cavity. Prognosis is poor. Bacterial pneumonia can also be seen in gerbils, but the disease is less common.

Various types of tumors are seen frequently in hamsters, more so than with gerbils. The more common tumors are carcinomas, cancerous tumors, or

adenomas, tumors involving the glands. Some tumors become so invasive that they are difficult to remove surgically due to the volume of blood supply nourishing the tumor. A catastrophic loss of blood volume may occur when the tumor is removed (**FIGURE 10-6**). **Lymphomas** occur fairly frequently in both species of hamsters. Tumors invade the lymph nodes, spleen, and liver. There may be few signs before the hamster is found dead.

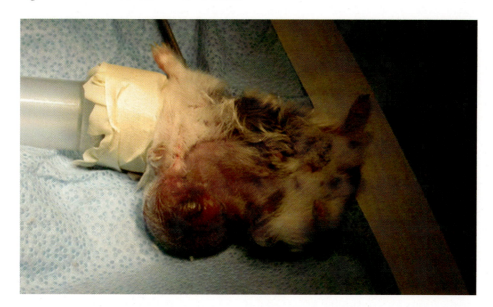

FIGURE 10-6 Large, vascular tumors are fairly common in hamsters. (*Courtesy of Eric Klaphake, DVM*).

Serious medical concerns are rare in the gerbil. Many gerbils are prone to seizure-like activity. The seizures are often triggered by the stress of handling or bouts of hyperactivity but rarely last longer than a few seconds. Owners should be cautioned to expect these incidents. The gerbil should be left alone and quietly observed during recovery. Attempts to assist in any way could result in a *latch-on* bite. Most gerbils recover without complications.

Red-nose occurs in gerbils that are incorrectly housed in wire cages. Constant chewing on cage bars irritates the end of the muzzle and nose, creating hair loss and inflammation. Changing the habitat will improve the condition. If not remedied, it may progress to facial **eczema** (a skin disease) with lesions around the nose, alopecia, inflammation of the skin, and moist dermatitis.

Rodents, including hamsters and gerbils, have a **Harderian gland** behind the eyes. This gland produces a lipid-based secretion when overstimulated. Stress and overcrowding can increase Harderian gland activity and secretion. One common, but easily remedied, reason is that the ambient humidity is too high. Humidity should not be greater than 50 percent, and moving the habitat to another location with a lower level of humidity will help remedy the condition. In severe cases, a veterinarian may recommend a topical antibiotic.

Parasites

Parasites are not commonly diagnosed in hamsters and gerbils. Both hamsters and gerbils, if in contact with rats and mice, may be infected with pinworms. If pinworm is suspected, it is usually diagnosed by placing a small piece of clear tape on the rectal area and examining the tape, mounted on a microscope slide, for the presence of ova. High parasitic infestations can cause lethargy, weight loss, rectal prolapse, and self-mutilation around the rectal area. Treatment includes ridding the environment of the pinworms along with treating the patient.

Ectoparasites are not a common problem in hamsters and gerbils. The demodectic mite has been reported in small numbers in hamsters and considered clinically normal. In large quantities, the mite can cause clinical signs such as alopecia and crusty skin along the back and neck of the hamster.

Clinical Procedures

A blood sample from a gerbil may be obtained from the lateral tail vein, with caution. It is usually best to quickly mask down the patient to help avoid tail degloving. The blood draw will be easier if the tail is warmed first by either dipping it in a warm water bath or wrapping it in a warmed cloth to dilate the vessels. This is not possible with hamsters because of their very short tails. Blood may also be drawn from the cranial vena cava. This is not without risk and the patient needs to be fully anesthetized. Injections may be given SQ or intraperitoneal.

Fluid therapy can be challenging in these small rodents. Vascular access is difficult and does not allow for large volumes of fluid to be administered. Most medications are considered extra label use in hamsters and gerbils because most medications are not currently approved for use in rodent species. Hiding medication in food can be a problem because most rodents are not tempted to eat something that has an unfamiliar taste or smell.

Anesthesia can be achieved by placing a cat mask over the animal with induction of either sevoflurane or isoflurane and maintained with a converted syringe case connected to the nonrebreathing anesthesia unit (FIGURE 10-7).

FIGURE 10-7 Converted syringe cases make excellent anesthesia masks for small rodents.
(*Courtesy of Eric Klaphake, DVM*).

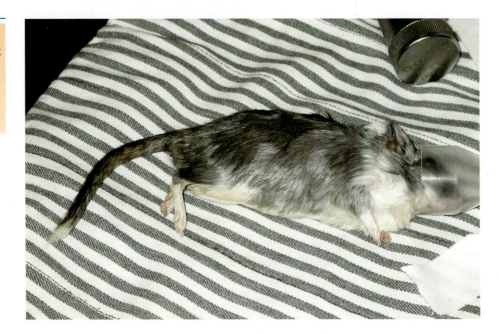

Anesthesia is required when taking radiographic images to obtain clear positioning and without harming the patient. Small tape stirrups around the limbs will help facilitate the extension of the limbs for appropriate positions. Radiographs are useful in detecting abdominal distention, abdominal masses, dilation of the cheek pouches, and gaseous dilation of the intestines.

Summary

Hamsters and gerbils are often first pets for children. There are many similarities between the species, but even a greater amount of difference in the housing, temperament, and sociability: Gerbils may be housed together while hamsters are solitary, gerbils are diurnal while hamsters are strictly nocturnal. Handling and restraint techniques, medical procedures, radiology, and anesthesia are similar in approach for both species.

fastFACTS

HAMSTERS/GERBILS

WEIGHT
- Russian hamster: 30–60 g
- Golden hamster: 40–70 g
- Gerbil hamster: 50–131 g

LIFE SPAN
- Russian: 1½–2 years (18–24 months)
- Golden: 2–2½ years (24–36 months)
- Gerbil: 2–3 years (24–39 months)

REPRODUCTION
- **Sexual maturity:** 5–8 weeks (Russian); 6–12 weeks (Golden); 9–12 weeks (Gerbil)
- **Gestation:** 16 days (Russian); 15–18 days (Golden); 23–26 days (Gerbil)
- **Litter size:** 4–8 (Russian); 4–12 (Golden); 3–8 (Gerbil)
- **Weaning age:** 4 weeks (Russian); 5–9 weeks (Golden); 3–4 weeks (Gerbil)

VITAL STATISTICS
- **Temperature:** 37.6°C (99.6°F)
- **Heart rate:** 300–460 beats/min (Russian); 200–400 beats/min (Golden); 260–600 beats/min (Gerbil)
- **Respiratory rate:** 60–80 breaths/min (Russian); 40–70 breaths/min (Golden); 85–160 breaths/min (Gerbil)

DENTAL
- Dental formula 2 (I 1/1, C 0/0, P 0/0, M 3/3)

ZOONOTIC POTENTIAL
Bacterial
- Salmonella

Viral
- LCM (lymphocytic choriomeningitis)

Fungal
- Dermatophytosis

Parasitic
- Giardia
- Coccidia

Review Questions

1. Why are hamsters not a good choice of pet for children?

2. What is wet tail?

3. Refer to Chapter 2. What is the significance of LCM?

4. Red-nose is a condition seen in gerbils. Describe this condition and its causes.

5. What are the most important things to remember about housing a hamster?

6. Seizures are common in gerbils. What events could trigger seizure activity?

7. What are the two species of hamsters?

8. Hamsters can be restrained with a scruff. What could happen if the scruff is maintained for too long?

9. Why should gerbils not be caught or held by the tail?

10. Gerbils are monogamous. What significance does this have with regard to a colony of gerbils?

Case Study I

History: A three-month-old hamster is presented by the new owner. She reports that she has had the hamster for only a few days but it seems very lethargic, has watery diarrhea, and smells really bad.

Physical Examination: The patient is dehydrated and feels cold to the touch. The hamster is reluctant to move, but when it does, ataxia is evident. There is a significant amount of wet fur around the caudal area and the hamster has a very noticeable odor. The veterinarian advises that the prognosis is grave and recommends euthanasia.

   a. What is the veterinarian's diagnosis?

   b. What is the causative agent for this condition?

   c. Would the owner be wise to return the hamster to the pet store for a replacement? Why, or why not?

Case Study II

History: A year-old male gerbil has been brought in by a concerned owner. The tail has been completely "skinned" and there is evidence that the gerbil has been gnawing at the tail. The gerbil is housed alone in an aquarium with a water bottle, hide box, and food dishes. The gerbil runs "constantly" in open-wire rodent wheel.

Physical Examination: The veterinarian confirms that the tail has been degloved and recommends amputation.

 a. Define the term *degloved*.

 b. What is the direct cause of this condition and how could it have been prevented?

For Further Reference

Brown, C., & Donnolley, T.M. (2004). Disease Problems of Small Rodents. In Quesenberry, K.E. and Carpenter, J.W. (Eds.), *Ferrets, Rabbits, and Rodents: Clinical Medicine and Surgery, 3rd Edition.* (pp 364–372). Philadelphia, PA: W.B. Saunders.

Ferris, N. *Hamsters: Syrian & Russian Dwarf Hamsters.* American Fancy Rat and Mouse Association, 1995–2006. http://www.afrma.org (accessed December 28, 2013).

Goodman, G. (2002). Hamsters. In Meredith, Anna & Redrobe, Sharon (Eds.), *BSVA Manual of Exotic Pets, 4th Edition* (pp. 26–32). Hoboken NJ: John Wiley & Sons.

HSUS. How to Care for Gerbils. In Meredith, Anna & Redrobe, Sharon (Eds.), *Animal Sheltering How-to Series,* March–April 2002. Hoboken NJ: John Wiley & Sons.

http://www.ocw.tufts.edu (accessed April 7, 2006), December 28, 2013.

http://www.merckmanuals.com/vet/index.html (accessed June 1, 2014).

Keeble, E. (2002). Gerbils. In Meredith, Anna & Redrobe, Sharon (Eds.), *BSVA Manual of Exotic Pets, 4th Edition* (pp. 34–45). Hoboken NJ: John Wiley & Sons.

Mayer, J.D. (2006). *Analgesia and Anesthesia in Rabbits and Rodents.* Conference Notes, Presented at the North America Veterinary Conference, Orlando, FL, January 7–11, 2006.

Nolen, S. *CDC on the Offensive to Stamp Out Rodent Virus (SIC).* http://www.avma.org/ (accessed March 28, 2006).

Quesenberry, K.E., Donnelly, T.M., & Mans, C. (2012). Biology, Husbandry and Clinical Techniques of Guinea Pigs and Chinchillas. In Quesenberry, K.E. and Carpenter, J.W. (Eds.), *Ferrets, Rabbits, and Rodents: Clinical Physiology and Clinical Techniques* (pp. 279–294). St. Louis, MO: Elsevier Saunders.

RATS AND MICE

OBJECTIVES

After completing the chapter, the student should be able to

- Correctly house a rat or mouse.
- Provide appropriate client education to new rat or mouse owners.
- Describe the correct diet for rats and mice.
- Provide basic nursing care to a rat or mouse.
- Assist in the anesthesia of a rat or mouse.
- Demonstrate appropriate restraint techniques.
- Understand common medical disorders in rats and mice.

Introduction

Rats and mice are two of the most successful and adaptive species of rodents in the world. They can be found in nearly every region in natural environments and following human habitation worldwide. They have been vilified, met with horror, and exterminated by the millions, yet they still survive in great numbers.

Both rats and mice are used extensively in research. Mice have contributed greatly to the decoding of the human genome (genetic tagging) and countless disease and reproductive studies. Rats are used in cancer and tumor research, developmental learning, and cognition studies, which measure awareness and intelligence. Rats and mice have not only caused human disease but also played an essential role in researching cures for human diseases (see Chapter 2).

Rats and mice are also bred as *feeders*, food for several species of carnivores kept in captivity. They are purchased to feed snakes and lizards, scorpions and tarantulas, and birds of prey. They are provided as food in zoos and wildlife rehabilitation centers. They can be purchased from any pet store or ordered frozen and vacuum packed. Feeders are available in different ages and sizes to meet any need. There are pinkies (hairless neonates), fuzzies (pups with a slight coat of hair) (**FIGURE 11-1**), hoppers (juvenile mice), and small, medium, large, and jumbo adults.

The Latin term *rodere* means to *gnaw*. The long chiseling incisors of rodents have also contributed to their success. Few things are impervious to gnawing rodents; they gain access to homes, feed bins, and barns, and they tunnel under roads and inhabit sewers.

References to rodents (*you dirty rat, like a rat in a maze*) pepper human language, and attempts to perfect solutions for complicated problems are dismissed as *trying to build a better mouse trap*. Rats and mice have also been personified in many children's books, nursery rhymes, and films as brave and heroic, cute and loving, wise and gentle. There is probably no other group of animals that elicits such varying and adamant reactions from humans.

The pet mouse, *Mus musculus*, and the pet rat, *Rattus norvegicus*, are very popular as alternative small-animal companions. These are domesticated strains that are healthier and have been bred specifically for the companion animal market. Mice are available in a variety of colors, referred to collectively as fancy mice. Some varieties of fancy mice are pied, Himalayan, satin, long haired, brindles, and seal points. There is a wide variety of colors in rats: solids of many colors, hooded (black and white), Siamese, blues, silver, rex, and agouti. There is also a hairless rat (**FIGURE 11-2**) and the newer Dumbo rat, with larger-than-normal ears.

People may join the American Fancy Rat & Mouse Association (AFRMA), an international organization that encourages rats and mice as pets. AFRMA organizes shows and exhibitions, complete with a rule book and standards for different varieties. It is a nonprofit organization with a focus on education and responsible ownership.

More exotic species of mice and rats have been imported, but aside from an initial limited interest, they have not become established in the pet trade. This is due in great part to legislation governing and controlling the importation and possession of exotic rodents; their zoonotic potential, known and unknown; and the devastating effects of introducing nonnative species that could quickly become established as viable populations. Giant Gambian rats, African Zebra mice (also traded as the Striped Grass mouse), and African Spiny mice have all been imported, many of them smuggled in by private collectors. One species that seems to maintain interest is the African Spiny mouse (**FIGURE 11-3**).

African Spiny mice, also called Egyptian Spiny mice, are occasionally available in pet stores. Legal ownership varies from state to state. It is from dedicated keepers and private owners that much of the information on Spiny mice is available.

Spiny mice are originally from the deserts and savannahs of Africa, India, and areas of the Middle East. They live in rock crevices and abandoned burrows of other species. As the name suggests, they have a unique coat with coarse, bristly, spine-like hairs. They are larger than the domestic mouse, are golden brown in color, and have a creamy-white underbelly. They have a long, narrow face with large ears and very dark eyes. Owners have found them to be more social than domestic mice, calmer to handle, and less inclined to bite.

Spiny mice are monogamous and have two litters a year, in the spring and summer. Gestation is more than twice as long as the domestic mouse (19 days) with as great a variance as 36 to 45 days. There is an average of two to three pups per litter. When pups are born, they are more developed than the common mouse. They do not have a full coat, but their eyes are open, they are active, and are able to follow the dam. They nurse vigorously and begin to nibble solid food within a

FIGURE 11-1 Rat and mouse pups are raised to feed other species. Feeder rodents are sorted by size and age. Pinkies (neonates) and a fuzzy (10-day-old mouse) are shown here.

FIGURE 11-2 A great variety of domestic rats are available. Illustrated is one of the hairless varieties. This rat has only black eyebrows, a patch of hair on the nose, and small tufts behind the ears. (*Courtesy of Eric Klaphake, DVM*).

FIGURE 11-3 One of the more unusual species is the African Spiny mouse. A dam is shown here with her 10-day-old pup.

few days of birth. The hair coat is complete by the time they are three to four days old. Breeding can be encouraged by feeding live crickets, grasshoppers, or mealworms once or twice weekly, not just during breeding seasons, but as a diet staple.

Spiny mice should not be picked up by the tail. They are delicate and easily damaged. Degloving wounds or complete amputation of the tail can occur in overcrowded cages. They are best housed as pairs or in small family groups.

Behavior

Generally, rats are considered to be better pets than mice. They are easily socialized as pups and rarely bite. Male rats tend to be more docile than females. Neutered male rats produce little odor. Rats are curious and intelligent and are capable of learning puzzle-solving skills, which they are able to retain and apply to new situations. They have been taught to play basketball, retrieve items, and indulge in social play with their owners. Because of the growing popularity of rats as pets, new behaviors and terms have been noted as owners spend more social time with them. **Boggling** is one such term. Boggling, or becoming *bug eyed*, can be somewhat alarming, but it is, apparently, a sign of contentment. Another often reported behavior in a contented rat is **bruxism**. In many other species, teeth grinding is usually associated with pain.

Mice are more active and will rarely sit quietly for long enough to allow much handling or interaction. They are also much quicker to bite and have a generally more nervous disposition. The odor of mice is much stronger than that of rats. They are less social with each other, often attacking or *ganging up* on a cage mate, inflicting severe injury or death. A group of mice is collectively referred to as a **gang** of mice. Rats rarely exhibit aggressive behaviors to one another.

Both rats and mice combine litters, care for, and nurse each other's young. Male rats are very tolerant of pups and will often hold them down and groom them. Male rats tend to the young when the female leaves the nest to feed. Both males and females pick up pups that have strayed and place them back in the nest with the others. Male mice will tolerate pups to a lesser degree and frequently cannibalize pups found away from the rest of the litter.

Sexing rats and mice is similar. In pups, sex can be determined by the ano-genital distance, the length between the anus and the genitalia. In a female, the distance is much shorter than in a male. In mature males, there is little room for doubt. Testicles are visibly large in lightly haired scrotal sacs. Testicles descend in juveniles prior to reaching sexual maturity.

Rats and mice reach sexual maturity between six and eight weeks of age. Once mature, females cycle every four to five days. Rats can be caged as a pair or set in small harems, with one male and two to three females. In large colonies of mice, some females may not cycle unless a new male is introduced. When a different male is placed in with a group of females, the females all begin to cycle at the same time. This is called the **Whitten effect**. While first recognized in mice, synchronization of reproductive cycles occurs with many other species.

Another peculiarity, called the **Bruce effect**, occurs in female mice that have been bred within 24 to 48 hours prior to being introduced to a different male. The first litter is frequently aborted or reabsorbed and the female will rebreed with the new male.

Average gestation time for mice is 16 to 18 days with litters averaging 8 to 12. Gestation time for rats is only slightly longer, with an average of 20 to 22 days. Rat litters are often larger and can produce as many as 18 pups.

Should a rat disappear from its cage, it is usually not a problem to capture it. The rat has probably curled up in a warm, soft spot to sleep. Many rats, on hearing their owners, will run up to them, expecting to be picked up and given a treat. When mice escape, they are very difficult to locate and recapture. Escaped mice may be heard somewhere in the house, but not seen. The use of a live trap is usually required. It is baited with familiar food and set at night in a location the mouse is most likely to be or where it was last heard.

Pet rats can be kept individually, in pairs, or in groups of one male and several females. Neutered male rats can be kept together. Introducing a young male rat to an established male can result in fighting and injury. Mice can be grouped similarly, but the introduction of a new mouse of either sex into an established colony will most likely result in death and cannibalization by the others. There is always one dominant male in a gang of mice. Other male mice are bitten and intimidated by the dominant male, and this may encourage the same bullying behavior from the females. Only the dominant male breeds with females. Younger males have been observed to form bachelor colonies, not even attempting to mate or showing interest in the females.

Housing

Cages for mice and rats should have good ventilation and solid floors. Without adequate ventilation, toxic ammonia fumes from urine will quickly build up and cause respiratory problems. Wire-framed cages for mice need to be chosen carefully by looking at the width of bar spacing. Many small-mammal cages would easily allow mice to escape through the bars. Long aquariums with fitted screen lids are appropriate for both species. Bedding substrates of recycled paper and aspen shavings are both suitable. Cedar should not be used because the aromatic fumes are toxic and will lead to respiratory problems and, if exposure is prolonged, death.

Cage furnishings should include a hide box, which is often shared (**FIGURE 11-4**), wheels for exercise, sections of PVC pipes, and empty toilet paper rolls for hiding in and chewing on. Any small dish that can easily be cleaned can be used for a food dish. Sipper tube bottles are used by both rats and mice. Water bottles should be checked and refilled twice daily. Bottles are frequently gnawed by rodents and damaged enough to cause the water to leak. If rodents run out of water, they have been known to cannibalize one another in order to obtain moisture.

Assorted wood blocks should be given for gnawing. Many sold in pet stores are flavored and dyed with nontoxic food coloring. Rats have been rushed to the clinic with bright pink and red blotches around the mouth, blue bottoms, and green ears. A closer look at the cage would probably produce a chew block under a dripping sipper tube.

FIGURE 11-4 Gangs of mice can be very social within an established group and will often share hide boxes.

Diet

Rats and mice have survived for so long and in such numbers because they will eat anything. However, with companion rats, the goal is not just survival but health and longevity. Commercial rodent blocks, which are nutritionally complete and fed exclusively to lab rodents, should also form the basis of companion rodent diets. They are hard and require gnawing, which helps keep the incisors worn down. Pet store shelves are full of various rodent mixes and special treats. Mixes should be evaluated for content, and those with excessive amounts of sunflower seeds should be avoided.

Rats particularly enjoy bananas, slices of apple, small pieces of cracker, and unsugared breakfast cereals. All of these add variety to the diet without contributing greatly to obesity. Often, a companion rat or mouse has the same *supper* as the owner. Feeding small amounts of scraps is not likely to harm the animal, but amounts should be limited and never substituted for a maintenance diet. At the top of the list of rat favorites is the all-American favorite, the hot dog. A hot dog will tempt any rat, and it is a good food item in which to inject medication that the rat might otherwise refuse. Pieces of hot dogs should not be offered on a regular basis. Very small pieces (a scraping), given occasionally, will condition the rat to accept medicated pieces without fuss. Both rats and mice have a very strong attraction to cheese. If cheese is given, it should only be a tiny piece, about the size of a single cottage cheese curd. Cheese has been known to cause constipation in pet rats.

Handling and Restraint

Mice, when restrained, will attempt to bite. A mouse can be picked up by the base of the tail and, with the tail hold maintained, placed on a rough surface. The mouse will attempt to run away, become stretched out, and allow the restrainer to collect it up in the other hand with a scruff. Once securely scruffed, the tail should be held with the little finger wrapped around it. This leaves one hand free for examination or treatment (**FIGURE 11-5**).

If a mouse is held by the tail for more than a brief transfer to another cage, it will quickly crawl up its tail and bite. Wearing leather gloves is not an acceptable approach to mouse restraint. Mice are easily injured or lost as it is impossible to secure them while wearing the bulk of a leather glove.

Rats can just be picked up with one hand. They should be grasped with one hand around the shoulders and the other hand placed underneath for support. Never carry a rat by its tail. It is not necessary and is one of the few instances that might provoke a bite. Rats, like mice, will run up the tail and bite the restrainer or, from the rat's perspective, the object or predator that has ahold of its tail. If the procedure is likely to cause pain, injections for example, place one hand around the shoulders as normal, but place the thumb under the mandible, preventing the head from turning. The other hand should be placed around the rat's rear limbs, stretching them out slightly.

FIGURE 11-5 A mouse is easily restrained for an examination with a scruff.

Medical Concerns

General signs of illness in rats and mice include weight loss, lethargy, and a poor hair coat. They sit hunched, separated from cage mates, and may exhibit **piloerection** (hairs standing upright). Piloerection is often a sign of pain in many species. There may be an accompanying red nasal and ocular discharge, **red tears** and **red sneezes**. This is not blood but **porphyrin** from the Harderian gland located behind the eye. Porphyrin may also be evident in overly stressed rats and mice. Rats and mice that are ill still attempt to groom and the secreted porphyrin can be transferred to other areas of the body, making it appear as if they are smeared with blood. Rats in excellent health sneeze frequently, and this alone should not be interpreted as disease.

Respiratory Disorders

Respiratory disease is frequently seen in pet rats and mice. In rats it is usually a mycoplasma bacteria that causes chronic upper respiratory infection (URI), while in mice it is more likely to be a virus causing acute URI. Signs include dyspnea, loud and raspy respiratory sounds, weight loss, lethargy, hunched posture, and most indicative of all, **heaving**, or abdominal muscle movements with each breath. Treatments may include antibiotics, **bronchodilators** (medications that expand the bronchial airway), subcutaneous fluids, and nutritional support. A **nebulizer**, or vaporizer, may also be used within an oxygen-infused isolation unit.

Gastrointestinal Disorders

Tyzzer's disease is seen in many rodents. It is a bacterial infection that attacks the GI tract and causes diarrhea, dehydration, and death. Most rodents are positively diagnosed postmortem. Rodents rarely survive Tyzzer's disease, even with the best supportive care, as they are usually presented in the late stages of the disease.

Nondigestive Disorders

Incisor malocclusion is fairly common in rats, but it is usually trauma induced rather than genetic. Lack of suitable gnawing material and chewing at the cage bars contribute to this condition. Chewing cage bars can loosen the roots of the continually growing incisors and cause them to stray from their normal position. A dental burr is the best tool to use in trimming abnormal rodent teeth. The teeth are open rooted, and trimming them will not cause pain. The use of standard nail clippers is likely to shatter the teeth, causing even more damage.

Rats and mice are both prone to **otitis interna**, a bacterial infection of the inner ear. Most present with a head tilt and reported behavior of constant circling. The direction of the circle is usually an indicator of which ear is affected. This condition is treatable with antibiotic therapy.

Abscesses occur from fighting or as a result of an injury from an unsafe area of the cage or furnishings. Abscesses are surgically opened, the pocket of pus removed, and the site flushed with a dilute betadine solution. The solution should be the color of weak tea. Surgical scrub should never be used to flush a wound of any type. The soapiness of scrub only compounds the problem. An abscess in a rodent is usually thick and similar to cottage cheese in appearance (**FIGURE 11-6**). It is recommended that all involved in removing the abscess should wear disposable gloves for their protection against bacterial infection and to avoid introducing other pathogens into the abscess cavity. Abscess sites are then allowed to heal as open wounds.

Tumors are by far the most prevalent problem found in both rats and mice. Tumors grow quickly and usually have a rich blood supply. Many times the question is raised, *Is the tumor removed from the rat, or the rat removed from the tumor?* Tumors can become so large that they restrict movement (**FIGURE 11-7**). The

FIGURE 11-6 Abscess material (pus) in rodents is thick and cream colored. It is often described as caseous, or cheese-like.

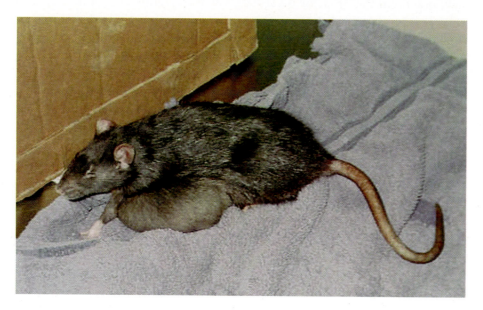

FIGURE 11-7 Tumors are by far the most prevalent problem found in both rats and mice. They can become so enlarged as to restrict movement. This is an example of a large, vascular tumor in a female rat. (*Courtesy of Eric Klaphake, DVM*).

FIGURE 11-8 A mammary gland tumor on a female mouse. Many tumors of mammary gland origin can be prevented by spaying the female when she reaches sexual maturity.

FIGURE 11-9 Mice will often gang-up on a cage mate. This is an example of a severely barbered female that has been bullied by cage mates. Barbering can also occur from boredom or overzealous grooming.

only treatment is surgical removal. There is also a predictability of another tumor occurring. Tumors are best removed when they are small. Many perioperative deaths can be attributed to the dramatic loss of blood and blood pressure directly related to the volume of circulating blood within the tumor. Mammary gland tumors can potentially be avoided by spaying a female before she reaches sexual maturity (**FIGURE 11-8**).

Barbering is a common problem in mice colonies. Mice normally groom each other, but the behavior is carried to excess by dominant or bored mice. There will be patches of alopecia and the damage may extend to skin abrasions (**FIGURE 11-9**). The traumatized sites should also be examined for the presence of mites. Initially, in severe cases of barbering, the conditions may appear similar.

Parasites

Common parasites in rodent colonies include pinworms and mite infestation. Pinworms are small roundworms of the GI tract. The ova are expelled with the feces. An easy diagnostic method for pinworm is to apply a transparent strip of tape to the rectal area. The tape will collect ova that can be visualized with a microscope. The tape is applied directly to the microscope slide and scanned using immersion oil at 100×.

Mites and lice can be diagnosed with clinical signs and a skin scraping. It is not unusual to see these ectoparasites with patients that have other medical problems or live in poor conditions. Anthelmintics including oral ivermectin are commonly prescribed.

Clinical Procedures

When collecting blood from a mouse or rat, not more than 10 percent of total blood volume should be taken. Blood volume in a healthy rat or mouse may be calculated from an average base volume of 70 ml/kg of bodyweight. Neither rats nor mice reach a bodyweight of 1 kg (2.2 lbs). Conversions must be careful and accurately performed with the use of a calibrated gram scale to obtain the exact weight just prior to drawing blood. The best choice for blood collection in a rat or mouse is a warmed lateral tail vein (**FIGURE 11-10**). Other options include the lateral saphenous vein, which is much smaller and more difficult to access and obtain an adequate sample. To collect blood samples, a 23-gauge needle is used on a tuberculin syringe or a heparinized microhematocrit tube collecting blood directly from the needle. The orbital sinus can be used as a collection site, but it is not recommended in companion rats and mice. Although it is a commonly used site in research animals, this method frequently causes severe damage to the eye.

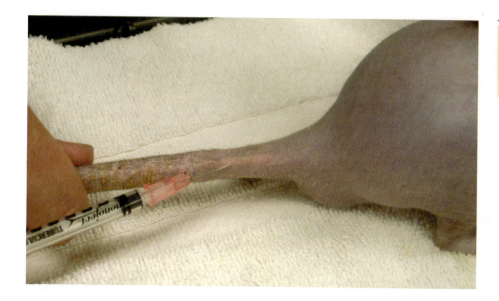

FIGURE 11-10 The approach for blood collection in a rat, accessing a lateral tail vein.

The rat can be manually restrained or placed head first into an empty 60 cc syringe case with only the tail protruding. If opting for syringe case restraint, the end of the case needs to be perforated from the inside or have the end tip cut away to allow air flow and avoid suffocation. Mice can be restrained in the same manner by using a 3 cc syringe case. Most 3 cc syringe cases are open ended. A tuberculin syringe with a 25-gauge needle offers the best chance of avoiding vein laceration or collapse. It is of adequate size for the small volume that will be collected.

Preferred routes of injection are SQ, between the scapula, and intraperitoneal (IP). If using the IP approach, injections should be given in the lower right quadrant to help avoid internal organs. Up to 5 ml of fluid can be given SQ or IP to rat for fluid therapy using a 25-gauge needle.

Oral medication and nutrition can be administered through a small tuberculin syringe, a dropper, or a small gastric feeding tube. Medication added to water or food is not recommended because the patient may not be eating or drinking due to illness.

Most rodents are masked down with an inhalant anesthesia. A large canine mask is placed directly over the patient so it sits flush with the table. Once a suitable plane of anesthesia is achieved, one without the requirement of further restraint, the patient may be transferred to a smaller mask, placed directly over the nose. Rodent-sized masks are commercially available, but they are easily improvised from syringe cases or small dispensing bottles that have been cut in half. The top end is connected to the nonrebreathing system with an elbow and taped securely in place to avoid leaks. The mask is formed by taping a latex glove over the large end and cutting a small slit in the glove, big enough only to accommodate the rodent's head, yet not putting pressure on the trachea (**FIGURE 11-11**). These *ready-made* maintenance masks should be disposed of after each use. Eye lubricant or artificial tears should be placed in the patient's eyes to prevent oxygen and anesthetic gas from drying out the eyes. Pulse oximeters can be applied to a foot or to the tail to monitor heart rate and oxygen saturation.

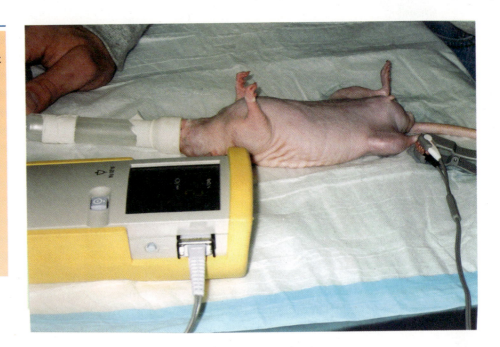

FIGURE 11-11 Converted syringe cases make excellent masks for inhalation anesthesia. The one used here is from a 35 cc syringe. The distal end has been removed and connected (taped) directly to the nonrebreathing unit. These *ready-made* maintenance masks should be disposed of after use. The patient is a hairless rat, about to undergo castration surgery.

Hypothermia is a concern with small rodents under anesthesia. They should be supported with additional heat in the form of a circulating water pad, preheated and ready. Small mammals can also be kept warm during surgical procedures by supporting them with disposable gloves filled with warm water. Gloves used in this manner must be monitored carefully for appropriate temperature. As they lose their heat they should be replaced as often as necessary to maintain warmth for the patient. The water-filled glove should also be checked for small leaks. Patients should be recovered in a prewarmed incubator (**FIGURE 11-12**).

To obtain diagnostic positioning for radiographs and prevent movement, anesthesia is required. Dental x-ray units have shown to be effective in taking radiographs of mice and small rats. These units are able to focus at short distances for small patients. Taping (cloth or masking tape) the patient's limbs to the cassette will prevent rotation of the body. Lateral and ventro/dorsal views of the whole body are taken to examine masses, fractures, and pulmonary edema associated with respiratory diseases.

Summary

Rats and mice have long been used extensively in research studies. They are also bred as food for a variety of carnivores. Generally, rats are better pets than mice. They are easily socialized and rarely bite. Rats are intelligent and responsive to their caregivers. Mice are quicker to bite and rarely interact with people and have a more nervous disposition. Rats and mice can be housed simply, often in large aquariums with screened lids to provide for adequate ventilation. Rats are easier to pick up and restrain than mice. Both species have similar medical concerns, the most common being the production of tumors that can affect any area of the body. Both rats and mice are susceptible to respiratory infections and bacterial infections of the inner ear. Blood collection can be difficult because of small veins that easily collapse. The preferred site is the lateral tail vein. General anesthesia is achieved by masking down the patient or with the use of an induction chamber. Both isoflurane and sevoflurane have been used successfully.

fastFACTS

RATS/MICE

MICE:
- Weight: 20–40 g

LIFE SPAN
- 1–3 years (12–36 months)

REPRODUCTION
- Sexual maturity: 8–12 weeks (2–3 months)
- Gestation: 18–19 days (average)
- Litter size: 8–12
- Weaning: 4 weeks

VITAL STATISTICS
- Temperature: 38.5–40°C (100.4–104°F)
- Heart rate: 320–760 beats/min
- Respiratory rate: 60–220 breaths/min

DENTAL
- 16 open-rooted teeth
- Dental formula 2 (I 1/1, C 0/0, P 0/0, M 3/3)

RATS:
- Weight: 250–450 g (males heavier)

LIFE SPAN
- 2.5–3 years average (30–36 months)

REPRODUCTION
- Sexual maturity: 8–12 weeks (2–3 months)
- Gestation: 21–23 days
- Litter size: 8–18
- Weaning age: 4–5 weeks

VITAL STATISTICS
- Temperature: 37.7°C (99.8°F)
- Heart rate: 250–450 beats/min
- Respiration rate: 70–115 breaths/min

DENTAL
- Total 16 teeth
- Dental formula 2 (I 1/1, C 0/0, P 0/0, M 3/3)

ZOONOTIC POTENTIAL
Viral
- Lymphocytic choriomeningitis
Bacterial
- Streptobacillus (rat bite fever)
- Campylobacter
- Leptospirosis
- Salmonella

Review Questions

1. When comparing mice and rats, why are rats considered better companion animals than mice?

2. Why should cedar not be used as bedding?

3. Rodents that are stressed or ill often have red tears. What is the cause?

4. How is sex determined in rats and mice?

5. What are the different types of feeder rodents?

6. How do restraint techniques differ between rats and mice?

7. What restrictions apply to owning African Spiny mice?

8. It is very common for rats and mice to produce large tumors. What is a common cause of death during tumor-removal surgery?

9. If rats and mice are deprived of water, how do they obtain moisture?

10. What is the Whitten effect?

Case Study I

History: A one-year-old neutered male rat has been brought in for examination because the owner has seen him "crying red tears." She also reports that "Winston" does not seem to have much interest in food or playing. He sneezes a lot and spends more time than usual curled up sleeping.

Physical Examination: The veterinarian also notices some red crusting around the nares and eyes. The hair coat is poor and there are areas of red stain on the coat. With quiet observation from a distance, he notes that the rat sits with a hunched posture and is slow to respond. There are audible respiratory sounds.

a. What is the veterinarian's most likely diagnosis?

b. What is causing the "red tears"?

c. What is the treatment he prescribes for Winston?

Case Study II

History: A six-month-old male mouse has been brought in for examination. The owner reports she noticed a small lump about two weeks ago, and it appears to be growing. At first the lump did not seem to bother the mouse, but now the patient has started to limp on a front leg. The other two mice in the cage seem fine.

Physical Examination: The lump is located in the axillary region of the patient's right foreleg. The veterinarian notes that the lump has interfered with the movement of the foreleg, but the foreleg itself does not seem to be directly involved. The veterinarian also notes that there is a small puncture wound on top of the reddened lump. He has ruled out a mammary gland tumor.

a. What is the most likely cause of this lump?

b. What is the recommended treatment for this problem?

c. Describe the contents of the lump.

For Further Reference

Brown, C., & Donnelly, T.M. (2004) Disease Problems of Small Rodents. In Quesenberry, K.E. and Carpenter, J.W. (Eds.), *Ferrets, Rabbits, and Rodents: Clinical Medicine and Surgery, 3rd Edition.* (pp. 364–372). Philadelphia, PA: W.B. Saunders.

Ferris, N. *Hamsters: Syrian & Russian Dwarf Hamsters.* American Fancy Rat and Mouse Association, 1995–2006. http://www.afrma.org (accessed June 12, 2006).

http://www.ocw.tufts.edu (accessed April 7, 2006).

http://www.alnmag.com/articles/2010/02/fundamentals-pain-assessment-rodents

http://www.ncb.nim.gov (accessed March 11, 2014).

http://www.purdue.edu (accessed March 11, 2014).

Joerg D. (2006). *Analgesia and Anesthesia in Rabbits and Rodents.* Conference Notes, The North America Veterinary Conference, Orlando, FL, January 7–11, 2006.

Johnson, R. Inside the Rat Pack, All Animals, The Humane Society of the United States, November/December 2013.

Quesenberry, K.E., Donnelly, T.M., & Mans, C. (2012). Biology, Husbandry and Clinical Techniques of Guinea Pigs and Chinchillas. In Quesenberry, K.E. and Carpenter, J.W. (Eds.), *Ferrets, Rabbits, and Rodents: Clinical Medicine and Surgery, 3rd Edition* (pp. 279–294).

SHORT-TAILED OPOSSUMS

12 CHAPTER

OBJECTIVES

After completing the chapter, the student should be able to

- Describe the correct habitat for a short-tailed opossum.
- Provide appropriate client education to new short-tailed opossum owners.
- Describe the correct diet for a short-tailed opossum.
- Provide basic nursing care to a short-tailed opossum.
- Assist in the anesthesia of a short-tailed opossum.
- Demonstrate appropriate restraint techniques for short-tailed opossums.
- Obtain basic knowledge of medical problems associated with short-tailed opossums.

KEY TERMS

eutherian
prehensile
hygrometer
neoplasia
hematuria
pyuria
melanoma
sarcoma
arthrosclerosis

Introduction

Marsupials are distinct from **eutherian** mammals. Eutherian mammals have a placenta, a temporary organ that forms and connects the developing offspring to the dam with an umbilical cord. At birth, the young have completed fetal development during a gestation period within the placenta. Marsupials do not form placentas, and the young are born in an immature, embryonic state where they continue to develop outside of the dam's body.

While most people think *pouched* when defining the characteristics of marsupials, not all marsupials possess a pouch, especially the smaller species. The pouch is a specially developed fold of abdominal skin in which to carry and nourish the developing young. Some marsupials, considered more primitive, do not have a pouch, but the skin around each nipple forms a donut-shaped cushion that keeps the neonates warm and secure in their attachment as they mature from their embryonic state outside of the dam.

All marsupials are born in an immature state, only partially developed at birth. The newly delivered young are able to crawl or wriggle up the mother's abdomen and locate a teat. The teat swells, completely filling the mouth and firmly attaching the neonate. Marsupials are so named because they have distinctive, modified bones that are absent in other mammals. The marsupial bones, *os marsupialia*, form part of the pelvic attachments for abdominal muscles.

There are approximately 260 species of marsupials. Many of them are native to Australia, but many of the smaller species are found in South America. The only representative in North America is the Virginia opossum. There are no species native to Europe.

Short-Tailed Opossums

The short-tailed opossum, known as the Brazilian short-tailed opossum or simply referred to as an STO, is related to several small opossums with approximately 17 subspecies. All are found in South America. They are native to fairly humid, forested areas of Brazil and Bolivia. They are ground dwellers, foraging on the forest floor for insects and small rodents. They are welcomed into countryside homes, and it is considered good luck to have a resident *possum*. They are very curious and unlikely to carry or transmit diseases, and they readily consume more unwelcome house guests such as large spiders and scorpions. While never considered pets in their native lands, the taxonomic name (*Monodelphis domestica*) reflects their long and amiable domestic relationship with humans.

FIGURE 12-1 Short-tailed opossums are appealing and social with their owners.

Short-tailed opossums are recent arrivals to the pet market, although they have been used extensively as research animals for many years. They are not only considered social and appealing but are also relatively easy to care for, making them an undemanding yet exotic pet (**FIGURE 12-1**). All STOs available in the pet trade and through private breeders are descendants of previously held research animals. They are now banned from being exported from their native habitats.

Short-tailed opossums are one of the smallest marsupials, with a body length of only four to five inches. The tail is slightly shorter than the body, hairless and without scales. They often use their tails, which are only minimally prehensile, and are able to grasp and carry small twigs and grasses back to the nest, usually in a hollow log. The coat is thick and very soft. Their color varies slightly from silvery gray to a more smokey-gray color. Some animals may have white feet or white markings on the chest. The underbelly is lighter in color. Each foot has five toes. The first toes on the hind feet are clawless and opposable (**FIGURE 12-2**). Although primarily nocturnal, short-tailed opossums are easily roused and become active during the day, seeking food items or just exploring their environment.

In the wild, STOs stalk and pounce on many smaller prey items. They eat a variety of insects, arachnids, small vertebrates, ground nesting birds' eggs, and even the occasional bird. They also eat small amounts of fruits, some vegetables, and grains; however, there is no report or evidence that they consume plant material.

Behavior

Short-tailed opossums are solitary by nature and will not tolerate the company of others. Attempting to place two or more together will result in severe injuries or death. In the wild, the only time they come together is when a female is ready to breed and will tolerate the presence of a male.

Short-tailed opossums may breed year round. Estrus varies from 3 to 12 days. Once mating has occurred, gestation is approximately 14 days. The tiny embryos

are barely visible as they crawl up the mother's abdomen to locate and attach to a nipple.

Short-tailed opossums have no pouch but instead have 13 teats arranged in a circle on their abdomen. There may be more fetal pups delivered, but only those that are able to attach to a teat will survive. The attachment occurs because the nipple swells and completely fills the mouth of the pup. Many may be lost, and the average viable litter size is 5 to 10. Pups remain attached to the nipple for approximately 30 days and are weaned completely at 50 days.

Under no circumstances should a pup ever be physically removed from a teat. The pup, once removed, will not be able to reattach, and it would be very difficult to save it. Prior to being completely weaned, pups climb up and ride on the dam's back as they begin to learn to search for food. A good time to start socializing the individual pup is when the little *hitch-hikers* begin to scamper off the dam and become fairly independent. Most pet short-tailed opossums are acquired around one or two months of age, by which time they are fully weaned and would normally leave the family group.

Prior to weaning, short-tailed opossums may nest together. Four- to six-week-old pups may still be in the maternal nest, but as soon as the young are independent of the dam, they will leave the company of their litter mates and begin their solitary lives.

Male anatomy is typical of other marsupials. The testicles are fur covered and cranial to the penis, located on the lower abdomen (**FIGURE 12-3**). The female has a very small genital opening that may be difficult to visualize because of the fur (**FIGURE 12-4**).

Sexual maturity is normally around four and a half months. STOs have an average life span of four years but are reproductively successful for approximately three years. Regardless of the size of habitat provided, it is not recommended to

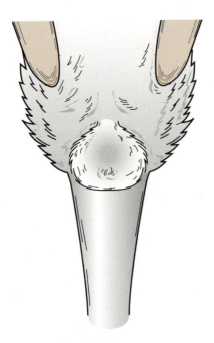

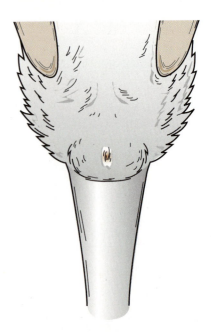

FIGURE 12-3 The external genitalia of a male short-tailed opossum. The scrotum is round and fully furred.

FIGURE 12-4 The external genitalia of a female short-tailed opossum. The female has a very small genital opening, which may be hard to visualize because of the density of the fur.

try to accommodate more than one animal per enclosure. Captive breeding can be difficult. Many times, even the introduction of a pair to each other will result in the death of one and serious injuries to the survivor. As for many exotic species, a USDA license is required to breed short-tailed opossums, even if only one pair is kept and the occasional litter produced.

Housing

Although the short-tailed opossum probably has a fairly large range of territory in the wild, they are easily kept in captivity without many special considerations. A 20-gallon-long (rather than high) terrarium tank with a screen lid works well. Basic set-up tanks are readily available in a variety of sizes, but the most important thing in choosing one is that the lid fits securely and can be latched. Even though this little opossum is a ground dweller, it can and will climb readily.

As with all exotic animals, natural environmental ranges of humidity and temperature should be provided. The humidity of South American forests is much higher than in the average household. The advantage of a glass aquarium is that it helps to maintain a slightly more humid environment. Recommended humidity is 50 to 60 percent. A small humidity gauge, or **hygrometer**, attached to the inside of the tank will help determine if the humidity is high enough. It may be necessary to add a small waterfall to the habitat. There are many self-contained waterfalls designed for use in reptile habitats that make an easy addition to the tank. The electrical cord should be checked regularly, but short-tailed opossums are not known to chew or gnaw at items. An additional source for raising the humidity is especially important during the winter months when ambient room humidity is far lower.

The ideal temperature range for a short-tailed opossum is between 72°F and 74°F. In a larger habitat, an additional source of heat is needed. Under-tank heating pads used for reptiles are safe and designed to provide small areas of greater warmth. Although live plants are attractive, lighting required for their benefit is not healthy for the short-tailed opossum.

All short-tailed opossums should be provided with a nest box and nesting material. Small rodent houses work well, as do half-round logs (**FIGURE 12-5**). Both sexes make sleeping nests, and material such as grass hay, shredded paper, or leaves should be provided so they can be gathered up by the opossum and put into the nest box. Bedding material should be nonabrasive, nontoxic, and at least two inches deep. Aspen shavings, shredded forest floor bark, or recycled paper bedding are all suitable. Cedar shavings should never be used as they contain oils known to be toxic and will cause severe respiratory problems and, if untreated, will lead to death.

Short-tailed opossums are fastidiously clean. They groom with their tongues, licking the fur clean. They have been observed *washing their hands* in the same manner as rats and mice and use their forepaws to clean their face and ears. They use a litter tray regularly. Small triangular plastic litter trays are available, but plastic food container lids work just as well as long as they are deep enough to contain the litter and are easy to clean and disinfect (**FIGURE 12-6**). Short-tailed

FIGURE 12-5 Short-tailed opossums should be provided with small rodent hide boxes. Both sexes will build a sleeping nest inside the hide box.

FIGURE 12-6 Short-tailed opossums are fastidiously clean and willingly use a litter box.

opossums do not cover or bury their droppings. Plain (nonclumping), unscented cat litter is recommended. A diet of meat, insects, and other meat-based protein causes their feces to have a stronger odor than that of herbivores. STOs often defecate or urinate while running in their wheels. It is important to routinely clean the wheels to prevent disease.

The addition of enrichment items not only makes the habitat visually appealing but also provides the opossum with extra areas to explore. Many items available for reptile habitats can be incorporated into the opossum habitat, providing interest for this curious little marsupial. Artificial plants, as well as artificial ledges and natural branches, may be hung from the sides of the tank. Short-tailed opossums will use a rodent wheel for exercise, both during the day and at night. For safety, solid or wire mesh wheels are safer than those with horizontal bars (**FIGURE 12-7**).

FIGURE 12-7 Exercise wheels should have a solid floor. Plastic is suitable as they are not known to chew on it and it is easily cleaned.

Access to fresh water is easily provided with a small rodent sipper bottle. The water bottle should be cleaned thoroughly and the water changed daily. The sipper tube needs to be checked regularly to be sure that the water droplets flow with the lightest touch. Never add any vitamin or mineral supplements to the water source. If supplements are needed, the best solution is to correct the diet. It may be inadequate, either in the amount fed or by not meeting nutritional needs.

Diet

The best captive diet is still being researched. Omnivorous by nature, short-tailed opossums have been kept successfully with a great assortment of food items. Most of the knowledge regarding captive diets has been derived from research colonies, where they are maintained for general and reproductive health. A mix of commercial carnivore and insectivore diets is always recommended, along with fresh fruits, meal worms, wax worms, and crickets. Food should be freshly prepared and fed in the evening.

Short-tailed opossums require a diet that is high in protein and low in fat. Many owners supplement the basic diet with low-fat cottage cheese, boiled egg (shell removed), live insects, and a weekly pinky or fuzzy mouse. Short-tailed opossums can be astonishingly quick when attacking live prey, and care needs to be taken when offering a pinky mouse to an eager short-tailed opossum. Small rodents are quickly grabbed and dispatched with a bite to the head (**FIGURE 12-8**).

FIGURE 12-8 Short-tailed opossums should be fed a live pinkie or fuzzy mouse once weekly. This female quickly dispatches her prey with a bite to the head.

A specifically formulated diet is available for the short-tailed opossum. It is a semi-soft crumble that does not need refrigeration, but it should be kept in a sealed container to preserve its moisture content. Commercial fox food diets have also been used successfully.

Handling and Restraint

The pet short-tailed opossum objects little to being handled. Many of them will scamper over and around an owner's lap, exploring pockets and clothing. They are rarely aggressive when being handled, but like any animal, they will bite if provoked. They may nip fingers if human hands smell like food or treats offered in the past.

Short-tailed opossums have little in the way of defense but will gape and hiss in alarm if they are frightened. Pups should be handled gently and allowed to explore without restraint. It is best to be seated on the floor to avoid any danger of a fall. However short the drop, they are easily injured.

Restraint for a physical examination should be as minimal as possible. As they become very trusting of people, a scruff may well alarm them. They can be held gently with one hand while being examined, the heart rate taken, and respirations observed. Should an examination require a closer evaluation, it would be better to use an inhalant anesthesia and prevent the stress that may occur.

Medical Concerns

Short-tailed opossums are generally hardy. The problems that do occur are usually the result of poor husbandry and an inadequate diet. Limited information is available on health and diseases. As they become more popular and are seen more frequently by veterinarians, greater knowledge will be gained. Often, there is reluctance on the part of the owner to seek medical advice when necessary as short-tailed opossums are illegal in many states.

Gastrointestinal Disorders

The cause of death most often reported is rectal prolapse due to aging, an incorrect diet, or the combination of both. It can also be secondary to bacterial enteritis. Rectal prolapse is seen most commonly in older females. If the initial cause of the prolapse can be identified and resolved, it may help to prevent a prolapse from recurring. It is critical that the opossum be seen by an exotic animal veterinarian as soon as possible so that the enlarged and inflamed rectal tissue can be reduced and replaced before tissue necrosis occurs. Any sign of straining to defecate is cause for concern.

Bacterial enteritis, implicated in some prolapses, can also be caused by poor diet and husbandry. The opossum becomes lethargic and has a distended abdomen due to a build up of gas in the gastrointestinal tract. Additionally, diarrhea and anorexia compound the problem. The patient can quickly become critically dehydrated because of the diarrhea and accompanying loss of appetite. Antibiotics and fluid therapy may be successful.

Nondigestive Disorders

Hepatic and pancreatic **neoplasias** (cancerous tumors) are frequently seen in short-tailed opossums two years old or older. Hepatic neoplasia produces lesions on the surface of the liver that can be readily seen on necropsy. If hepatic neoplasia is suspected, a definitive diagnosis can be determined with a chemistry panel to determine liver function. Both aspartate aminotransferase (AST) and alanine aminotransferase (ALT) will be elevated. The patient will most likely present with abdominal enlargement and weight loss. Signs may be slowly progressive or appear in a very short period of time.

Pancreatic neoplasia may be primary or secondary to hepatic neoplasia. It is most often diagnosed postmortem. There may be signs similar to hepatic neoplasia, or the owner may find the short-tailed opossum dead in the cage with no prior signs of illness. Both types of neoplasia have a very poor prognosis.

Older opossums frequently develop nephritis in association with a bacterial infection. Signs may include polyuria/polydipsia (PU/PD), evident when the patient drinks more water than normal and urinates frequently. Hematuria and pyuria, blood and pus in the urine, may also be present. A routine urinalysis should be performed, along with a bacterial culture of the urine to identify the causative agent. Radiographs may show kidney enlargement. Treatment usually involves antibiotics and fluid therapy to ensure hydration and support for kidney filtration.

Pituitary adenoma can occur in the short-tailed opossum. It is usually first evaluated as a skin disorder because the primary sign is a rather large area of alopecia on the rump, which may progress up the lower back (FIGURE 12-9). The hair loss is not associated with inflammation or irritation, and many owners report that the bald patch appeared suddenly. With a pituitary adenoma, blood serum levels of both estradiol and prolactin (luteotropic hormone [LTH]) will

FIGURE 12-9 Pituitary adenoma is not unknown in short-tailed opossums. One of the primary clinical signs is an area of alopecia across the rump, which may progress up the back.

be elevated. Prior to a blood draw, it is important to decide which exotic specialty laboratory will be used and the specifics of the samples required. All labs will provide information regarding the type of samples that should be submitted for specific tests, serum or blood volume required, preferred preservative, and whether the sample should be frozen or cooled.

Dermatitis is another condition that may be caused by an inadequate diet and poor nutrition. Hair loss is common with flea or mite infestations. The alopecia may be isolated or appear patchy anywhere on the body. There may be other causes, such as allergies, skin infection, or neoplasia. A skin scraping will help to determine the exact cause.

Benign skin lipomas have also been reported and, as with any small mammal, these should be surgically removed. Excessive ultraviolet light can induce melanoma (skin cancer) and sarcoma (soft tissue cancer). The history should include whether or not UV lighting is used and the number of hours of exposure. Habitats are often set up with live plants, and although beneficial to the plants, short-tailed opossums should not be exposed to this type of light.

Long-term dietary deficiencies and age can cause heart and coronary artery disease in short-tailed opossums. They can develop high levels of cholesterol leading to arthrosclerosis (hardening of the arteries). Signs are typical of heart and circulatory problems as seen with other small animals, that of exercise intolerance, cold extremities, and ascites.

Parasites

Incidental reports of parasites in short-tailed opossums are rare. They could potentially be exposed to and infested with fleas or mites. If suspected, intestinal parasites could be diagnosed with standard fecal floatation methods or by direct smear. To date, data cannot be found regarding the use of specific anthelmintics in the short-tailed opossum.

Clinical Procedures

The veins are very small and fragile. It will be difficult to safely obtain a blood sample without the use of general anesthesia. Both isoflurane and sevoflurane have been used successfully. The patient can be placed in an induction chamber or contained in a dog mask placed flat against the table. Induction is rapid and they are easily monitored. General anesthesia will reduce stress on the patient and help to prevent a hematoma or laceration of a vessel in a struggling and very small patient. For recovery, a clear plastic container with a ventilated lid provides a secure container and the patient is easily observed. A soft cloth should cover the bottom of the enclosure to prevent injury during the recovery process.

A 25-gauge needle with a tuberculin syringe is adequate for the small amount of blood required, which should not be greater than 1 percent of bodyweight by volume. Blood may be collected from the ventral coccygeal vein, femoral vein, or saphenous vein. Attempts with a direct cardiac puncture or accessing the vena cava often results in disaster and death for the patient.

Radiographs are often taken to evaluate the respiratory system and fractures. Sedation is required to obtain correct positioning without harming the patient. Ultrasonography is not commonly used in these small animals.

Summary

Short-tailed opossums are small marsupials native to the jungle floors of South America. Unlike some of the larger marsupials, STOs do not have a pouch but instead, the tiny neonates attach to a donut-shaped cushion that forms around each nipple. The teat swells, completely filling the mouth and securing the tiny infants during their development. These little opossums are omnivorous, eating a variety of small prey including insects, the eggs of ground nesting birds, and even the occasional small bird or rodent. They are solitary by nature and will not tolerate the company of others. They are very curious and social with human caregivers and are not known to bite. They are fastidiously clean, and will use a litter box and an exercise wheel with a solid floor. Generally, STOs are hardy but can develop pituitary adenomas and hepatic and pancreatic neoplasia, and with aging, rectal prolapse is common. UV lighting should not be used in the habitat. Excessive ultraviolet light can induce melanomas and sarcomas for these small marsupials because they have adapted to living in the dense growth of the forest floor with low levels of natural sunlight.

fastFACTS

SHORT-TAILED OPOSSUMS

- Brazilian short-tailed opossum (STO) (*Monodelphis domestica*)
- Diurnal/nocturnal ground-dwelling marsupial
- Solitary, aggressive with others
- Omnivore/insectivore diet
- Five toes on each foot; first two of each hind foot is clawless and opposable

WEIGHT
- **Males:** 100–125 g (average)
- **Females:** 75–95 g (average)

LIFE SPAN
- Approximately four years (48 months)

REPRODUCTION
- **Sexual maturity:** 16–20 weeks (4–5 months)
- **Gestation:** 14 days

- **Litter size:** 5–10
- **Weaning:** 48–50 days

VITAL STATISTICS
- **Temperature:** 89.6°F (31.6°C)
- Marsupial body temperature is normally much lower than that of placental mammals
- **Heart rate:** 239–300 beats/min
- **Respiratory rate:** 52–65 breaths/min

DENTAL
- Total 48 teeth total
- Dental formula 2 (I 4/3, C 1/1, P 4/4, M 3/3)

ZOONOTIC POTENTIAL
- None reported

Review Questions

1. Why are short-tailed opossums classified as marsupials if they do not have a pouch?
2. How do short-tailed opossums use their tails?
3. Describe the typical diet of the short-tailed opossum in captivity.
4. How long do neonates stay attached to the nipple before exploring independently?
5. What happens when the tiny embryos are expelled?
6. What is the most common cause of illness or disease in a short-tailed opossum?
7. What can cause rectal prolapse in a short-tailed opossum?
8. What is the best method of anesthetizing a short-tailed opossum?
9. List the veins that may be used for blood collection.
10. What types of bedding should be used in the habitat?

Case Study I

History: A three-year-old intact male opossum is brought in for an examination. The owner has noticed that the bedding has become wetter than normal and complains that she has to fill the water bottle every night when, normally, the water bottle doesn't need to be changed for two or three days. Questioned further, the owner reports that there might have been some small spots in the bedding that could have been blood.

Physical Examination: The veterinarian finds nothing significant during the initial physical examination but recommends a radiograph and a urinalysis. The owner declines the radiograph due to expenses but agrees to the urinalysis.

Laboratory Findings (from the urinalysis): There is blood and pus in the urine sample.

a. What problem does the laboratory result indicate?
b. What is the recommended treatment for this condition?
c. What husbandry concerns arise from the owner's comment regarding the water bottle and the frequency it is being refilled?

Case Study II

History: A two-year-old female opossum has been brought in by the owner. He is concerned because he has noticed that the opossum has been losing some of the fur on the top of the rump. The opossum is eating and seems to be behaving normally, but his concern is that the rodent wheel might be rubbing off the fur and says that "she probably runs five miles a night in her wheel."

Physical Examination: There is an area of alopecia on the rump that appears to be spreading to the lower lumbar region. The skin is visible and devoid of all fur, but there is no evidence of inflammation or flaky skin. The hairs that remain are undamaged in any way.

    a. What is the suspected cause of this regional alopecia?

    b. What is the recommended treatment?

    c. Is the exercise wheel implicated in any way?

For Further Reference

Johnson-Delaney, C.A. (2005). *The Marsupial Pet: Sugar Sliders, Opossums, and Wallabies*. Presented at the Dallas Veterinary Medical Association, St. Paul, MN.

Holz, P. (2003). Marsupialia (Marsupials). In Fowler, M.E. and Miller, R.E. (Eds.), *Zoo and Wild Animal Medicine, 5th Edition*. St. Louis, MO: Elsevier/Saunders.

http://www.rainbowwildlife.com (accessed January 4, 2014).

SUGAR GLIDERS

KEY TERMS

arboreal
syndactylism
volplane
patagium
joeys
OOP
glider mills
detritus
seasonal feeders
osteodystrophy

OBJECTIVES

After completing the chapter, the student should be able to

- Describe the correct habitat for sugar gliders.
- Provide client education to new sugar glider owners.
- Discuss the correct diet for a sugar glider.
- Provide basic nursing care to a sugar glider.
- Assist in the anesthesia of a sugar glider.
- Demonstrate the appropriate restraint technique for a sugar glider.
- Understand common medical disorders in a sugar glider.

Introduction

Sugar gliders are unique for many reasons, not the least of which is the way they move. Rather than jumping, running, or walking on the ground, they launch themselves from high branches and glide from place to place, rarely leaving the treetop canopy and the safety of the forest.

Sugar gliders belong to a family of arboreal (tree living) marsupials native to Australia, Tasmania, Indonesia, and New Guinea. They are members of the same order that includes possums, kangaroos, and wombats. Within this order, there are three species taxonomically grouped as families, one of which is the Petauridae and includes the sugar glider, *Petaurus breviceps*.

Sugar gliders are small, with a body length of only five to seven inches. The tail is as long as or slightly longer than the body. They have five clawed toes on the front feet and four on the hind feet, with an opposable grasping thumb. The middle two toes of the hind feet are fused, an anatomical adaptation called syndactylism. Sugar gliders have a total of 46 to 48 teeth with long, forward-pointing lower incisors designed for gouging.

Sugar gliders are strictly nocturnal. During the day, they sleep huddled together in small colony groups. The colony nest is high in the treetops and lined with leaves. Their natural range (territory) covers an area greater than two acres.

The ability to glide, or volplane, is due to a flap of fur-covered skin on each side that connects from carpus to tarsus and opens somewhat like a parachute when the limbs are extended. Sugar gliders are able to guide themselves while airborne by moving their limbs and altering the tension of the skin flap. The tail is used as a rudder, much like the tail of an aircraft or the steering rudder of a boat, which helps them to change horizontal direction and assist in landing (**FIGURE 13-1**). Sugar gliders do not fly; they glide from place to place by launching themselves in the air, often from great heights. They are able to glide for distances as great as 300 feet. This anatomical *parachute* is called the patagium.

Sugar gliders are very vocal with one another and make a variety of sounds. A surprisingly loud bark serves as a warning or alarm call to other gliders. They also produce a soft, churring sound that has been interpreted as contentment because it is often heard when the animals are settling into the nest together. The most common noise heard by owners occurs when sugar gliders are disturbed during the day. This unique sound has been likened to an electronic pencil sharpener, or to a *toy chain saw with a run-down battery*. The term *crabbing* is used to describe this sound. Regardless of how it is described, it is a sound produced when sugar gliders are annoyed or disturbed.

The sex of an adult sugar glider is easily determined. Males have a very large scent gland on the top of the head, which is absent in females (**FIGURE 13-2**). The gland becomes more pronounced with sexual maturity. The scrotum is located more cranial than in placental mammals. It is fur covered, pendulous, and attached by a stalk (**FIGURE 13-3**). The penis is bifurcated, caudal to the scrotum, and difficult to visualize under normal circumstances. There are bilateral anal sacs near the penis, and these are often mistaken for testicles in young male gliders.

FIGURE 13-1 Sugar gliders do not have the ability to fly, but glide from place to place by launching themselves into the air and extending the patagium. The fur-covered flap of skin acts somewhat like a parachute.

FIGURE 13-2 Male sugar gliders have a large scent gland on the head. The gland becomes more pronounced with sexual maturity.

Males also have a scent gland on the thorax, visible as a small area of discolored fur (**FIGURE 13-3**). During the breeding season, the saliva of males has also been reported to produce a strong scent.

The female has a bilobed uterus with laterally paired vaginas and one central birth canal. The birth canal is temporary. It forms just prior to birth and disappears

FIGURE 13-3 The external genitalia of a male sugar glider. The scrotum is attached by a stalk; it is round, fur covered, and, in this photograph, positioned just lateral to the abdomen, near the right leg.

soon after delivery of the embryos. Because of the uterine anatomy, it is fairly common to have separate pregnancies with developing joeys of differing ages. The pouch is vertical and contains only two teats. In addition to anal scent glands, females have additional scent glands located within the pouch. The frequency of female scent marking increases with her readiness to breed. Once mating occurs, gestation is approximately 16 days. Parturition usually takes place mid-morning with an average of two joeys per litter. The joeys are born in an embryonic state, being about the size of a grain of rice. Their forelegs are developed only enough to allow them to crawl from the birth canal up into the pouch. An observable sign that birth is imminent is when the dam begins to lick the fur leading to her pouch. The licking prevents the joeys from becoming stuck in the dam's fur and helps the neonates to migrate into the pouch so they do not die of exposure. Crawling from the birth canal to the pouch usually takes five or six minutes.

Once in the pouch, the embryonic joey attaches to a nipple, which swells to prevent the newborn from being dislodged. Joeys should never be forcibly removed from the nipple at any stage in their development as it is not possible for them to reattach. Joeys will die if forcibly removed from the nipple.

Joeys stay attached to the nipple for two months and during this time their internal organs and limbs develop and fur appears. Around eight weeks, they begin to venture out of the pouch. They continue to nurse by sticking their heads into the pouch as, by now, they are too large to fit inside the pouch. Joeys usually travel on the back or underbelly of either parent as they begin to learn to forage and eat solid food (**FIGURE 13-4**).

Both parents contribute to the care of the offspring. Most joeys become independent at approximately 117 days, but they may stay in the parental nest for up to one year. Males are driven from the family nest once they reach sexual maturity. Sexual maturity is usually reached at eight months for a female, but may not be

FIGURE 13-4 An unweaned joey traveling on the back of the dam. (*Courtesy of Laura DeVries, CVT.*)

FIGURE 13-5 The age of a joey is usually stated by the number of days the joey has been out of pouch, or OOP. (*Courtesy of Laura DeVries, CVT.*)

reached for as long as a year. Males mature slightly later, with an average of 12 to 14 months. Pouch-independent young are described (their ages given) by the number of days OOP (Out Of Pouch) (**FIGURE 13-5**).

No one knows for certain when sugar gliders first arrived in the United States, but it is probable that they were brought in by a private collector who found them appealing. They first appeared in the pet trade in the early to mid-1980s and very quickly became popular as *pocket pets*. Sugar gliders breed readily in captivity, and there is now a population glut.

Even Though the Market Seems to Have Been Saturated, thousands of sugar gliders are still being bred by **glider mills**, the term that refers to the indiscriminate breeding of sugar gliders for profit. With the constant supply of gliders and products connected to this industry, Internet web pages abound. One must be extremely cautious when evaluating the content of many of these sites, forums, and discussion boards. Misinformation not only expands but somehow becomes *fact*. Advice is offered on everything from disease diagnosis and treatment to commercial carnivore diets that are being repackaged, relabeled, and then sold as *secret recipes*. All medical concerns should be discussed with an exotic animal veterinarian.

Behavior

Behavior in captivity is often aberrant from natural behaviors observed by field researchers. Regardless of the number of generations that are bred in captivity, the sugar glider is still far from being a domesticated species. Sugar gliders respond and react to their captive state similarly to other wild animals, that is, by exhibiting abnormal behaviors and disorders not seen in the wild.

Sugar gliders need the company of other gliders for social interaction and mental stability. A solitary sugar glider will self-mutilate, causing trauma that often leads to cage death or euthanasia. They have been known to chew off limbs, scrotal sacs and stalks, and their own tails. There may also be gouging wounds on their body caused by their lower incisors and trauma from excessive, compulsive licking. This behavior occurs regardless of the amount of time a caregiver provides for an individual animal.

Many Dealers Offer an Item Called a Bonding Pouch.
The idea is that the owner carries a single glider around with the pouch tucked inside a shirt or pocket, to allow the estranged and isolated little animal to become familiar with the scent of the owner. This may well accustom the glider to a specific scent, but the glider is still a captive wild animal and it is doubtful that any real bonding occurs. Bonding pouches may, in fact, increase the anxiety of the lone sugar glider. Being held in the company of another species (human) is not a substitute for natural, normal interaction with its own kind. Knowledgeable people and those who truly care about the welfare of their animals would never support this *bonding pouch* idea and would never sell or obtain a single sugar glider as a *pocket pet*, dooming it to a short and very stressful life.

Sugar gliders are colony animals, with small family groups living and nesting together, both male and female and young, pouch-independent joeys. A *minimum* of four gliders should be housed together to provide a social environment that is as normal as possible. Sugar gliders are usually very good parents, but too often the dam is allowed to breed indiscriminately, resulting in close inbreeding that intensifies many of the problems already seen in captive glider colonies.

Housing

With an understanding of natural behaviors and the issues of captivity, housing can be problematic and needs serious consideration before choosing and caring for a sugar glider colony. Ideally, there should be a *glider room* that is large enough to allow the gliders to glide, to interact with one another, and to behave in a manner somewhat resembling their behaviors in the wild. As they are nocturnal, bright daylight and artificial light are both very stressful to them. They need to be fed in the evenings and will come out to eat only when it is dark. One should never attempt to change a nocturnal species into a diurnal species; it will not work and the interruption of normal cycles of sleep causes a great deal of physiological stress to the animal.

Understanding the natural behavior of a species is vital to its well-being in captivity. Sugar gliders need a great amount of space and the usual cage setup for other small animals is totally inadequate. Simply put, gliders cannot be housed in cages designed for other small animals. Moreover, it must be reiterated that it is extremely important that they are not kept alone with *pocket pet* status. For the average pet owner, providing a healthy habitat that meets the animals' needs is not easy.

Considering all these things, it is possible to house sugar gliders safely and successfully. Some owners have adapted two or more large avian flight cages. If constructing a cage, the choice of material is important. Sugar gliders will chew on items that may be toxic to them, and gouge at the wood of the enclosure in a natural attempt to find food items. They are especially adept at using their fore-paws to open cage doors. All doors should have a secure latch to prevent escape.

Minimum Cage Requirements as Set by Individual States' Departments of Agriculture do not even come close to actual requirements. For the most part, sugar gliders are not understood as a species and are grouped together with other small animals such as rodents, chinchillas, and cavies. The author of *Cage Size Matters: The Current Industry Standard Borders on Animal Cruelty* (Glider University) points out that there is a vast difference in one U.S. state's minimum housing requirement of 216 square inches for one animal and the 81 square feet (11,644 square inches) per two animals as required by Australian regulations.

The enclosure should be equipped with a sleeping pouch near the top of the cage, one that most mimics the sleeping nest of the colony. Sleeping pouches designed for ferrets work well and can be easily attached to the upper part of the cage. There should also be an assortment of branches, perches, and food dishes placed to accommodate their arboreal nature. Dishes placed in the lower part of the cage collect urine and feces, contaminating the food. Access to fresh water must be easily available. Several separate water dishes should be provided. Sugar gliders do not readily use sipper bottles as do rabbits and rodents. The habitat should be set up to meet the needs of the gliders, yet should also provide easy access for the caregiver.

Diet

Field researchers including veterinarians, zoologists, botanists, and others have spent countless hours observing wild gliders in an effort to determine their dietary needs in captivity. Dietary needs are not just what they eat, but how and when they feed, their behaviors, and their social interactions. Researchers examine fecal samples, browsing and foraging sites, and the detritus or litter around nest sites and on the ground. Anatomical features of a species offer clues as to what an animal eats: for example, the lower incisors of the sugar glider are elongated and protrude forward. They are very sharp, and are used for gouging and peeling tree bark to search for insects or to gain access to tree sap.

Like many species, gliders are seasonal feeders; that is, their diet varies depending on natural food availability and abundance. They are omnivores, consuming insects, arachnids, a great deal of plant-derived material, and, occasionally, small vertebrates. Observations and studies like this are also very important in determining what sugar gliders do not eat. Many breeders of captive gliders erroneously recommend diets that include seeds, nuts, and cereal grains.

For captive gliders, the common name prefix *sugar* has contributed to the mistaken belief that they eat only sweet foods. While it is true that gliders seem to enjoy foods with a sweet taste, this in no way begins to meet their nutritional needs. Sugar cubes and various other candies, referred to as *lickey treats*, have been promoted by breeders and sellers as a method of taming a sugar glider. Misinformation such as this, however, well intended, contributes greatly to one of the major concerns in captive gliders: malnutrition and premature death.

Table 13-1 lists food items consumed in the wild. This natural diet is not easily replicated for captive sugar glider colonies, but good alternatives have been developed. There is no one easy formula that meets all their nutritional needs; however, there are several recommended diets that are not only adequate, but are proving to be successful.

Analysis of food items for nutritional content of proteins and carbohydrates, including those derived from plant fiber, fats, minerals, and vitamins, is essential in providing a diet that is not only adequate to promote an animal's health and well-being but also allows the animal to exhibit normal, natural behaviors when foraging and selecting food items. Additionally, consideration must be give to seasonality (the variety and type of food available) in the wild within their normal geographic range. The sugar glider inhabits the Southern Hemisphere, which is seasonally opposite to the Northern Hemisphere. Food sources change dramatically and seasonally. This difference in natural food availability contributes to the

TABLE 13-1

NATURAL DIET OF *PETAURUS BREVICEPS*	
Natural Diet	**Ingredients/Source**
Eucalyptus Sap	Sap is the fluid produced by plants and trees. It is usually sticky to the touch and provides the circulating nutrients to the tree. Eucalyptus trees are native to Australia and grow to great heights. Their leaves produce an aromatic oil, often used for human medicinal purposes.
Manna	Manna is produced from sap that leaks from a tree at the site of insect damage. It can be found on the leaves and tree bark.
Honeydew	Honeydew is produced by insects that feed on large quantities of sap. It is the excess sugar content of sap excreted by these insects. It is usually white and crusty.
Nectar	Nectar is a liquid produced by flowering plants and trees; it has a sweet taste and attracts pollinating insects.
Gums	Gums are multiple sugars produced by some plants and trees. It is thicker than sap and becomes gelatinous when mixed with water. It is produced in abundance by many Australian and African trees, in particular, the acacias, which are a variety of tropical tree. It is also referred to as gum Arabic, a substance used in the production of many human food products such as candy, chewing gum, and other products that require non-artificial thickening agents.
Insects and Arachnids	Many species are consumed by the sugar glider, depending on their availability. Gliders, both wild and captive, will pull off the legs and wings before they are consumed.

notion that they are *picky eaters*. Many owners report that their gliders will suddenly refuse food that they previously consumed readily.

Periods of decreased consumption may be normal, and attempts to encourage greater intake usually involve random substitutions of food items selected for palatability rather than items that are nutritionally appropriate and meet metabolic needs. Obesity in captive gliders, coupled with malnutrition, lack of normal exercise (room to glide), and human expectations of amounts that should be eaten, all contribute to the health problems seen in captive gliders.

Benjamin Leadbeater, an English Victorian naturalist and taxidermist, traveled the world collecting specimens for zoos and museums. He is credited with discovering many species and for his observations and writings of their natural history. Leadbeater's Mix of nutrients may be dated to 1834, when he recommended suitable diets for new specimens delivered to the London Zoo.

Leadbeater's Mix
150 ml each of warm water and honey
1 boiled egg, shell removed
25 g of high-protein baby cereal
vitamin/mineral supplement

Mix water and honey and gradually add to blended egg. Add vitamin/mineral supplement and baby cereal, blend until smooth.

The mixture needs to be refrigerated. Portions may be frozen in ice cube trays and thawed completely before feeding. Leadbeater's Mix should equal 50 percent of the daily diet. The other 50 percent can be an assortment of chopped fruits, frozen mixed vegetables (thawed), and live insects such as crickets and mealworms.

Another diet that is used with considerable success is based on the Taronga Zoo diet (Sydney, Australia). There are slight variations to this diet, but none seem to affect the nutritional balance. For example, the diet calls for *once weekly, day-old chicks*. Aside from the inherent unpleasantness of feeding day old-chicks, they are not readily available to the average caregiver. Many breeders have recommended substituting a feeder mouse. This substitution needs careful thought as there is a great potential for injury to the sugar gliders and there is the possibility of parasite transmission (toxoplasmosis, which is zoonotic). Also, it has been noted by many owners that citrus fruits (oranges) contribute greatly to the odor produced by sugar gliders. Their urine becomes stronger smelling and more pungent.

Taronga Zoo Diet.
Equal portions of apple, banana, fresh sweet corn (frozen corn may be used if thawed), grapes, and kiwi fruit. Fresh oranges with the skin on, melon, papaya, and sweet potatoes
Leadbeater's Mix 50 percent of total volume
Dog Kibble (which appears in a few variations of this formula)
Once weekly, day-old chicks

Many owners capture insects, grasshoppers, and moths, but care must be taken as they may have been exposed to insecticides and fertilizers. Fireflies should not be fed as they are known to be toxic in many species. Insects should be gut-loaded prior to feeding (*feed the food*), or they may be dusted with a vitamin/mineral supplement just prior to feeding. There are commercial diets for sugar gliders, as well as a nectar powder that is mixed with water. Gliders will also readily lap natural fruit juices. Several veterinarians recommend small amounts (teaspoonful) of exotic feline diets as an additional protein source.

Handling and Restraint

Owners of sugar gliders allow them to scamper around on their hands and bodies without much handling. Well-handled sugar gliders often leap from the cage top or other high areas in a room to land on the owner. Should one unintentionally escape from the cage, it is easier to recapture by turning on all the lights or with the use of a flashlight. A sudden light source will temporarily immobilize a sugar glider and a thick towel can be placed over it so that it may safely be picked up. Another method is to put the sleeping sack close to its location. Most sugar gliders will enter the sleeping sack. Bare-handed attempts to capture the sugar glider may result in a bite that can be deep and painful.

It can be very difficult to examine a sugar glider with regard to its health, potential illness, or injuries without the use of general anesthesia. In assessing general health, mobility, coat condition, and attitude, it often works well to keep one's distance and observe the glider's interactions with the owner while obtaining all relevant history. History should include how and where it was obtained, number of gliders in the colony, housing, and diet. If it is kept as a *pocket pet*, that is, a solitary glider in a bonding pouch, information should be obtained as to how often and when the animal is most disturbed, and the initial reason for the examination.

Sugar gliders should be observed in subdued lighting and that alone can make a more complete examination difficult. Bright, artificial lights startle sugar gliders and reactions may be misinterpreted. It is possible to scruff a glider for a very quick cursory examination, but be prepared and warn the owner that the glider will produce loud vocalizations, most of which they have never heard before. Even a relatively simple procedure, such as nail trims, often requires general anesthesia, as sugar gliders are difficult to restrain safely.

Medical Concerns

Most medical problems seen in the sugar glider can be directly related to inadequate husbandry, cage requirement, and dietary needs that are not being met. Malnutrition leads to weight loss, generalized weakness, muscle atrophy, lethargy, ataxia, and hind limb paralysis. These conditions may develop slowly, over a period of months, before an owner is able to detect a real problem. Once these conditions become apparent, the glider may be already hypothermic, dehydrated, and exhibiting neurological signs. Because of their nocturnal nature, it is sometimes difficult for owners to carefully observe their sugar gliders for abnormalities or signs of illness.

Gastrointestinal Disorders

Gastrointestinal problems are seen with an incorrect diet that may also contribute to rectal prolapse. The muscles become weakened and lose tone to the point where the rectum protrudes, exposing the tissue. Veterinary care is necessary and anesthesia is required to reduce and restore the prolapsed tissue. Success depends on how long the tissue has been exposed and the overall health of the glider. Prolapse often becomes a chronic, recurring problem regardless of dietary changes made.

Enteritis can be associated with bacterial and protozoal problems. Diarrhea is the most common sign with both of these conditions. Recommended veterinary treatment includes administration of antimicrobial and/or anthelmintic therapy for possible parasites, bismuth subsalicylate to control the diarrhea, and an analgesic and anti-inflammatory medication to calm the gastrointestinal track.

Nondigestive Disorders

Cataracts have occurred in sugar gliders and are linked to nutritional deficiencies and genetics. Cataracts seem to occur more frequently in closely related, inbred sugar gliders. Vitamin A deficiency can also cause cataracts to form over a period of time. When the diet is incorrect, the glider can become hyperglycemic, causing cataracts to form more readily.

Urinary tract disorders can occur in the sugar glider. Dietary-related nephritis, renal disease, and male urinary tract blockages are common. Bladder rupture, prolapse of the penis, and ensuing necrosis may occur with trauma. Instances of penile necrosis are also seen due to the accumulation of fur, human hair, or bedding material that has tightened and formed a tourniquet around the penis, cutting off the blood supply.

Reproductive problems may be caused by an inappropriate diet or insufficient amounts of food. There may also be infectious agents within the pouch or reproductive tract. Joeys that are weak and fail to thrive can be attributed to an insufficiency problem of the dam, as she may not be producing an adequate volume of milk or may have developed mastitis. Human interference will also produce weak joeys and will kill them if they are pulled from the pouch early. Ill-advised attempts are made to hand-feed joeys in order to more easily imprint the baby glider. Occasionally, joeys die in the pouch from an undetermined cause.

Neurological and limb problems are often seen together. These conditions occur most frequently in sugar gliders that are confined to a bonding pouch and receive little opportunity for exercise and are not being given an adequate diet to meet their nutritional needs. Affected sugar gliders are ataxic and may exhibit muscle tremors when they attempt to move. **Osteodystrophy**, decalcification of the bone, may also be present with accompanying hind limb paralysis (**FIGURE 13-6**). Depending on the duration and severity of these problems, there may be seizures and permanent damage to the nervous system and bone with pathological fractures. Radiographs will help evaluate the severity of the condition and diagnosis of fractures. Therapy consists of supportive care with dietary and husbandry changes but treatments are not curative. A veterinarian may prescribe injections of calcium gluconate and appropriate vitamin/mineral supplements.

FIGURE 13-6 Neurological and limb problems are often seen together in captive sugar gliders. These conditions occur more frequently in sugar gliders confined to a bonding pouch. The male sugar glider shown here was barely able to crawl from the bonding pouch.

Parasites

The fur of a healthy sugar glider is dense and very soft. Poor hair coats can be a result of dietary insufficiency, barbering, or fighting. If the glider is kept in a cage with wood shavings as bedding, mites can be a problem. Mite infestations, more common in wild sugar gliders, are treated by a veterinarian with injections of ivermectin and a recommended change in the cage substrate.

Sugar gliders may also have giardia and coccidia, usually evidenced by diarrhea. Fecal flotations and direct smears will determine if these organisms are present. The microscopic slide should also be carefully scanned for ova of other intestinal parasites. If present, ova may not be easily identified as there is little information or reports of intestinal parasitism in sugar gliders.

Clinical Procedures

Either isoflurane or sevoflurane can be used for anesthesia in the sugar glider. Inhalant anesthesia and induction with either agent can safely be achieved with an anesthesia induction chamber (**FIGURE 13-7**). Prior to anesthesia induction,

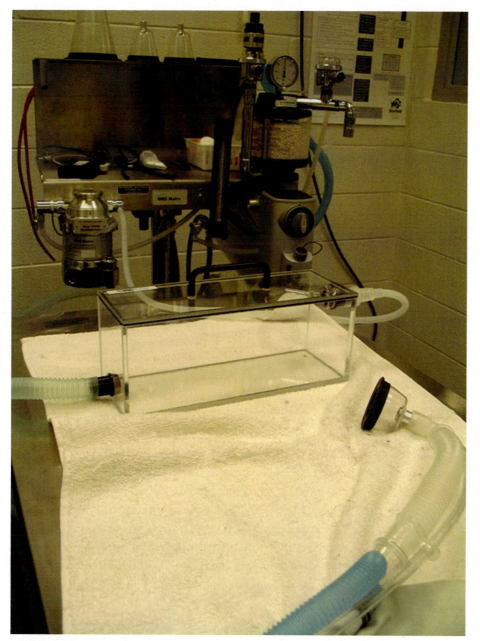

FIGURE 13-7 The setup of an anesthesia induction chamber in preparation for surgery.

FIGURE 13-8 When a suitable plane of anesthesia has been achieved, the patient is transferred directly to the nonrebreathing system. The elbow forms a perfectly fitted mask and is easily detached after use for cleaning and cold sterilization.

preanesthetic agents are administered. Either atropine or glycopyrolate works well for sugar gliders. When the sugar glider reaches a suitable plane of anesthesia within the chamber, it can be maintained by transferring the patient to the examination table so that it breathes directly from a nonrebreathing system (**FIGURE 13-8**). As with any small mammal under general anesthesia, a decrease in body temperature may be expected. Examination gloves filled with warm water can be used to maintain body temperature and provide positioning support, as can a circulating warm water pad. Normal marsupial body temperature is approximately 30 percent

FIGURE 13-9 A recovering sugar glider in a stockinette jacket. (This is a young male. The scrotal stalk and scrotum is evident between the rear legs.) (*Courtesy of Carol VandenAkker.*)

lower than that of placental mammals and care must be taken not to overheat the anesthetized glider while compensating for body heat lost.

Sugar gliders are very dexterous and quickly remove sutures unless they are subcutaneous. A *jacket* of stockinette may help in many instances (**FIGURE 13-9**). Sugar gliders object less to this and seem to find comfort by being in the stockinette. Adequate analgesia and the injection of a local anesthetic at the surgical site can help prevent the patient from chewing at the site and creating an even greater wound. Gliders do best when allowed to recover in a soft, dark pouch.

Injections may be given to administer fluids, antibiotics, or other life-saving interventions. Most commonly, subcutaneous fluids are administered to a dehydrated glider. Fluids may be given subcutaneously between the scapulae. It is important that the volume is administered slowly and the site massaged to prevent pooling in the patagium. Isotonic fluids are administered for the treatment of dehydration and shock. Critical instances of shock are often related to severe self-mutilation or from trauma by another household pet. The volume of fluids administered need to be carefully calculated and should not exceed 10 percent of bodyweight.

Intramuscular injections are used to administer antibiotic therapy. Preferred injection sites are the epaxial muscles of dorsal neck or the ventral thorax. Because of the short-term pain of the injection, a firm but gentle scruff is necessary and the patient is likely to respond with loud vocalizations. The patient should be released into a sleeping sack where it is less likely to attempt escape.

Obtaining a blood sample can be challenging due to the restraint required and the size of vessels in the glider. The easiest and least traumatic method for obtaining a sample is to mask down the glider with gas anesthesia. This is not only less stressful but will help prevent laceration of the blood vessel. The amount of blood that can be collected should be no more than 1 percent of bodyweight by volume. The patient should be weighed on a gram scale just prior to the blood draw and calculations performed accurately. The jugular vein or the cranial vena cava provides a larger amount of blood. Because only small amounts of blood are obtained, microcontainer collection tubes are adequate, and a 25-gauge needle attached to a tuberculin syringe will help prevent vein collapse. The coccygeal vein of the tail is recommended when collecting blood in capillary tubes, and this is made easier by first warming the tail to dilate the vein. This is achieved by dipping the tail in warm water or applying a warm cloth around the tail. Most tests can be run on blood collected in tubes containing either EDTA or heparin, but if an outside specialty diagnostic laboratory is used, always check with them for their sample requirements prior to collecting the sample.

Radiography can be used to detect soft tissue and orthopedic disease. Most sugar gliders will require anesthesia to obtain quality radiographs. Radiograph techniques and positioning are similar to other small-mammal radiographs. Ultrasonography can be useful to examine the abdomen or heart.

Summary

Sugar gliders are small, arboreal marsupials native to Australia, Tasmania, Indonesia, and New Guinea. The name *glider* is derived from their ability to volplane (glide) due to a flap of fur-covered skin on each side of the body that opens rather like a parachute, allowing them to glide from treetop to treetop. In the wild, sugar gliders live in small colonies and will not thrive as isolated individuals. They are strictly nocturnal and require a specialized diet and an enclosure that is large enough for them to be able to glide and interact with one another—a habitat that mimics, as closely as possible, their natural arboreal environment. Medical and behavioral problems are often directly related to an inadequate diet and a habitat that does not meet their specialized needs. Sugar gliders are difficult to examine with regard to health, injuries, and illness without the use of general anesthesia. Sugar gliders should be examined under subdued lighting and that alone can make a more complete examination difficult. Inhalation anesthesia is induced and maintained with either isoflurane or sevoflurane and the use of an adapted small face mask. A complete physical examination and clinical samples should be obtained quickly to avoid further stress to the patient.

fastFACTS

SUGAR GLIDERS (PETAURUS BREVICEPS)

- Nocturnal, arboreal marsupials
- Native to Australia and New Guinea
- Live in colonies of 7–12 animals, usually family groups
- Defend individual trees and territories of up to 2.5 acres
- Gliding membrane (patagium) begins at the fifth digit, continues from carpus to tarsus
- First and second digits of hind feet partially fused
- Will enter a torpid state with adverse temperatures (ideal environmental temperature: 64 to 75°F)

FEMALE ANATOMY

- Bilobed uterus with lateral vaginas and central birth canal
- Pouch with two teats, two offspring common
- Scent glands contained within the pouch; secretion and urine marking increase with breeding receptivity

MALE ANATOMY

- Bifurcated penis with mid-ventral pendulous scrotum
- Scent glands on the forehead, chest, and perianal region
- Both sexes scent mark territory, nest site, and each other

WEIGHT

- **Male:** 115–160 g
- **Female:** 96–135 g

LIFE SPAN

- **Wild:** 4–7 years (48–84 months)
- **Captivity:** 12–14 years (144–168 months), dependent on diet and husbandry

REPRODUCTION

- **Sexual maturity:** males: 48–56 weeks (12–14 months); females: 32–48 weeks (8 to 12 months)

- **Gestation:** 16 days, followed by fetal migration to pouch
- **TIP (time in pouch):** approximately 70 days
- **Litter size:** 1–2
- **Weaning age:** approximately 117 days; will remain in parental nest

VITAL STATISTICS
- **Temperature:** 32°C (89.6°F)
- **Heart rate:** 239–300 beats/min
- **Respiratory rate:** 52–65 breaths/min

DENTAL
- Total 46 to 48 teeth
- Dental formula 2 (I 4/3, C 1/1, P 3/3, M 4/4)

ZOONOTIC POTENTIAL
Bacterial
- Salmonella
- Leptospirosis

Parasitic
- Giardia

Review Questions

1. What does OOP refer to?
2. When are sugar gliders active?
3. What is the name of the anatomical parachute that allows the sugar glider to glide?
4. What is the diet of a captive sugar glider?
5. What are some of the problems seen when a sugar glider is kept as a solitary pet?
6. What are the recommended methods of restraint to examine a sugar glider?
7. How is sex determined in a sugar glider?
8. What is the most common gastrointestinal disease in sugar gliders?
9. Explain how a male sugar glider develops penile necrosis.
10. Which veins are used for blood collection?

Case Study I

History: A two-year-old intact male sugar glider is presented by the owner. His main concern is that the glider "gets hurt in his cage all the time." He has checked carefully for the source of these injuries but cannot find any evidence of what could be causing the injuries. He has brought the glider in today because there is a fresh laceration on the abdomen, approximately 1½ inches long. Questioned further, the owner describes the cage as a two-story condo with an "old suspended sleeping sack" he had for his ferret. He has just this one sugar glider. He originally had two, but the other one died approximately six months earlier. The owner agrees to wound repair and complete examination under general anesthesia.

Physical Examination: The patient has several old and fresh new injuries on the thorax and abdomen. The right pinna has been shredded and the scrotal sac appears to have some mild mutilation.

    a. What is the most likely cause of these injuries?

    b. How are these injuries best treated?

    c. What further recommendations and advice should be offered to the owner?

Case Study II

History: A young (age unknown) female sugar glider is presented by the owner. She has brought the patient to the clinic in a small soft pouch she wears around her neck. Her chief complaint is that the glider seems weak. She also thinks it has lost weight but there is no record of a previous weight: she is "just guessing" because it isn't eating very much.

Physical Examination: The patient is removed from the owner's pouch and evaluated as a mature female. The patient is weighed on a gram scale and the weight is well below normal for an adult female. The glider is reluctant to move and when gently prodded exhibits muscle trembling and ataxia.

    a. What further questions need to be asked of the owner?

    b. How are these signs and brief history associated with the diagnosis?

    c. What treatment(s) is the veterinarian likely to recommend?

For Further Reference

Barnes, M. (2005). Sugar Gliders. In Gage, L.J. (Ed.), *Hand-Rearing Wild and Domestic Mammals* (pp. 55–62). Ames, IA: Iowa State Press.

Holz, P. (2003). Marsupalia (Marsupials). In Fowler, M.E. and Miller, R.E. (Eds.), *Zoo and Wild Animal Medicine, 5th Edition* (pp. 288–303). Philadelphia, PA: W. B. Saunders.

http://www.merckmanuals.com (accessed June 3, 2014).

http://www.nessexotics.com (accessed June 3, 2014).

Johnson-Delaney, C.A. (2002). Other Small Mammals. In Meredith, A. and Sharon, R. (Eds.), *BSAVA Manual of Exotic Pets, 5th Edition* (pp. 102–115). Quedgeley: British Small Animal Veterinary Association.

Lightfoot, T.M. (1999). Clinical Examination of Chinchillas, Hedgehogs, Prairie Dogs and Sugar Gliders. In Orcutt, C.J. (Ed.), *The Veterinary Clinics of North America Exotic Animal Practice Physical Examination and Preventative Medicine* (2: pp. 262–269). Philadelphia, PA: W. B. Saunders.

Lightfoot, T.M., & Bartlett, L. (1999). Sugar Glider Orchietomy, an Excerpt from Small Mammal Clinician's Notebook. In *Exotic DVM. Florida* (1–4: pp. 11–13). ICE Proceedings.

Ness, R.D., & Johnson-Delaney, C.A.(2012). Sugar Gliders. In Quesenberry, K.E. and Carpenter, J.W. (Eds.), *Ferrets, Rabbits, and Rodents: Clinical Medicine and Surgery, 3rd Edition* (pp. 393–410). Philadelphia, PA: W. B. Saunders.

Rivera, S. (2003). Other Species Seen in Practice. In Ballard, B. and Cheek, R. (Eds.), *Exotic Animal Medicine for the Veterinary Technician, 2nd Edition* (pp. 269–272). Ames, IA: Iowa State Press.

Shapiro, Daniel, M.D., Director, Clinical Microbiology Laboratory, Lahey Clinic, Burlington, Massachusetts; Adjunct Associate Professor of Medicine (Infectious Diseases), Boston University School of Medicine.

UNIT III

14 AVIAN

OBJECTIVES

After completing the chapter, the student should be able to

- Describe the correct housing for a companion bird with species-appropriate cage size, perches, and toys.
- Provide client education to new bird owners.
- Describe the correct diets for different species of companion birds.
- Provide basic nursing and emergency care to a bird.
- Provide nutritional support to an ill bird through tube feeding.
- Assist in the anesthesia of a bird.
- Demonstrate appropriate restraint techniques for birds.
- Understand common medical disorders in birds.

Introduction

Aviculture, the keeping of companion birds, has been a part of human culture for centuries. Birds were kept for their songs, beautiful colors, intelligence, and, of course, the wonder of flight. The feathers of exotic birds have been used to decorate ceremonial dress, birds of prey were trained as hunting companions in the ancient art of falconry, pigeons were employed during war time to deliver messages behind enemy lines, and then there was the wonder of speech. Today, birds are kept for all these reasons and more. They are the third most popular companion species, following only dogs and cats (**FIGURE 14-1**).

There is such great variety in the types of companion birds, from the smallest singing finches to the largest macaws, that there can be no one volume that could even begin to cover or keep up with current information.

In general, companion birds are limited to a few taxonomic families: those commonly and collectively called parrots, finches, softbills, pigeons, and doves. Within each of these groups is an enormous variety of species, each with its own distinct characteristics, behaviors, and personalities. Diet and husbandry requirements that meet their needs are just as specific.

Birds are classified according to their anatomical features. Psittacines are birds with powerful hooked beaks. They have four toes, two forward and two back. In addition to perching, they use their feet to pick up and hold food items,

FIGURE 14-1 Birds have been kept for centuries. The feathers of many species, such as these Scarlet macaws, were used to decorate ceremonial dress.

to play, and to climb. This group is collectively called *parrots* or *hookbills*. Popular species include budgerigars, cockatiels, cockatoos, conures, amazons, macaws, and the African parrots. Psittacines are known for their ability to talk, their intelligence, and their longevity. Some hookbills may live 60 years or more.

Passerines are the largest group and include the song birds. They have small straight beaks and four toes; three toes are directed forward and one toward the back. Passerines do not use their feet like psittacines and are often called *perching birds* (**FIGURE 14-2**). They are usually too small to become hand tamed but are enjoyed for their variety of songs and color. Included in this group is an enormous variety of finches and canaries. They have a much shorter life span of four to eight years.

Many of the finches are also softbills to distinguish them from the seed eaters. Softbills require different diets and generally do not eat seeds but feed mainly on insects and a variety of fruits.

A separate group, Rhamphastidae, includes toucans, toucanetes, and aricaris. These are often kept in display aviaries (enclosures for birds) rather than as

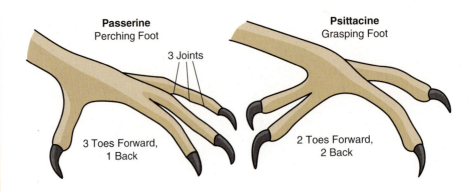

FIGURE 14-2 While both passerines and psittacines have four toes, the foot of the passerine (or perching bird) has three toes pointing forward and one pointing back. A psittacine's foot is adapted for grasping and has two toes pointing forward and two toes pointing back.

companion birds. They have very specific dietary requirements and are not usually hand tamed. They eat a variety of fruits, insects, small mammals, and lizards. They are kept for their beauty and relative rareness. This group has very large beaks for catching prey, which they toss into the air and swallow whole.

Pigeons and doves, from the group Columbiformes, are very popular and are usually housed outdoors in **lofts.** They are known for their homing instincts and gentle nature. Flocks of *homers* may often be seen in the sky as they fly back to the loft, returning after morning exercise.

A Homing Pigeon is a domesticated rock pigeon that has been selectively bred to be able to find its way home over long distances. While it is not certain how these birds navigate, it is thought that they are able to follow landmarks, but even so, this does not explain how birds are able to identify landmarks and return to the home loft when transported in a vehicle for a first return flight. Some of the longest flights recorded have been greater than 1600 miles. Many of these birds compete in races. Pigeon racing is a popular sport the world over.

There are many varieties of fancy pigeons. They are not only raced, but are exhibited in specialty shows, released at functions, and kept by millions of people who just enjoy watching their flocks whirl and swoop though the air.

Pigeons are larger with squared tails. Doves are much smaller with tail feathers that end in a point. They are unique in that both parents produce **crop milk** to feed their young, known as **squabs.** Crop milk is produced by specialized cells within the crop of parent birds. The milk is high in fat and protein and regurgitated directly into the crop of the nestlings.

Respiration in birds is quite different from that in other species. The lungs are fairly rigid and lay flat against the dorsal vertebrae and ribs. Depending on species, birds have between seven and nine air sacs that function in respiration. The air sacs are anatomically located as two posterior thoracic, two abdominal, two anterior thoracic, and one interclavicle. There are also two cervical air sacs that may be absent in some species. Movement of the breast muscles, keel bone, and ribs provides for lung respiration, which in turn inflates the air sacs. Inspired air moves from the lungs and through the air sacs before returning to the lungs and exhaling exchanged gases. Birds do not have a diaphragm to assist in respiration (**FIGURE 14-3**).

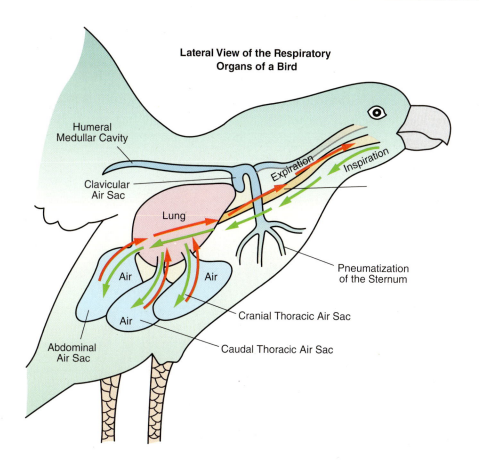

Lateral View of the Respiratory Organs of a Bird

Humeral Medullar Cavity

Clavicular Air Sac

Lung

Expiration

Inspiration

Air

Air

Air

Pneumatization of the Sternum

Cranial Thoracic Air Sac

Caudal Thoracic Air Sac

Abdominal Air Sac

FIGURE 14-3 The respiratory system of birds functions in a different way from that of mammals. Birds do not have a diaphragm to assist in respiration. Inspired air moves through the lungs and air sacs assisted by the keel bone, breast muscles, and ribs.

Avian digestion also differs compared to other species. Food enters the oral cavity and travels down the esophagus to the crop. The crop is an expandable food storage pouch (**FIGURE 14-4**). After leaving the crop, the food passes to the **proventriculus**, or true stomach. The proventriculus contains digestive enzymes that start to break down the food. From the proventriculus food travels to the **ventriculus**, commonly called the **gizzard**. The strong, tough muscles of the

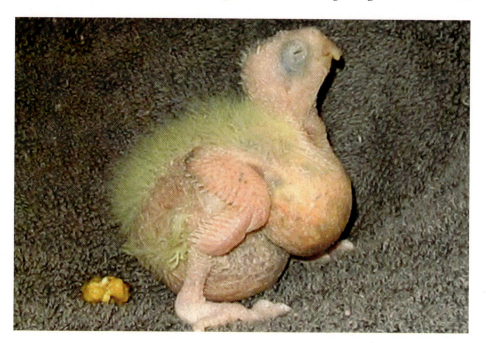

FIGURE 14-4 A young, parent-fed cockatiel chick with a full crop. This chick is approximately eight days old.

ventriculus grind the food before it enters the small intestine, where nutrients are absorbed. Waste products are expelled from the cloaca. Not only is the cloaca the terminal collecting sac for the digestive system, but the urinary tract also empties into the cloaca before waste is expelled (**FIGURE 14-5**).

Males have two testicles enclosed within the abdominal cavity (**FIGURE 14-6**). During breeding season, the testicles enlarge as they produce sperm. In many birds, only the left testicle is functional. Most of the parrot species have only one functional ovary, located on the left side (**FIGURE 14-7**).

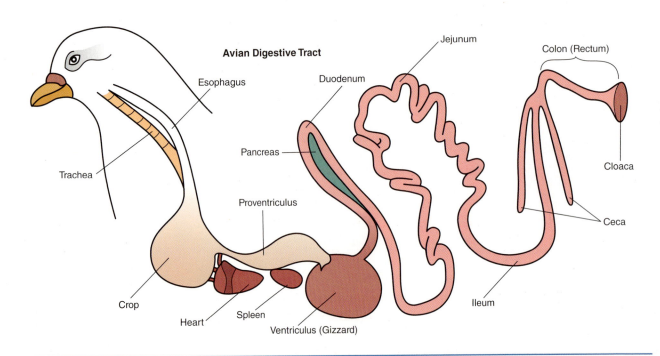

Avian Digestive Tract

Esophagus
Trachea
Proventriculus
Crop
Heart
Spleen
Ventriculus (Gizzard)
Pancreas
Duodenum
Jejunum
Colon (Rectum)
Cloaca
Ceca
Ileum

FIGURE 14-5 The digestive system of birds also differs from that of mammals. Digestion begins in the crop before moving to the proventriculus (or true stomach) and the ventriculus, commonly called the gizzard.

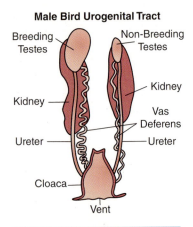

Male Bird Urogenital Tract

Breeding Testes
Non-Breeding Testes
Kidney
Kidney
Vas Deferens
Ureter
Ureter
Cloaca
Vent

FIGURE 14-6 The urogenital tract of a male bird. Usually, only the left testicle produces sperm during the breeding season.

FIGURE 14-7 Newly hatched cockatiel chicks. Not all eggs produced by the ovary will be fertilized by the male as shown here with the remaining egg being nonfertile.

With some species, sex determination is fairly simple and can be done visually. Species with distinctly different physical appearances between males and females are called **sexually dimorphic**. For example, the **cere** (the fleshy tissue between the beak and the head) of a male budgerigar is blue whereas the female has a tan or pale pink cere. The Eclectus parrot has an even greater difference: males are bright green and females are red. In some species, sound may be identifying; for example, only the male canary sings. In other species, it is not possible to determine sex visually. Positive identification is usually determined through DNA testing. A single feather sent to a lab is all that is required. Previously, before DNA testing, birds were examined surgically. This procedure and newer endoscopic techniques require the bird to be anesthetized and the reproductive organs examined internally. Because anesthesia potentially puts the bird at risk, invasive procedures used just for sexing are not usually performed in companion birds.

The **uropygial gland** is located at the base of the tail and produces oils that the bird uses when **preening** (cleaning feathers). Birds rub their beaks over the gland, picking up the oils, which are transferred and groomed into the feathers. Not all birds have a uropygial gland. It is inconsistent in psittacines and absent in Amazon parrots, Hyacinth macaws, and Palm cockatoos. All birds preen to clean and maintain their feathers.

Birds have six different types of feathers. Contour feathers are found on the wings and tail; they cover the body and outline its shape. The wing feathers, both primary and secondary, are called **remiges**, and the tail feathers are called **rectrices**, and they are the longest feathers in all birds (**FIGURE 14-8**).

Down feathers are close to the body, provide warmth, and are the first to appear in newly hatched chicks. Semiplumes lay under the contour feathers and over the down feathers. These also help insulate and keep the bird warm. Some

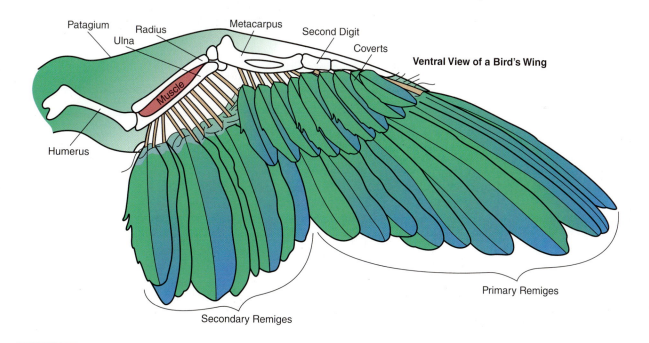

FIGURE 14-8 The ventral view of a bird wing showing feather attachment and skeletal anatomy.

down feathers produce a fine white powder that provides waterproofing. These feathers are called **powder down** feathers to distinguish them from those that only provide warmth. Some species such as cockatoos and African Greys produce a greater amount of *feather dust* than other birds. **Filoplumes** are small, hair-like feathers with barbs along the shaft that help guide the bird in flight and in repositioning the feathers. **Bristle feathers** are on the head and around the eyes and nares. These are sensory feathers (**FIGURE 14-9**).

New and growing feathers have a rich blood supply that nourishes the feather as it matures (**FIGURE 14-10**). As the feather grows, the blood supply recedes and the shaft of the feather becomes transparent. If a large growing flight feather is broken or damaged, it will bleed profusely as these feathers are directly attached to bone and nourished with a rich blood supply. To stop the bleeding, the broken **blood feather** needs to be completely removed. In a small bird such as a cockatiel it should be grasped firmly with forceps, close to the skin, and pulled straight out in the same direction as the feather growth. The blood vessel will seal off under the skin. Direct pressure on a blood feather will usually not stop the bleeding. Traumatized birds have been known to bleed to death from broken blood feathers. Extraction of a blood feather is painful and anesthesia is recommended, and necessary in large birds such as the macaws.

Molting, the natural replacement of feathers, occurs in birds once or twice a year, depending on species. Molting is also influenced by season, health, and

FIGURE 14-9 External anatomy of a bird and feather differences.

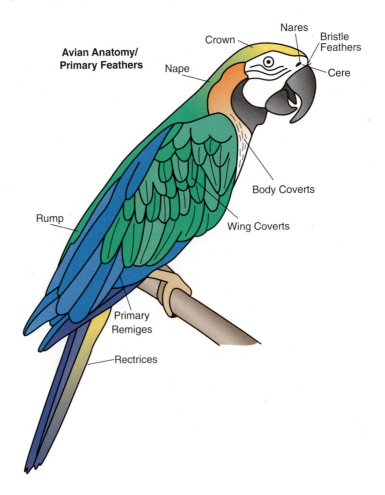

Avian Anatomy/
Primary Feathers

Nares
Crown
Bristle Feathers
Nape
Cere
Body Coverts
Wing Coverts
Rump
Primary Remiges
Rectrices

nutrition. Most molts take several weeks to complete as feather replacement is a gradual process. Molting can be a stressful time for a bird. Birds that are normally vocal often become subdued and quiet. Care should be taken with molting birds, as blood feathers are coming in and they are more vulnerable to damage.

Behavior

Flock status is very important in establishing hierarchy and social order. When people acquire companion birds, they become part of the social order. In the wild, the birds on the highest branches are the dominant birds. With companion birds, the flock status is no different. The height at which the bird is placed, either by cage location or while socializing with a human, determines rank and social status. Many behavioral issues with birds are a direct result of placing the bird above the owner's head or allowing the bird to sit on a shoulder, giving the bird priority at the top of the perceived hierarchy.

Birds are very intelligent. They have the ability to use tools, problem-solve, and transfer learned information. Because of this level of intelligence, they require daily interaction with humans or other birds. They need to be provided with toys appropriate for their mental abilities, toys that interact, stimulate, and entertain. Many behavioral problems seen in birds are created by boredom. Some birds become aggressive and cage territorial and start to bite. Others begin to feather-pick out of boredom and from little or no social interaction with people (**FIGURE 14-11**). Birds that are hand-raised by humans imprint and bond with people, not other birds, and need daily quality time with their human flock.

Species that are known for their ability to talk or use human speech also use appropriate vocabulary for their situation. African Greys in particular are very vocal about their predicament and will voice an opinion or make a request (or complaint) very clearly.

FIGURE 14-11 Feather-picking is a common behavioral problem in many companion birds. As with any behavioral change, a medical cause should be ruled out. This otherwise healthy Green Wing macaw has become a chronic feather picker. The exact reason, or reasons, is undetermined.

Stop It! Don't Do It! Want To Come Out, Cookie? Birds are **cognizant**. They know and understand exactly what they are saying and use the vocabulary they have learned to make their feelings very clear. *It needs to be remembered that birds will use the words they hear and learn, and in a clinical situation, staff need to be aware of their own word choices. Occasionally, just one harsh inappropriate word used even casually could send an avian patient home with an unpleasant surprise for the owner.*

Correcting undesirable behavior in psittacines is similar to correcting a young child's bad behavior. A *time out* is frequently used. The owner does not interact with the bird until the unacceptable behavior stops, for example, screaming for attention. Redirecting and positive reinforcement are also excellent techniques to correct behavioral problems.

For Many Years, Dr. Irene Pepperberg has studied the learning ability of parrots. Thanks to her research studies with Alex, an African Grey parrot, Griffin, and other parrots in comparable studies, we know that psittacines are cognizant, intelligent, and possess a greater ability to communicate than previously thought.

Housing

All birds need to be provided with as large a cage as possible. A cage is not just a place to contain a bird, but it becomes a safe place, a playground, the source of food and water, and a quiet sleeping area. Cages should be wider than they are tall. Birds do not fly straight up and down. Their flight and movements are more lateral or at a gradual incline. Round cages provide no place for a bird to settle into and feel safe.

Cages should not be coated with lead-based paint or made of galvanized metal. Cage bar spacing needs to be of the appropriate size for the bird so it cannot squeeze through the bars or become stuck between them.

Perches in the cages should vary in dimension with sizes appropriate for grasping and resting the feet. Natural wood, such as manzanita, makes ideal perches. They are available in varying widths and shapes and attach easily to cage bars. Manzanita wood is easy to clean and disinfect. Dowel rods and sand-covered perches should not be used. The constant abrasion from standing on sandpaper-covered perches creates sores on the bottoms of the feet. The foot wraps three-fourths of the way around a correctly fitted perch, and the nails do not come into contact with the surface of the perch, so the use of sandpaper-covered perches is not a method of keeping nails trimmed. Perches should not be placed directly over food and water bowls where they can become contaminated with fecal material. Most birds have a sleeping perch, used only at night or when resting. It should be placed toward the back of a cage. Some birds, especially conures, use a sleeping tent that hangs from the cage top.

A variety of materials are available for bird cage floors. The easiest, cheapest, and healthiest is still plain newspaper. Sawdust and wood shavings create dust that can cause respiratory problems and possible crop blockages if ingested. Ground corncob, walnut shells (which are toxic), and recycled paper products promote mold growth when they become wet and should not be used. Birds pick up and play with many items, including cage substrates, when allowed to play on the cage floor. Shredding newspaper is a fun activity; newspaper can not only be shredded but can also be turned into a *tent* or a good place for finding hidden toys and treats, and these fun activities are safe and stimulating.

The cage should be located in an area of the house where the bird is able to interact with family members, but not near the kitchen. All brands of nonstick cookware, when overheated, release PTFE (polytetrafluoroethylene) fumes, and because of the uniqueness and efficiency of the avian respiratory system, even low levels of PTFE are toxic to birds. Burning food, such as microwaved popcorn, can also release enough toxic fumes to kill a bird. Common household items that are toxic to birds include fumes from self-cleaning ovens, aerosol sprays, concentrated cleaning products, and oil-based paint. New carpet often contains formaldehyde, some candles contain lead, and scented air fresheners have also been implicated. These products can potentially cause respiratory distress and possibly death in birds that are exposed to them.

Placing the cage near a window is not recommended. Birds are easily frightened by things they see outside and panic. Dramatic temperature changes, magnified by the glass, can overheat the bird, or drafts from winter weather can chill the bird. Companion birds are most comfortable at room temperatures. Birds that are housed entirely indoors also benefit from full spectrum UVB lighting. It provides the bird with the option of sun-bathing and aids in the absorption of vitamin D. The light should be on during normal daylight hours and turned off at night (**FIGURE 14-12**)

FIGURE 14-12 Birds that are housed indoors can receive the benefit of sunlight with the addition of a full spectrum bulb to the exterior of the cage.

Diet

Not all birds eat seeds, and birds that do should not eat them exclusively. Feeding a seed-only diet contributes to malnutrition and shorter life spans. Seed mixtures should be limited and measured based on the size of the bird and the suitability for the species. Smaller species feed on a variety of plant seeds, including millet, rape, and linseed, while mixes for larger birds contain safflower, sunflower, pumpkin, and cereal grains. Peppers, shredded coconut, and whole nuts are often found in the higher quality mixes.

Diets high in sunflower seeds can cause obesity. With the advances in understanding the nutritional needs of companion birds, formulated, pelleted diets have been developed. Pellets can be mixed in with seeds to allow the bird to become familiar with the new food. Neither seeds nor pellets exclusively will meet all the bird's needs.

Food also becomes enrichment and a part of *family life* in the human flock. Birds frequently share small portions of the family dinner. For the most part, it is safe to feed birds *people food*. Food items that should *never* be offered are avocados, onions, rhubarb, garlic, and some uncooked squashes that contain enzymes that can be toxic. Chocolate and caffeine are both on the *NO* list, as well as highly salted items, alcoholic beverages, and all food taken from the human mouth.

Although it is not a sound dietary practice, white meat and dairy products can be offered in limited amounts. Many psittacines enjoy small pieces of cheese, a teaspoon of natural yoghurt with fruit, or a piece of diced cooked chicken. They will grasp and totally destroy an occasional cooked chicken drumstick when offered as a bone, stripped of most meat and the small, sharp fibula bone removed for safety.

> **Aside from the Possibility of Zoonotic Disease Transmission,** feeding a bird from the human mouth may send the *wrong message* to the bird. As birds bond with owners, they can also view them as mates. As a part of bird courtship behaviors, they regurgitate food to feed their chosen mate.

A wide variety of fresh fruits and vegetables should always be a part of the bird's daily diet. Other foods that can be added to the diet are cooked pasta, potatoes, and rice; legumes that have been soaked and then cooked; cooked scrambled eggs; and slices of hard-boiled eggs. Smaller birds, cockatiels, finches, and canaries seem to prefer vegetables over fruit. It can be difficult to tempt birds to try different foods if they were not introduced to them when they were chicks. The sense of smell is not very well developed in birds, but many companion birds have come to associate the sounds of certain noises in the kitchen with certain foods that they enjoy.

Lories and lorikeets require a specialized diet of nectar and do not eat seeds or vegetables. Powdered nectar is mixed with water and offered as a liquid. The tongue of a Lory has bristles that hold the nectar as it is lapped. They also eat very ripe and juicy fruit.

Pigeons and doves consume seeds whole. They eat a variety of cereal grains and pod seeds, such as peas and beans. Pigeons are fed different combinations of proteins and fats, which better enable them to compete and race. Many pigeon breeders develop their own mixes of grain and legumes to enhance performance in racing or in specialty breeds, which are bred for show. Grit (bird gravel) was often recommended for all companion birds. It was scattered on the cage floor or offered in a small dish. It was thought to be necessary for birds to digest seeds. However, grit eventually ends up in the gizzard (ventriculus), where it grinds the food and breaks down the shell hull. Therefore, not all birds should have grit as it can harm those that do not need it. It is now recommended only for birds, like pigeons and doves that consume whole, intact seeds. Most other birds crack the seed hull and cast it aside before consuming the small kernel inside.

Cuttlebones are recommended for small birds, but they are of benefit to all birds. Finches, canaries, cockatiels, and *budgies* all enjoy pecking on them. Cuttlebones provide a source of calcium and trace amounts of iodine. Nibbling on them also helps groom the beak. Larger birds will break off pieces, and a cuttlebone may last only a day or a few hours; it is offered more as a treat/toy. Cuttlebones are often erroneously referred to as *cuddle* bones, but they are actually the endoskeleton of the salt water cuttlefish, which are related to squid (**FIGURE 14-13**).

FIGURE 14-13 Cuttlebones are given to small birds as a source of calcium and trace amounts of iodine. A cuttlebone is the endoskeleton of this salt water cuttlefish.

Restraint and Handling

Most companion birds are hand tamed and willingly step up to an extended pair of fingers. Hookbills often use their beaks to grasp something prior to stepping onto it. This should not be taken as an attempt to bite. Jerking the hand away will earn immediate mistrust from the bird. It is the equivalent of offering someone a seat, then pulling it out from under them. If a bird steps toward an extended hand, it is not likely that it will bite unless something frightens it or the person attempts to grab the bird.

By necessity, restraint for an examination or procedure requires a different approach from social interactions. Birds should never be restrained from the front. Even slight pressure on the keel bone, pectoral muscles, and ribs will restrict respiration. Birds should always be restrained from the back, either cupped in the hand or wrapped in towel (**FIGURE 14-14**).

The owner should be requested not to hand the bird to the restrainer, but allow the restrainer to safely catch up the bird. When an owner passes the bird to a staff member, it can damage the trust the bird has with its human companion. The bird should be placed on a perch for the restrainer to retrieve and restrain without owner involvement.

The use of a towel as a restraint device prevents oils and disinfectant soap residue from the restrainer's hands from damaging the feathers, and it also acts as a distraction for the bird. Gloves should never be used with companion birds. They are frightening to the bird and can result in trauma to the patient. When wearing gloves, the restrainer is unable to feel the amount of pressure being applied to the bird. Birds should be approached from behind and the towel wrapped around them, including the wings. The patient should be held with one hand under the lower beak, avoiding pressure on the cheeks, which will result in bruising. This is especially important with macaws that do not have feathers on their cheeks (refer to Figure 14-14). The other hand grasps both feet and supports the bird. To prevent regurgitation, the bird should always be held upright. Holding the bird close to the restrainer offers security to the bird. Birds that arrive in their cages can also be retrieved with a towel. To avoid injury to the bird, toys, dishes, and perches should be removed prior to attempting to towel the bird inside the cage.

Most large companion birds know exactly what the dreaded towel means and some will attempt avoidance at the very sight of it. If a bird panics with the approach of a towel, the lights should be turned off. Diurnal birds have limited vision in the dark, and this will give the restrainer the advantage of retrieving the bird without difficulty. Toweled birds can become very vocal. Macaws scream loudly enough to cause ringing in human ears, and many clinic staff have learned to be prepared and use ear plugs.

Medical Concerns

Birds are susceptible to a variety of diseases. Some diseases are chronic while others are acute, and there may be different forms of the same disease, with varying clinical signs. Because they are a prey species, birds are very good at hiding signs of illness, and not only does the appearance of weakness attract hunting predators, but members of their own flock have also been known to attack an ill

FIGURE 14-14 This Blue and Gold macaw is correctly and safely restrained with the use of a towel. The restrainer's right hand is under the towel and has the bird's head comfortably under control without the danger of bruising the bird's cheeks.

or injured bird. Often avian patients are presented by concerned owners because they have noticed a change in behavior or feeding. It is important for veterinary staff to obtain a detailed history that includes open-ended questions about the bird's home environment and any recent changes in the home or the bird's cage.

Bird droppings (feces, urates, and urine) can indicate the health status of the patient. A normal dropping consists of three parts: fecal material is normally green or brown, the urates are white, and the urine is clear and watery. **TABLE 14-1** provides a summary of changes and conditions associated with bird droppings.

TABLE 14-1

EVALUATION OF BIRD DROPPINGS	
Dropping Consistency or Color	**Causes**
Red dropping: Bright red	Fresh blood from the lower GI tract
Dark red	Old blood from the upper GI tract
Tomato soup- or chocolate-appearing dropping	Suggests lead poisoning
Chartreuse green diarrhea	Suggests septicemia, true diarrhea, chlamydiosis
Undigested seeds in the dropping	Suggests a digestive disorder
Popcorn-looking stools	Suggests pancreatic insufficiency
Very large droppings in the morning	Suggests the bird is not defecating overnight
Polyuria (excessive urine)	Suggests fruits/veggies in the diet, medication, diabetes
Yellow or yellow-colored urates	Suggests a liver disorder

General signs of illness in birds include decreased appetite, discharge from the nares or oral cavity, decreased activity, abnormal droppings, and a huddled or fluffed appearance. Birds with respiratory disease often exhibit a *tail bob*. The rectrices (tail feathers) move up and down with each breath. Birds that are very ill may sit with their eyes closed and/or exhibit open-mouth breathing (**FIGURE 14-15**).

Bathing or showering at least twice weekly helps maintain feather health. Most birds enjoy having a bath; they can be misted with a spray bottle, put into a sink with a sprayer, or taken into the shower with the owner. The water should be lukewarm and controlled to gently spray over the bird with no risk of soap contamination.

For Whatever Unknown Reason, the sound of a household vacuum cleaner will trigger a bathing response in most birds, and they will usually try and bathe in their water dish if there is no other option.

Smaller birds while also enjoying a light spray will normally bathe in a small bowl of water placed on the cage floor.

All companion birds need routine nail trims, and some birds may also require beak trims. (Refer to Clinical Procedures.) Leg bands placed on birds by breeders are for breeder identification purposes only. These bands may include initials that identify the aviary name and a chick number. Breeder bands on young birds are fully closed and slipped onto the leg when the chicks are newly hatched. Quarantine bands placed by the USDA are *open* bands. Breeder leg bands can be removed by a veterinarian using a pin cutter if it is too loose and presents a danger of becoming caught in the cage or on cage furnishings or if it has become too tight. Care has to be taken that the veins and arteries are not nicked in the process of removing the band from the leg. USDA bands should not be removed. Breeder leg bands are not proof of ownership. One way to positively identify a bird and prove ownership is with a microchip implant. This is placed in the pectoral muscle on either side of the keel bone.

Respiratory Disorders

Respiratory tract and air sac problems are common in companion birds. Normal respiration in birds should be relaxed and regular. Tail bobbing, open-mouth, and labored breathing, with an increase in the respiratory rate, are signs for concern and always indicate a medical emergency. A bird with glottis and tracheal problems (possible obstruction) may present with an increased rate of inspiration and, in severe cases, open-mouth breathing. With small airway disease involving the bronchi, there may not be any initial signs, or the patient may be presented with severe dyspnea and an expiratory wheeze. Lung or air sac disease may present with mild to severe dyspnea and an increased respiratory rate.

Aspergillus is a fungal organism that can cause rhinitis, tracheitis, and air sacculitis when birds inhale the fungal spores. Initially, clinical signs most often resemble pneumonia; however, aspergillosis may become systemic, invading other body systems and cause osteomyelitis and encephalitis. Malnutrition and an excessively humid environment, with poor husbandry and hygiene practices, will increase the bird's susceptibility to an aspergillus infection. A positive diagnosis is complicated and may involve several types of testing because routine blood work alone is not diagnostic. Radiographs may show signs of changes in the airway and lungs and further endoscopic examination may be warranted. A fungal culture of the upper respiratory tract may also aid in a positive diagnosis. Treatment includes oral antifungals and nebulization.

Gastrointestinal Disorders

Sour crop, or crop stasis, is not a disease but is usually a sign of other conditions. The cause may be a bacterial, fungal, or yeast (candida) infection. It also occurs in hand-reared chicks. Food that has been left out, reheated, or fed at the incorrect temperature is a common cause of sour crop. Not allowing the crop to completely empty before another feeding can also lead to crop stasis (**FIGURE 14-16**).

FIGURE 14-16 Correct hand-feeding methods, including temperature of the formula and scrupulous hygiene practices, are essential to prevent sour crop and aspiration in baby birds. This young conure is being hand-fed with a syringe.

Tumors may also invade the crop and become so invasive that they prevent the bird from filling the crop. **FIGURE 14-17** shows a bird with a large tumor of the crop. The tumor took up three-fourths of the crop capacity. Surgery was not an option due to the rich blood supply feeding the tumor.

A serious gastrointestinal disease of birds is proventricular dilatation disease (PDD). This disease is thought to be caused by a virus that attacks the smooth muscle of the crop, proventriculus, ventriculus, and small intestine. PDD causes delayed motility in the gastrointestinal tract and dilatation of the organs. Common signs include weight loss, regurgitation, crop impaction, and eventual death.

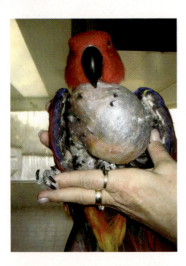

FIGURE 14-17 Tumors may develop anywhere along the digestive tract. This female Eclectus parrot has a large, rapidly growing tumor of the crop. Due to the invasive and systemic involvement of this tumor (only the small bulge in the upper left area of the crop was functional), the bird was humanely euthanized.

Nondigestive Disorders

Beak deformities can be either congenital or acquired. The most common cause is trauma. In young birds, trauma-related injuries are due to aggression among cage mates and aggressive feeding of chicks by parent birds. Scissor beak is a common congenital deformity where the upper beak is bent to one side.

Psittacine beak and feather disease (PBFD) is caused by a virus that can be transferred from an infected hen to the egg or directly to feeding chicks. It is also transmitted by inhalation or ingestion of infected feather dust and dry, powdered fecal material. It can be carried on clothing, hands, feeding utensils, and nest boxes.

There are three forms of PBFD. The peracute form occurs in chicks. There may be no signs and it is usually diagnosed postmortem. The acute form appears in young birds that are beginning to grow feathers to replace neonatal down. There may be several days of depression and changes in the appearance of developing feathers. New, growing feathers may die, break off or bend, and have a clubbed appearance, or they may be molted prematurely, before they are fully grown.

The chronic form of PBFD develops in mature birds. During each molt, more and more abnormal feathers appear. There may be bleeding in the feather shaft or within the pulp cavity of new feathers. Short, clubbed, and abnormally colored feathers may appear. Beak deformities usually occur following feather changes. The beak may become elongated and there may be fractures and necrosis of the beak. Many infected birds die within a few months of the appearance of clinical signs and a positive diagnosis. Others have been known to survive for a few years. At present, there is no vaccine or cure for PBFD.

PDD (proventricular dilatation disease) or avian bornavirus (ABV) was previously known simply as macaw wasting disease. Researchers have identified a bornavirus that has been linked to PDD, and it is now known that this virus can affect not only macaws but also many other psittacines. With the identification of the causative agent the disease is now referred to as ABV.

Potentially infected patients present with weight loss, the presence of whole seeds in the droppings, and chronic regurgitation. In some birds there may be neurological signs including seizures and ataxia as a bornavirus primarily targets the nervous system. Blindness and overall weakness may also be present with or without GI signs. Radiographs can be diagnostic in determining the presence of a dilated proventriculus. PCR is the preferred method of testing to help confirm the diagnosis, but this cannot be relied on entirely as the virus is shed intermittently and may not be present in a singular blood sample and PCR testing should be repeated to confirm results. There is no cure for ABV but supportive treatment includes providing easily digestible soft foods and administrating NSAIDs. Affected birds should be isolated to prevent further outbreaks.

Chlamydiosis is caused by a gram-negative bacteria and is of concern to bird owners because it has zoonotic potential. Human transmission occurs by inhalation of infected particles in feather dust and/or dried powdery droppings. Birds can be asymptomatic carriers and can transmit the disease to other birds and people, and there are no specific signs that a bird may be infected. Many birds can be chronic intermittent shedders, carrying the disease for several years with no signs of being infected and become clinically ill only after a period of stress. Some birds may show general, nonspecific signs including a lack of appetite and weight loss, lethargy, and general disinterest. Bright green, watery droppings may also be apparent and there may be a discharge from the eyes and periorbital swelling. Extreme cases can cause sudden death.

C. psittaci has been found in many other species of birds, not just parrots, and the disease is sometimes also referred to as ornithosis. Pigeons are assumed carriers; birds of prey, shore birds, and pet doves have also been implicated and, to a lesser extent, canaries and finches.

In humans, the disease is called *psittacosis*. This is to distinguish it from other species of chlamydia bacteria, which is spread from person to person. Transmission to humans occurs in the same manner as bird-to-bird transmission: inhalation of the organism. This can come from feather dust and dried fecal material, but also from contact with respiratory secretions, handling infected birds, and mouth-to-beak contact.

Diagnosis of the disease in birds can be difficult. Blood, serum, and swabs from the choana (the opening between the nasal and the oral cavity) and cloaca (vent) swabs are submitted at the same time to a diagnostic laboratory for different evaluations. Treatment may begin if the disease is suspected, even pending laboratory results. *Chlamydiosis* (in birds) and *psittacosis* (in humans) can be treated with the prescribed use of a tetracycline such as doxycycline. Because this disease is zoonotic, confirmed cases of psittacosis are reportable to the CDC (Center for Disease Control). The largest at-risk groups of people are bird-fanciers and companion bird owners. Other groups include veterinarians and veterinary staff, wildlife rehabilitators, pigeon fanciers, and city sanitation engineers (pigeon clean-up duty). There is no over-the-counter prevention or cure and no one is immune.

> **Tetracycline Tablets Are Readily Available for Aquarium Treatments.** These tablets should *never* be used in an attempt to prevent or self-medicate. They are ineffective for anything other than fish aquaria.

Polyoma is a contagious, fatal disease of birds. Some birds die without developing any clinical signs and others die within 48 hours of developing clinical signs. There may be delayed crop emptying, regurgitation, diarrhea, and subcutaneous bleeding.

Polyoma is readily transmitted to other birds through shared air space and contaminants (fomite) brought in on clothing, skin (from handling infected birds), jewelry, and hair. People may bring the virus into the home when visiting bird fairs or pet stores. Virus particles can survive in the environment for several months. Preventing the spread of any disease is important; this can be done by not visiting multiple bird/pet stores on the same day. People should always disinfect themselves and their clothing before handling their own birds. When purchasing a bird, never accept a seller's remark that the bird *had all his shots*. Vaccinated birds are issued a certificate by the manufacturer of the vaccine and it must be completely and accurately filled out. Vaccination will protect a bird from polyoma. Initially, the bird should receive two vaccinations given at a two-to three-week interval and boosted annually thereafter.

Cloacal prolapse in birds is caused by many variables, depending on the sex of the bird. In hens, it is commonly associated with straining to lay eggs. Some cases are idiopathic, with no known cause. The recommended treatment is surgical replacement of the prolapsed tissue (**FIGURE 14-18**).

FIGURE 14-18 This cockatiel has a cloacal prolapse that was successfully repaired. A prolapse is a medical emergency and should be attended to by a veterinarian immediately to prevent tissue necrosis. Surgical intervention is often required. *(Courtesy of Eric Klaphake, DVM)*

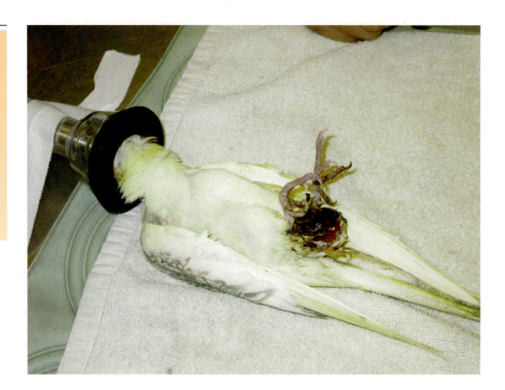

Hens that become eggbound can potentially die from this condition if left untreated. Most birds usually lay eggs at 24- to 48-hour intervals during a laying cycle. Cockatiels, lovebirds, budgerigars, and finches are more commonly affected than the larger psittacines. Straining, fluffed feathers; matted vent; anorexia; the presence of blood in the dropping; dyspnea; and sudden death can all be signs that the hen is unable to lay the egg. Poor nutrition, obesity, and chronic egg laying can predispose a bird to becoming eggbound. Diagnosis includes palpation of the lower coelomic cavity and radiographs to determine the presence of an egg. Initial treatment includes supportive care. Fluids are usually administered along with providing an increase in environmental temperature and humidity, and the hen should be placed in an incubator unit to monitor and control the immediate environment. Supplemental feedings and analgesia are often administered. If the egg is visible in the vent, lubrication with a water-soluble jelly can often assist in the delivery of the egg. Once the hen is stable, the veterinarian may administer a hormone to induce contractions, attempt a manual expulsion of the egg, or perform an ovacentesis to collapse the egg so that it can be extracted. There is a high risk of egg yolk coelomitis with these procedures.

Cloacal papillomas are often diagnosed in birds. A papilloma is a benign wart that originates from epithelial cells, and they are associated with a Herpes virus infection. It can cause **tenesmus** (painful straining while defecating), bloody droppings, and staining of the feathers around the vent. Birds kept in unclean environments are particularly susceptible. A simple diagnostic test can be done to determine if it is papilloma by placing a dab of white vinegar on the affected tissue. If the vinegar turns the growth white, it is indicative of papilloma because the acetic acid in the vinegar reacts with the papilloma tissue. Treatment may consist of surgically removing the papilloma, but recurrences are common (**FIGURE 14-19**).

Hypovitaminosis A is usually associated with birds that are fed only a seed-based diet. A swollen and narrowing of the choana, chronic upper airway disease, hyperkeratosis of the feet and legs, poor feather quality, and anorexia can all be signs of vitamin A insufficiency. The recommended treatment is administering vitamin A through injections and correcting the diet.

Heavy metal toxicity should be suspected in birds that are exposed to metals within their environment. Lead and zinc are the most common sources for companion birds. Lead is found in paints, toys, drapery weights, fishing weights, stained glass solder, plumbing materials, batteries, and linoleum. Zinc is a component of galvanized hardware, wire, costume jewelry, and most imported bird cages. Clinical signs are nonspecific and variable but include lethargy, weakness, polyuria, hematuria, anorexia, ataxia, regurgitation, and seizures. Along with a detailed history, a complete blood panel and radiographs will help confirm the diagnosis. If pieces of ingested metal are visible on the radiograph, surgical removal is necessary; however, not all toxicities involve ingestion of a solid object, and blood testing is also required. The course of treatment includes the administration of a **chelating agent**, such as calcium EDTA, which binds with toxic metals in the blood and other tissues. The bonded metals can then be excreted. Supportive care includes fluid therapy, supplemental feedings, additional warmth, and seizure control.

Birds are susceptible to a variety of medical problems unique to them. It is important to establish a good relationship with an avian veterinarian prior to needing emergency care.

FIGURE 14-19 The exact cause of avian papilloma is undetermined, but there is a greater incidence in birds that are kept in dirty conditions. This Blue and Gold macaw was rescued from serious neglect. When the caked-on fecal material was carefully removed from the vent area, the papilloma became evident. With recommended veterinary care and cleanliness, the bird's condition has improved greatly.

Avian Blood Samples Are Usually Collected in Microtainer Tubes. The rubber stoppers in red-top tubes contain trace amounts of zinc, which could affect the results. Additionally zinc is present in low levels because it is an essential trace element of several enzymes. When specifically testing for zinc, royal blue–top tubes are preferred.

Parasites

Parasites are rarely a concern in companion birds. If birds have parasites, they are usually ectoparasites. The Knemidocoptes mite is the most common mite seen in budgerigars and passerines. This mite attacks the tissue around the cere and legs. In budgerigars, hyperkeratosis is also present around the cere. In passerines, the legs are attacked, the feathers look ragged around the legs, and the nails grow excessively. Ivermectin is the treatment of choice to address most mite infestations. Lice are not a problem in well-cared-for birds kept indoors in a clean environment. Lice can be seen along the feather shaft, close to the skin. A bird cannot contract lice from other species, including humans.

Clinical Procedures

In smaller birds, the nails can be clipped with tiny nail trimmers. In larger birds, nails can be trimmed with a Dremel-type tool. The advantage to using a powered rotary tool is that it also cauterizes the nail if it is inadvertently cut too short and into the quick. Beaks can also be trimmed and shaped using the Dremel® (**FIGURE 14-20**). If a rotary tool is not used or bleeding from a cut nail or beak doesn't stop, a styptic powder can be applied. Silver nitrate sticks should not be used on a bird's beak or nails. While not toxic to dogs or cats, silver nitrate is toxic to birds if ingested. Correct trimming of nails and beaks requires knowledge of the tools used to avoid injury to the bird (**FIGURE 14-21**).

Wing trims are often a personal preference. Some owners prefer their birds to have free flight while others would like feather trims but do not want the bird to appear to have incomplete wings. The purpose of trimming flight feathers is to prevent the bird from flying too far or too high yet allow the bird to glide safely down should it decide to fly. Most veterinarians recommend cutting the flight feathers just under the second row of contour feathers, and the number to cut varies with different species. The feathers are cut from the distal end of the wing, trimming toward the body. Both wings need to be trimmed in the same manner. Feathers should be examined from the underside of the wing and cut one at a time. Blood feathers should never be cut. If a blood feather is present, one mature

FIGURE 14-20 The use of a Dremel® tool to trim a bird's beak is an acquired and exact skill. The veterinarian has both hands free to control the Dremel® and the beak. Serious trauma could result if the bird is not restrained properly for this procedure.

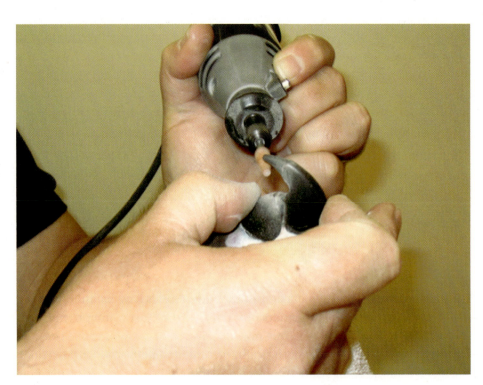

FIGURE 14-21 A variety of grooming tools used for beaks, nails, and wings. These include a Dremel® tool, small side cutters for feathers, two sizes of nail clippers, an emery board, and a buffer for smoothing and polishing the beak and nails.

feather on each side of the blood feather should be left to protect it. Scissors should not be used to cut straight across and through all of the feathers at once.

Blood samples are most frequently collected from the right jugular vein because in most birds the left jugular vein is much smaller. The right jugular vein is superficial and easy to visualize. The crop should be empty prior to obtaining a blood sample to prevent regurgitation and possible aspiration. Occasionally, an air sac may sit on top of the vein. To avoid inadvertent puncture, the air sac should be gently pushed aside. A 25- to 27-gauge needle connected to a tuberculin syringe is used to obtain the sample. To restrain for a jugular venipuncture, the patient is wrapped in a towel and positioned on its left side. With one hand holding the head (thumb under lower beak and the rest of the hand around the back of the head), the other hand lowers the right wing down to expose the jugular vein (**FIGURE 14-22**). Direct pressure is applied to the puncture site to prevent hematoma formation or death from blood loss. The patient should not be released until all bleeding has stopped.

Blood may also be taken from the medial metatarsal vein. Collection from this site should be approached with caution as the vein is in very close proximity to the medial artery. The artery can be easily lacerated if the patient is allowed to move. Accessing this site may require anesthesia. Another potential site is the cutaneous ulnar vein, on the medial surface of the wing. Because this vein is superficial and the skin is thin, the area is easily bruised and requires a longer period of direct pressure to the puncture site to prevent hematoma formation.

It is never recommended to obtain a blood sample by clipping a toenail so short that it bleeds. This is very painful for the bird and the injured toe is easily contaminated. It may also cause a permanent disfigurement of the toe and create great distrust for future visits.

FIGURE 14-22 The correct restraint for obtaining a jugular blood draw. The patient is always placed on its left side with the head extended and the right wing lowered to easily visualize the jugular vein.

Most routine blood samples are collected in heparinized tubes. Blood volume in birds is normally 10 percent of bodyweight. Only small amounts are required, and for most patients, 0.5 to 1.0 ml of blood will be an adequate sample. All patients should be weighed on a gram scale prior to a blood draw to determine the maximum volume of blood that can be safely taken at that time (FIGURE 14-23). In small birds such as canaries and budgerigars, the use of a 27-gauge needle on a tuberculin syringe is recommended. With medium to large psittacines, a 25-gauge needle attached to a tuberculin syringe works well. Microtainer tubes containing EDTA are used for hematologic evaluations and lithium heparin for biochemical profiles. Blood films need to be prepared immediately after blood collection because anticoagulants can alter some cell morphology.

FIGURE 14-23 A young Sun conure sitting quietly on a gram scale with a perch. An accurate weight should always be obtained just prior to a blood draw.

Intramuscular injections are given in the pectoral muscle. Drugs administered IM need to be considered carefully, as many drugs can cause muscle necrosis at the injection site. When multiple injections are being administered, alternate sides should be used.

Fluid therapy can be administered through intravenous or intraosseous catheters or by subcutaneous injections. If giving subcutaneous fluids, the bird is restrained and one leg is pulled slightly forward and lateral. A 25- to 27-gauge needle is inserted through the thin membrane between the leg and the body and the fluids are injected into the triangle formed. The fluids should appear as a small bubble forming at the site. If the needle is pushed too deep into the skin pocket, fluids may be deposited in the air sac and the patient could, in effect, be drowned (**FIGURE 14-24**).

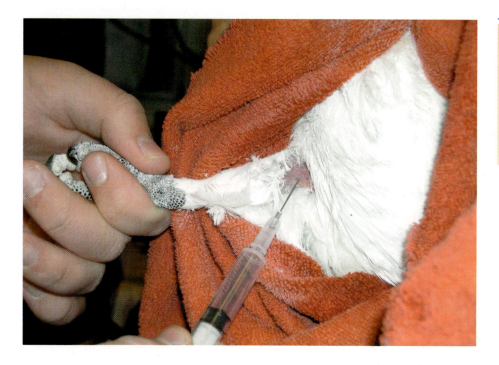

FIGURE 14-24
Subcutaneous fluids may be given in the web of skin between the body and the leg. It is important to gently pull the leg slightly lateral to visualize the correct area for injection.

Intraosseous catheters are placed more routinely than IV catheters because the veins are so fragile. In small birds, a 25- to 27-gauge spinal needle or injection needle is used. A 20- to 22-gauge spinal needle or injection needle can be used in larger birds. Common sites for placement include the distal ulna, proximal ulna, and the proximal tibiotarsus. Patients are anesthetized and the area is prepared with the same care as a surgical site. The catheter needle is secured by tape or sutures.

Occasionally, following surgery, trauma, or self-mutilation due to excessive feather plucking, variations of the E-collar may be necessary to prevent the patient from removing sutures or further traumatizing an injury. Collars should be placed only under the supervision of a veterinarian to address individual case needs. Various kinds of collars are available for avian use. Some are made from soft foam rubber tubes that gently extend the neck and prevent the bird from flexing its neck but still be able to bend down to eat and drink. Stockinettes can also be used to cover an area and protect the site from injury by the bird's beak or nails.

An E-collar of any type must be carefully placed and adjusted so that it does not interfere with the crop. To avoid further stress to the patient, the collar is fitted

FIGURE 14-25 Chronic obsessive feather-picking, for whatever reason, can lead to serious self-mutilation. This cockatoo has opened a large wound in the lower abdomen.

FIGURE 14-26 This is the same patient with an E-collar that has been applied to prevent the bird from further self-injury. The wound has also been bandaged.

under general anesthesia. Upon recovery, the bird should be placed in a small and well-padded enclosure without perches. Birds take a little time to adjust to the presence of an E-collar and will often tumble and fall as they attempt to regain their balance. During this time they should be observed carefully but be allowed to adjust to it themselves (**FIGURE 14-25** and **FIGURE 14-26**).

There are many differences in avian anesthesia compared to other species. Their unique physiology affects the way they are anesthetized and monitored. Air sacs function in respiration and are directly involved in the inhalation, uptake, and metabolism of anesthetic gas.

Birds should not be fasted for more than three hours prior to anesthesia. They have a high metabolic rate and little glycogen storage ability; withholding food for longer periods can result in hypoglycemia. If the crop is full, it needs to be manually emptied prior to anesthesia induction to avoid regurgitation and possible aspiration.

Their unique respiratory system also makes birds prone to hypothermia. Heat loss occurs during anesthesia because of the constant need for air flow across the parabronchi. It is important to keep the bird warm while under anesthesia and maintain steady, positive pressure ventilation (PPV).

Inhalant anesthesia is preferred for use in birds. Injectable anesthetics are available but not used often because of the prolonged recovery and possible muscle necrosis at the injection site. The advantage of inhalant anesthesia is a quick induction and recovery time with minimal cardiopulmonary effects. Isoflurane and sevoflurane are both used successfully with a nonrebreathing system.

African Greys Usually Have Lower Blood Calcium Than Other Species. Administering an appropriate dose of calcium gluconate by injection prior to anesthesia may help prevent seizures during anesthesia and recovery.

Birds are intubated by gently pulling the tongue forward and placing the endotracheal tube (ET) into the glottis, which is easily visualized at the base of the tongue. An uncuffed ET should be used as the tracheal rings of birds are complete and fragile. If there is an upper airway obstruction, an experienced avian veterinarian may perform an air sac intubation. It is not uncommon to experience mucous plugs in the ET tube during long procedures. If the avian patient is having difficulty breathing or the anesthetist is having difficulty ventilating the patient, the tube needs to be carefully checked for a mucous plug and immediately replaced with a new tube.

Careful monitoring during anesthesia is extremely critical for birds. With induction, voluntary muscle activity decreases, but corneal and pedal reflexes remain. Dopplers are used to monitor heart rate and heart strength. One of the best sites to place the Doppler probe is over the superficial ulnar artery or the radial artery on the medial side of the ulna. A respiratory monitor, an esophageal stethoscope, and an ECG are additional devices that can be used to monitor the patient. Monitoring blood pressure is difficult and not accurate in smaller species of birds.

As the depth of anesthesia increases, respiration begins to stabilize, but vital signs continue to decrease and respiratory arrest may occur. Respiratory rate under anesthesia is two to seven breaths per minute. Respiratory arrest is not uncommon in an anesthetized avian patient, and with close monitoring this can be reversed or prevented. *Cardiac arrest will quickly follow without immediate corrective measures.* Cardiac arrest may not be reversible. Manual ventilation and adjustment of the flow of the anesthetic agent can help reverse respiratory arrest and establish a normal respiratory rate.

During recovery from anesthesia, the bird should be wrapped in a towel and held upright until fully recovered. Extubation occurs only when the bird is alert enough to attempt to bite the tube. Keeping the bird intubated will help prevent possible regurgitation. Most patients recover from inhalant anesthesia within five to ten minutes after the gas anesthetic is turned off. Because of their high metabolic rate, food should be offered shortly after recovery to prevent hypoglycemia.

Avian patients are usually anesthetized prior to positioning for a radiograph. This not only removes stress on the patient but also allows for better positioning, without movement and the potential of injury. Debilitated birds with a high risk for general anesthesia can be placed in a paperbag, plastic container, or glass tank, or lightly rolled up in towel for restraint when taking radiographs. The two views are lateral, with the wings fully extended and placed over the bird's back (**FIGURE 14-27** and **FIGURE 14-28**). The legs are extended caudally to expose the abdomen. The other view is ventral/dorsal (V/D). The patient is positioned dorsally with the wings spread out to either side. The legs are pulled down and away from the body.

Birds are held in position by using *masking* tape. White, or zonas (cloth tape) tape should not be used. When these tapes are removed, feathers are pulled out and there may be skin tearing.

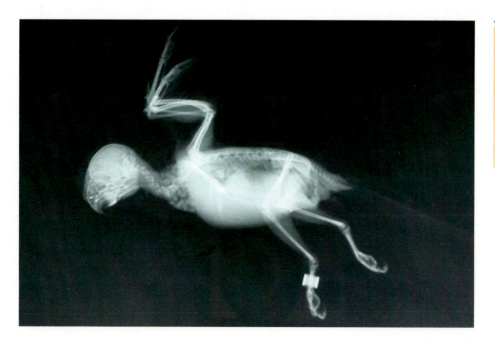

FIGURE 14-27 Position for a lateral radiograph. The patient is anesthetized; the wings and legs are held in place with masking tape. *(Courtesy of Martin G. Orr, DVM, Bird and Exotic All-Pets Hospital)*

FIGURE 14-28 Lateral radiograph of a budgerigar with a full crop.

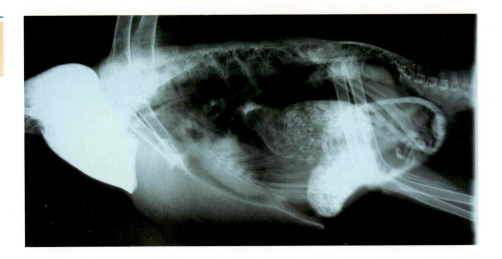

FIGURE 14-28 Lateral radiograph of a budgerigar with a full crop.

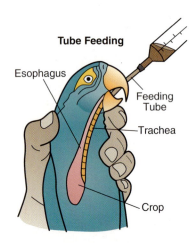

FIGURE 14-29 It is extremely important to confirm that the gavage, or feeding tube, is correctly placed within the crop to avoid aspiration and death of the patient.

Endoscopy is frequently used in birds to visualize internal structures, obtain biopsy samples, to remove small objects in the upper airway or GI tract. To avoid damage to the tissues, most endoscopes used in birds will have a diameter ranging from 1.9 to 2.7 mm. The patient is anesthetized and a small incision is made without the need for a major invasive surgical procedure. A common site for endoscopic imaging includes the coelomic cavity to view the cloaca, caudal air sac, sinuses, trachea, and choanal slit. Accessing the caudal thoracic air sac will allow the visualization of the internal organs including the gonads, kidneys, liver, intestinal tract, and heart.

Birds can be given nutritional support by tube feeding. The amount given is based on weight and need and placed directly into the crop. The bird is restrained with a towel and held upright. The gavage tube is gently maneuvered past the tongue and down the right side toward the esophagus and into the crop. If the placement is correct, the feeding tube can be seen to slide down the esophagus into the crop. To ensure correct placement, the metal feeding tube should be felt or visualized in the crop *before* administering the food. If the tube cannot be felt in the crop, it is in the trachea. If the food is deposited into the trachea, it will enter the lungs and kill the bird instantly (**FIGURE 14-29**).

Fecal gram stains provide information on the percentages of different types of bacteria present in the gastrointestinal tract. Normal intestinal flora are grampositive rods and cocci with 1 percent gram-negative rods. Changes in the amount of bacteria, the presence of yeast, and changes in the types of bacteria indicate disease. Enterobacter, *Escherichia coli*, Proteus, Klebsiella, and Pseudomonas are gram-negative bacteria that can cause disease in birds. Skin, air sac, crop, and gastrointestinal infections are often caused by overgrowth of gram-negative bacteria. Microbiologic cultures provide information on possible infections that affect the intestinal tract, reproductive tract, and urinary system. Culture sites include the crop, cloaca, choana, and skin. Culturing the crop is indicated when the bird shows signs of crop stasis or regurgitation.

Sinus flushing and nebulization treatments are performed on birds with respiratory diseases. Both of these treatments involve adding medication to saline or sterile water to administer to the patient. Ultrasonic nebulizers are used to provide daily treatments that last an average of 10 to 30 minutes per session.

Bandages in birds are used to immobilize fractures, protect wounds, and prevent self-mutilation. If bandages are applied incorrectly, this can make the condition worse. External coaptation devices involve the placement of splints and bandages. The figure-of-eight bandage is a frequently used external coaptation bandage applied to the wing and can be used to immobilize fractures distal to the ulna or maintain the wing position for IO/IV catheter placement and stabilization. A nonstick wrap (Vetwrap®) is preferred for this type of bandage (FIGURE 14-30).

FIGURE 14-30 Detailed process of the application of a figure 8 bandage.

Tibiotarsal fractures are often seen in the avian patient. Lateral splints are used to stabilize this type of fracture. A soft padded wrap reinforced with an aluminum rod, tongue depressor, syringe case, moldable plastic, or paper clip is used to make the splints. Tape splints are used on smaller avian patients.

Birds with external coaptation devices require cage rest and close monitoring. The patient is placed in a smooth-sided enclosure to prevent climbing with no or low perches. Bite-size food should be offered to avian patients with leg injuries. Bumblefoot may develop in patients with leg injuries on the weight-bearing foot.

Summary

Companion birds have been kept for many centuries. Today, most of the species are either psittacines or passerines. Each species has specific diet and husbandry requirements. Avian digestion and respiration are quite different from those of mammals and these differences need to be fully understood to care for the avian patient. Birds are highly intelligent and cognizant; they can and do use tools and solve problems. Many behavioral problems develop because of a lack of understanding and awareness of avian perception and intelligence. Housing and husbandry practices are important in keeping a companion bird healthy, both mentally and physically. Because of the uniqueness of the respiratory tract, mastering avian restraint techniques is extremely important in order to perform an examination or any clinical procedure. Companion birds are susceptible to many diseases not seen in other species. Testing, diagnostics, and clinical procedures for the avian patient require a different approach with additional skills. Owners expect veterinary staff to not only be familiar with their companion birds but also be proficient in the knowledge and skills required to care for them.

fastFACTS

TABLE 14-2

AVIAN WEIGHT

Species	Weight
Finches	10 to 18 g
Canaries	15 to 40 g
Budgerigars	30 to 70g
Cockatiels	70 to 110 g
Small parrots	90 to 130 g
African Grey parrots	350 to 600 g
Amazons	400 to 550 g
Cockatoos	200 to 950 g
Macaws	1000 to 1800 g

TABLE 14-3

AVIAN LIFE SPAN

Species	Life Span
Finches	4 to 5 years average
Canaries	8 to 10 years average
Budgerigars	6 to 8 years average
Cockatiels	8 to 20 years average
Small parrots	20 to 30 years
African Grey parrots	40 to 50 years
Amazons	70 to 80 years
Cockatoos	30 to 40 years
Macaws	50 to 60 years

TABLE 14-4

AVIAN REPRODUCTION

Species	Egg Incubation Time (days)
Finches	18
Canaries	18
Budgerigars	18
Cockatiels	21
Small parrots	22 to 24
African Grey parrots	24 to 28
Amazons	26
Cockatoos	24 to 29

TABLE 14-5

AVIAN HEART RATE/RESPIRATORY RATE

Species	Heart Rate (beats/min)	Respiratory Rate (breaths/min)
Finches	274	60 to 70
Canaries	274	60 to 70
Budgerigars	206 to 225	35 to 50
Cockatiels	190 to 215	35 to 50
Small parrots	190 to 215	35 to 50
African Grey parrots	147 to 163	20 to 40
Amazons	147 to 154	20 to 30
Cockatoos	130 to 178	5 to 40
Macaws	110 to 127	20 to 30

ZOONOTIC POTENTIAL

Bacterial

- Mycobacterium (tuberculosis)
- Psittacosis (avian chlamydia)
- Salmonella

Fungal

- Aspergillosis

Review Questions

1. How does avian respiration differ from mammalian respiration?
2. What is the purpose of the crop?
3. What are the methods of determining the sex of a bird?
4. What is a blood feather?
5. Discuss the drawbacks of allowing a bird to ride around on the owner's shoulder.
6. List human foods that can be toxic to birds.
7. Describe the methods of restraint used with the avian patient.
8. Why is it so important to confirm that a feeding tube has accurately been placed in the crop?
9. List the various types of bird feathers and the purpose of each.
10. What is the purpose of the uropygial gland?

Case Study I

History: A 15-year-old Blue and Gold macaw has been brought in by the owner for "some kind of test." The owner states that his physician said to "take the bird to the vet." The owner has been ill with flu-like symptoms for the past two months. While taking the history, the staff member observes the close bond between the bird and its elderly owner, who exchanges several *kisses* with the bird as he comforts it in the strange surroundings. The owner comments sadly, "He's my best buddy, all I have since my wife passed away some months back. He even shares my dinner. It was just us, the three of us."

Physical Examination: The bird appears to be healthy and has a bright attitude. The feathers and beak are normal and the weight is within normal range for a Blue and Gold macaw. The bird's droppings are normal.

a. What disease does the physician suspect this bird may have?

b. Discuss the connection to the owner's health and his beloved companion. Consider the history, observations from the staff, and comments from the bird's owner.

c. What specific test will determine if the physician's concerns are supported or ruled out?

Case Study II

History: A 12-year-old Meyer's parrot is presented by a concerned owner. She has noticed that during the last molt, the new feathers look "weird." She is also concerned that the beak is growing at an odd angle. The owner comments that she enjoys attending bird shows and visiting pet stores but she has never seen anything like this in other birds.

Physical Examination: There are several small fractures along the length of the beak. The beak is malformed and appears to be twisting. The new feathers are not growing straight from the feather shaft but curl and appear clubbed. Several of the new feather shafts are necrotic.

a. Given all the relevant history of this patient, what tests would the veterinarian recommend to confirm what she suspects?

b. What is the prognosis for this patient?

c. What extra precautions need to be taken by staff in contact with the patient and in cleaning the examination room, lab, and treatment areas?

For Further Reference

Bays, T.B., Lightfoot, T., & Mayer, J. (2006). *Exotic Pet Behavior*. St Louis, MO: Saunders.

de Matos, R., & Morrisey, J. (2005, April). *Emergency and Critical Care for Small Birds. Seminars in Avian and Exotic Pet Medicine* 14(2):90–105.

Frunefield, S. (2010). The Goal: Quality Avian Medicine. *Journal of Exotic Pet Medicine* 19:4–21.

Gunkel, C., & Lafortune, M. (2005, October). *Current Techniques in Avian Anesthesia. Seminars in Avian and Exotic Pet Medicine* 14(4):263–275.

Harcourt-Brown, N. (2002). Avian Anatomy and Physiology. In Meredith, A. and Redrobe, S. (Eds.), *BSAVA Manual of Exotic Pets, 4th Edition* (pp. 138–148). British Small Animal Veterinary Association.

http://www.avianbiotech.com (accessed June 2, 2014).

http://www.cdc.gov (accessed June 5, 2014).

http://www.thegabrielfoundation.org (accessed January 18, 2014).

http://www.merckmanuals.com (accessed January 3, 2014).

http://www.michigan.gov (accessed May 26, 2014).

http://www.usyd.edu (accessed May 26, 2014).

http://www.veterinarypartner.com (accessed January 18, 2014).

Judah, V., & Nuttall, K. (2001–2003). *Tech Talk: A Look Beyond, Session I*. Ronie's for the Love of Birds, Educational Seminars for Bird Owners, 2001–2003.

Orosz, S., PhD, DVM, AVBP, http://www.lafeber.com (accessed June 5, 2014).

Ritchie, B., Harrison, G.J., & Harrison, L.R. (1994). *Avian Medicine: Principles and Application*. Wingers Publishing, Inc.

Stanford, M.C. (2002). Aviary Birds. In Meredith, A. and Redrobe, S. (Eds.), *BSAVA Manual of Exotic Pets, 4th Edition* (pp. 157–167). British Small Animal Veterinary Association.

UNIT IV

©iStockPhoto.com/CathyKeifer

REPTILES

OBJECTIVES

After completing the chapter, the student should be able to

- Describe the housing requirement for common reptile species.
- Provide appropriate client education to new reptile owners.
- Provide basic nursing care for reptile species.
- Demonstrate appropriate restraint techniques with each species of reptile.
- Describe the correct diets for different reptile species kept as pets.
- Determine the appropriate temperature and humidity (POTZ) for different reptile species.
- Identify potential problems with inappropriate housing, restraint, and diet.
- Understand common medical disorders in reptiles.

Introduction

Reptiles have been kept in captivity for many years. However, the knowledge necessary to keep them healthy was severely lacking. There was no understanding of the importance of UVB lighting, temperature and humidity requirements, and diet. Most, if not all, reptiles were wild caught and kept or traded by hobbyists.

For example, in the 1950s, small turtles could be purchased for a quarter and came with flowers painted on their very tiny shells. Their habitat was a small plastic bowl with an island and a green-topped palm tree. They were fed dried ants and flies in a fish food shaker, and the only advice available was to feed them raw hamburger and put eye drops in their swollen eyes. It is now illegal to sell hatchling turtles as pets with a carapace (upper shell) less than four inches. This regulation gives some protection not only to the turtles but to young children as well. After an outbreak of Salmonella, it was discovered that young children would put these tiny turtles into their mouths, exposing themselves to Salmonella.

Today, many of these remarkable and ancient animals are captive bred and are protected from being collected in the wild. New information is forthcoming almost on a daily basis, and with new information comes improvement in diet and husbandry practices.

Reptile medicine has become a highly skilled and specialized field. Advanced medical diagnostics and procedures are available specific to the uniqueness of reptiles. Specifically designed habitats have become household focal points because of their beauty and appeal, recreating natural habitats that are quickly disappearing. Reptiles are bred for beauty, color variations, and health. Their variety seems endless. The attraction to reptiles is in their differences, not their similarities.

Lizards inhabit a wide variety of ecosystems from deserts to tropical rainforests. They are divided geographically between Old World species and New World species. There are approximately 4450 different species. The iguanids comprise the largest number of New World species. Within this group are the green iguana, anole, basilisk, horned lizard, and spiny lizard. The Agamidae family contains the largest of the Old World lizards. Examples from this group include the bearded dragon, agama, frilled lizard, water dragon, and uromastyx (**FIGURE 15-1**).

FIGURE 15-1 A member of the *Agamidae* family is the Chinese water dragon.

Chameleons form a separate classification of Old World lizards, Chameleonidae (**FIGURE 15-2**). A great variety of captive-bred chameleon species are commonly available: Veiled, Panther, Jackson's, and Mueller chameleons are all popular and under the correct conditions are fairly easy to maintain. Another group includes the geckos, skinks, and monitor lizards. Many of these species are noted for their aggressiveness in captivity, especially the monitors and Tokay geckos. Care needs to be taken when handling these species.

FIGURE 15-2 Chameleons belong to a separate classification of Old World lizards, *Chameleonidae*. This Veiled chameleon has just caught a cricket with a very long and sticky tongue.

Lizards have keen eyesight with excellent color vision. Their binocular vision causes them to tilt their heads to one side when focusing on a prey item or movement in the habitat. Chameleons are able to focus each eye independently, with one eye on one object and the other eye looking around at something completely different.

Chelonians (turtles and tortoises) are facing many threats to their survival. They are used as food, their nests are robbed of eggs, and many species are illegally traded by the thousands for use in *traditional medicine*. Approximately 257 species of turtle and tortoises are remaining, divided into 12 families.

In very general terms, tortoises differ from turtles in that tortoises are terrestrial vegetarians while turtles are aquatic, or semiaquatic, and omnivorous. Collectively, they are called chelonians. The most common cause of disease in chelonians is poor husbandry. Tortoises and turtles differ greatly in their nutritional and husbandry requirements.

All chelonians have an upper shell, the carapace, and a lower shell, the **plastron**. The two halves are connected by bony bridges. The carapace is formed from the ribs and spinal column that are fused to form the **dermal bone**, and the plastron is a bony fusion of the clavicle and abdominal ribs. The dermal bone grows directly from the dermal (skin) layer. The shell is covered in a keratin layer, with distinctively shaped **scutes** (the small plates of the shell). Scutes are capable of regeneration if damaged (**FIGURE 15-3**).

FIGURE 15-3 Chelonians, turtles and tortoises, are similar in anatomy. Both have an upper and lower shell connected by bony bridges to protect internal organs.

Ventral View of a Turtle

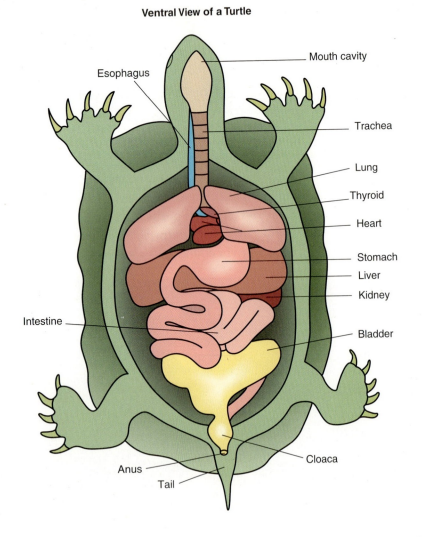

Esophagus — Mouth cavity — Trachea — Lung — Thyroid — Heart — Stomach — Liver — Kidney — Bladder — Intestine — Cloaca — Anus — Tail

flicking its tongue for sensory input. Infrared receptors that sense heat are located between the nares and eyes. These sensory adaptations allow snakes to strike with great accuracy.

Snake skulls are very flexible and mobile. Snakes are able to move the top jaw independently from the bottom jaw, which allows them to open their mouths wide enough to swallow whole prey. Snakes need to be fed a prey item that is large enough that it forces the jaws to open normally and to be able to expand further to accommodate the body growth of the snake. Feeding one appropriate-sized rat that allows normal feeding behavior and jaw movement is always preferable to offering two or three mice. The correct size of prey to feed is one that is as large as the largest part of the snake's body, not the size of its head. A hungry snake will sometimes yawn widely. Yawning is also seen after swallowing prey and serves to reposition the jaws (**FIGURE 15-4**).

FIGURE 15-4 The anatomy of a snake, showing the elongation of internal organs.

Anatomy of a Snake

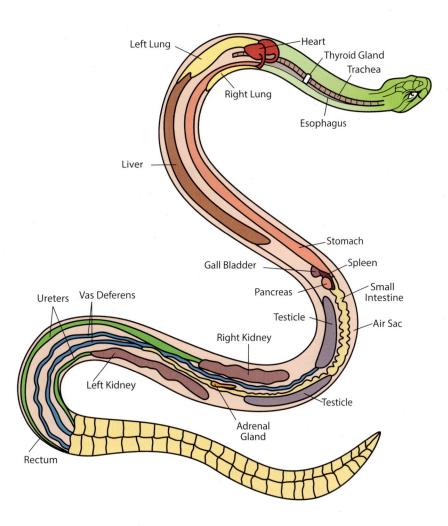

Tracheal rings are incomplete in most lizards and some species of snakes but complete in chelonians. In snakes, the trachea does not bifurcate as with most

Owners frequently ask if they should allow their turtles and tortoises to hibernate. There are many differing opinions from veterinarians and herpetologists. The most common problem seen by veterinarians is that animals are allowed to hibernate when they are not nutritionally prepared. Posthibernation anorexia can be caused by disease prior to hibernation, poor nutrition, cold weather shortly after coming out of hibernation, and the owner not feeding the chelonian sufficiently during recovery from hibernation. They are usually presented to the veterinary hospital with respiratory problems and a dramatic weight loss. Many chelonians, doing poorly, die during hibernation. All owners should discuss hibernation options with their veterinarian and have a complete physical examination prior to hibernating a turtle or tortoise.

> **There Is No Truth in Counting the Lines within the Scutes to Determine Age.**
> Lines and patterns of the scutes are determined by species, diet, overall health, and growth rate.

Sexual determination in chelonians is somewhat easier than in other reptiles. In the mature chelonian, and depending on the species, the plastron of the male is often generally more concave than the plastron of the female. Some males have a longer and thicker tail than females. In aquatic species, males have long claws on the front limbs to grasp the female during copulation. In some Box turtles, males have a red iris and females a yellow iris.

All chelonians are oviparous, or egg layers. Most species of Box turtles lay eggs beginning in May and continuing through July. They lay an average of two to eight eggs at one time. Box turtles can store sperm for approximately four years. Female chelonians must be in prime condition before they start egg production.

There are more than 2500 species of snakes. Of these, approximately 1700 are colubrids, which include some of the most popular snakes kept in captivity. Examples are corn snakes, rat snakes, king snakes, milk snakes, and garter snakes. Colubrids are aquatic, arboreal, or terrestrial.

Another popular group is the boids, with approximately 63 species. This group includes some of the largest species of snakes, the boas and pythons, such as the Burmese python and the Columbian red-tailed boa. Boids are native to North, Central, and South America. The native North American boids are represented by the Rubber boa and the Rosy boa. New World boids are viviparous—that is, they give birth to live young—while the species native to Africa, Asia, and Australia are oviparous, egg-layers.

Snakes have poor eyesight and their hearing is limited to low frequencies. They rely on olfactory input to hunt and evaluate their environment. The Jacobson's organ (vomeronasal organ) is a highly developed sensory organ that enhances an animal's sense of smell. The forked tongue of the snake picks up scents that have filtered through pits at the sides of the mouth and into the oral cavity and are transferred to the Jacobson's organ. A healthy, active snake is continually flicking its tongue for sensory input. Infrared receptors that sense heat are located between the nares and eyes. These sensory adaptations allow snakes to strike with great accuracy.

Snake skulls are very flexible and mobile. Snakes are able to move the top jaw independently from the bottom jaw, which allows them to open their mouths wide enough to swallow whole prey. Snakes need to be fed a prey item that is large enough that it forces the jaws to open normally and to be able to expand further to accommodate the body growth of the snake. Feeding one appropriate-sized rat that allows normal feeding behavior and jaw movement is always preferable to offering two or three mice. The correct size of prey to feed is one that is as large as the largest part of the snake's body, not the size of its head. A hungry snake will sometimes yawn widely. Yawning is also seen after swallowing prey and serves to reposition the jaws (**FIGURE 15-4**).

FIGURE 15-4 The anatomy of a snake, showing the elongation of internal organs.

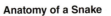

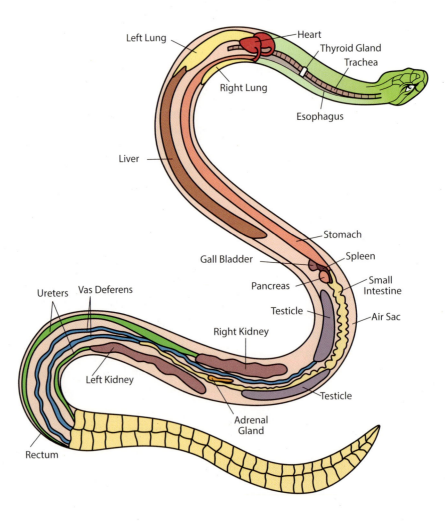

Tracheal rings are incomplete in most lizards and some species of snakes but complete in chelonians. In snakes, the trachea does not bifurcate as with most species. Chelonians have a long, flexible trachea that bifurcates close to the heart. Most reptiles breathe **episodically**, with periods of apnea. Reptiles do not have a diaphragm to assist with respiration, but instead use intercostal muscle movement. Snakes and lizards breathe with rib movement, while turtles and tortoises breathe by using shoulder muscles to change the pressure in the pleural cavity.

Reptile teeth are **polyphyodontic**; that is, they are reabsorbed or shed and replaced at a rapid rate throughout life. Not all reptiles have teeth. Some have a hard bony plate along the mandible and maxillary bones.

The digestive tract of reptiles is shorter than that of mammals but slower in digesting and assimilating food. Herbivorous reptiles have a slower digestive metabolism than omnivores and insectivores.

Male snakes and most lizards have paired reproductive organs called **hemipenes**. Gender determination of some lizards and snakes can be done by locating the hemipene bulge, which is just cranial to the vent. The hemipenes can be everted manually, or they may prolapse. Chronic prolapse needs to be investigated for a medical cause. In chronic conditions, one or both hemipenes may be surgically removed. Male chelonians have a single penis.

The sex of a mature iguana (and many other lizard species; **FIGURE 15-5**) was thought to be determined by femoral pores. If the femoral pores were pronounced, the lizard was assumed to be a male. This is no longer accepted as there are many female lizards with large femoral pores.

FIGURE 15-5
Reproductively mature iguanas often undergo dramatic behavioral and color changes during breeding season. This mature male green iguana has turned a deep reddish orange and has become very aggressive toward its owner. (*Courtesy of Eric Klaphake, DVM*)

Some species of snakes have small spurs, located on either side of the vent. They are short bones covered in keratin. Spurs are used in courtship and mating. Examining the relative size of spurs is not a reliable method for sex determination. The sex of snakes can be determined with the use of a probe.

Probing requires a slender, blunt-ended metal rod that is inserted into the vent directed caudally and carefully moved along the ventral wall of the tail. If the probe can be passed only two to three scale lengths caudally, this indicates the snake is a female. In a male, the probe can be advanced six to seven scales. The probe should slide in easily, and force should never be used. A sterile lubricating jelly is applied to the tip of the probe before insertion into the vent (**FIGURE 15-6**).

FIGURE 15-6 The sex of a snake may be determined with careful use of a probe inserted into the vent.

Since DNA sexing in reptiles is not always accurate, many laboratories have abandoned this method. Ultrasound may reveal ovaries in a reproductively active female, but so will a radiograph with the distinctive outline of multiple eggs.

Reptile reproduction is regulated by the pineal gland, the hypothalamus, and environmental stimuli. These, in turn, stimulate the release of reproductive hormones. Many species of reptiles from the Northern Hemisphere will not breed unless they have been in hibernation for a period of time to mimic the natural hibernation period. Most breeding hibernations begin in November and continue through February.

Incubation temperature determines the sex of the offspring in more than 70 species of reptiles, including 90 percent of turtles and some lizards. The exception to this is snakes. In some reptile species, sperm can be stored in the oviduct. Fertilization is triggered when the ova enter the oviduct, which may be months later. In some species, the sperm can be stored for up to six years. The female is able to produce fertile eggs over a period of several years without coming into contact with a male. The majority of reptile eggs are incubated at around 28°C (82°F) to 32°C (86°F). To determine if an egg is viable, an egg candle can be used several days after the egg has been laid. The egg is taken into a dark room and a bright light source is held directly against the shell. One to seven days after the egg has been laid, the yolk drops and a blood spot will appear. During stages of the incubation period, it may be possible to observe development of the fetus and movement within the egg (**FIGURE 15-7**).

FIGURE 15-7 A reptile incubation setup. These Bearded Dragon eggs were all hatched successfully. (*Courtesy of Bev and Dan Ring*)

Reptiles shed their skin as they grow. A healthy snake sheds its skin in one piece, starting with the head. It should be complete all the way down to the tail. Chelonians shed their skin in small pieces from the limbs, neck, and tail over a short period of time. A chelonian that appears to be constantly shedding needs to be examined for malnutrition or a skin infection. Aquatic turtles shed not only their skin but also layers of scutes as the turtle grows. Lizards shed their skin in patches of small pieces over a period of several days. Shedding completed in a healthy reptile is referred to as ecdysis. Failure to shed completely, dysecdysis, can cause problems. As the dead skin dries out, it constricts around toes and the end of the tail, cutting off blood supply. Ecdysis occurs throughout the lifetime of the reptile. Young snakes shed more frequently than adults because of their rapid growth rate. Moisture from the snake's body enters the space between the old and new layers of skin, which aids in loosening and lifting the old skin up and off. Snakes beginning to shed appear opaque and their eyes appear blue (**FIGURE 15-8**).

Ecdysis takes approximately five to seven days, during which time the snake is unable to see and should be handled carefully or, preferably, not at all. A snake *in the blue* is more likely to strike. A good, healthy shed comes off in one piece, complete with eye caps. If the eye caps are retained, it could cause ocular damage. Retained eye caps need to be removed very carefully to avoid injury to the eye. A slow trickle of warm water will moisten the eye cap. Using *only* cotton tip swabs or cotton balls, gently and carefully roll the layer of dead skin away from the eye. Never attempt to pull eye caps away with forceps as doing this could also remove the lens of the eye. The most common reason for a poor shed is lack of humidity

FIGURE 15-8 The blue eye of a snake that is about to shed.

or poor nutrition. If the snake needs assistance, it should be soaked in warm water and the old skin pulled off caudally in the direction of the scales, starting from the head and moving toward the tail (**FIGURE 15-9**).

FIGURE 15-9 The recently shed skin of a healthy snake, complete with eye caps.

Behavior

Several studies have been done on reptile behavior, especially with the green anole. Most of the studies have been related to reproductive behavior. For reptiles in captivity, it is important to know the social behavior of each species so as not to stress the reptile. Many reptiles suffer from environmental stress in captivity.

If danger is near, most reptiles will try to avoid detection by hiding, as this requires the least amount of energy to avoid being detected. If this does not work, some species fake a bite or strike. The Eastern hog-nosed snake, for example, performs a *death display* by turning upside down and lying motionless. While playing dead, this species can produce a bloody, smelly liquid that imitates blood and the odor of decay. Other reptile species may self-inflate by taking in air, causing them to appear bigger to a predator, or they may wedge themselves into a small space and self-inflate so that they are tight within the crevice.

Different species of lizards have various ways of protecting themselves in their natural habitats. Many species of lizards have the ability to release their tails and escape unharmed. Species with autonomy, the voluntary release of the tail, also have the ability to regenerate it. The regrown tail has a distinctly different appearance and color. It may take several months to regenerate. The head-bob, a quick up and down movement, is always a warning or a threat and should not be copied by owners. It is not a friendly greeting.

Tail flicking and tail whipping are behaviors that indicate that the reptile is agitated. It is seen in snakes when they vibrate their tails when a predator or threat approaches. The display threat as demonstrated by the well-known rattle snake species also occurs in nonvenomous species when threatened. Lizards often use their tails to strike out at the predator or if agitated during an examination in a veterinary hospital.

> **Aggression in Male Green Iguanas Is Well Documented.**
> Women owners need to be especially careful when male iguanas become sexually mature, as they can become aggressive toward women during their menses. There have been several instances of male iguanas attacking and causing severe lacerations that required emergency room treatment, sutures, and, in some instances, reconstructive plastic surgery.

Snakes have the ability to empty their musk glands when provoked or frightened. These glands are located near the cloaca. Milk snakes are known for this behavior. Chelonians do not have this gland, but they will urinate often during handling with the same effect.

Chelonians retract their head and legs into their shells for protection against predators. This can pose a problem when trying to examine or medicate them. Several species of lizards, for example, the Horned lizard, can spurt blood from their eyes as far as three feet or nearly a meter. They are able to do this by restricting blood flow from the head until the small blood vessels rupture due to the rise in blood pressure around the eyes.

The ability of some reptile species to change colors helps them with camouflage. Color change can also indicate courtship, stress, illness, or interaction with other species. Chameleons are known for their ability to change color. Other species of reptiles have the ability to change according to changes in their environment, including heat, cold, stress, and perceived threats.

Housing

Reptiles are ectotherms; that is, they require an environment that provides the heat necessary for metabolic activity and health. Ectotherms do not have the ability to generate and maintain their own body heat. Each species requires a different temperature range to achieve optimum ability to digest prey, fight off infection, and perform day-to-day functions. These temperatures are referred to as POTZ (preferred optimum temperature zone). A reptile enclosure should take into consideration all of the variants and provide areas to accommodate reptile movement from one area to another, allowing the reptile to obtain optimum temperature. Reptiles that are ill usually seek out the warmest part of the enclosure. Ambient humidity influences a reptile's ability to eat, to shed normally, and to eliminate waste. Humidity can range from 30 percent to 100 percent, depending on the natural habitat of the species.

Correct lighting is a critical component in keeping reptiles healthy. Ultraviolet light assists in the assimilation of vitamin D3, which is necessary for the absorption of dietary calcium. Natural sunlight is the best source of ultraviolet light, but most reptiles are kept indoors. Reptiles should not be housed in direct sunlight from windows. The full spectrum rays of sunlight cannot penetrate through window glass and habitat temperatures will dramatically increase, potentially killing the reptile.

Full spectrum lighting is provided by bulbs and fixtures designed specifically for reptile habitats. Varying degrees of ultraviolet wavelengths or ultraviolet light are generated. UVA bands affect the behavior of reptiles. Full spectrum lights (UVB and UVA) should be changed every six to eight months, or according to the manufacturer recommendation. While the bulb may still produce a light, the ability to produce a *full spectrum* may have dropped dramatically. The lights should cycle for 13 to 15 hours during the summer months and 9 to 12 hours during the winter months to mimic the natural photoperiod of the species.

Lizard habitats should be spacious enough to allow the lizard to stretch lengthwise and to turn around completely, including the full length of the tail, without touching the sides of the enclosure. Terrestrial lizards require more floor space than height. Aquatic and semiaquatic species need access to an area with water and a dry area for basking. Adequate ventilation while keeping heat and humidity within the POTZ range needs to be considered. Glass or acrylic habitats help maintain appropriate levels of heat and humidity (FIGURE 15-10).

Chameleons require a cage that allows for plenty of air circulation, such as a wire mesh or screen, yet is able to maintain the correct temperature and humidity range for the species. Chameleons should *not* be housed in aquariums, as they do not provide adequate ventilation. Housing chameleons in aquariums, however large, is often a main cause of death.

Humidity is difficult to maintain for some lizards. Tropical species need 80 percent to 100 percent humidity, while the desert species need only around 30 percent to 40 percent. Misting the cage daily will help provide some of the humidity. Adding waterfalls, water bowls, and live plants can also increase the humidity.

Cage substrates vary for different species. Substrates should mimic the natural environment of the species. A variety of commercially prepared substrates are available that provide for species variation. These include sand, shredded bark,

FIGURE 15-10 When designed correctly, reptile habitats not only are beneficial to the inhabitant but can also become a focal point and an area of great beauty and interest. (*Courtesy of CagesByDesign.com*)

soil, and soft reptile litter. Some are digestible, others are supplemented with calcium, but all are manufactured for specific reptile requirements and feeding behaviors.

Some species inadvertently swallow substrate material when catching prey. This can cause gastrointestinal blockage. Wood shavings of any type, silica sand, corncob, and ground walnut shell should not be used. An example of using an incorrect substrate, and its consequence, is keeping the Leopard gecko on sand. Sand impaction is common in this species because of the way it catches live food items with a quick *pounce and swallow*, ingesting sand along with the prey.

Most lizards need climbing objects as a part of the cage furnishings. These include various sized branches, driftwood pieces, cork bark, plastic or live plants, and rocks. There are also artificial vines, realistically created to mimic jungle vines. All items should be placed at varying levels to provide ease of access to basking areas.

Maintaining the POTZ for each species is very important in keeping reptiles healthy. The basking lamp should be placed at one end of the cage, keeping the temperature in that area of the enclosure constant. Levels of furnishings provide for temperature gradients. Several methods and products can be used to provide heat. Heat rocks should not be used for any reptile, especially lizards. There is no way to regulate the amount of heat produced, and these products can malfunction, either by not producing enough heat or by becoming excessively hot. Iguanas, for example, will lay on a heat rock until it becomes so hot that the skin starts to burn. Normal basking activity is occurs in the heat from the sun, dorsally, and reptiles do not have the ability to detect excess heat from the ventral surface.

Basking lights placed outside the cage or enclosed within the habitat at a safe distance from the reptile are more easily monitored and controlled. Any heating elements placed within the environment need to be *caged* so the reptile has no direct access to them. Many species of lizards are excellent at leaping and have been found fatally burned by unprotected heating elements.

If the nighttime temperature drops too low for the reptile, an additional source of heat needs to be provided. This includes ceramic heat bulbs, red incandescent bulbs, or reptile heating pads placed underneath the cage. In the morning, lizards often appear darker than normal. Dark pigments absorb more heat and light. By mid- to late afternoon, the lizard may be lighter in color because of absorbed heat and light.

Some species of tortoises can grow quite large and require an area that is big enough for exercise and has easy access to a water source for soaking. Tortoises absorb some amount of water through their skin but also need to be provided with fresh drinking water to prevent dehydration. Substrates that are digestible include alfalfa rabbit pellets or a layer of grass hay placed over a rubber mat. Rabbit pellets bought in bulk are relatively inexpensive, and the area can be cleaned of fecal material similar to cleaning a cat litter tray. Loose grass hay is raked up and disposed of in a sealed trash bag. It should not be used as garden mulch at the risk of introducing Salmonella. The mat underneath is simply hosed off and replaced. Nondigestible floor coverings commonly used include indoor/outdoor carpet and reptile soil mixes. Tortoises also need a source of UVB light. Many tortoises are housed outdoors during the summer months and allowed to graze and wander at will. Still, they need a shelter and protection from predators, especially curious dogs.

> **In One Tragic Instance,** a newly purchased Panther chameleon was left inside a box in the car while the owners stopped for *just a minute* to run an errand. The chameleon turned pure white in an attempt to dissipate excessive heat. It did not survive.

Aquatic turtles need a body of water for swimming and feeding. They need to be provided with a basking platform that allows them to easily climb completely out of the water. Water quality is maintained with an underwater filter system. Water should be dechlorinated, and depending on the species, the water may need to be heated. Suitable substrates for semiaquatic turtles include peat moss, commercial reptile soil, or cypress mulch. Clay, walnut shell (which is toxic), rocks, and gravel should not be used as they are too abrasive.

Chelonians are heliotherms, meaning that they actively seek out sunlight for heat. In captivity, the POTZ range for tortoises, depending on species, is between 26°C and 37.7°C (79°F and 100°F). POTZ for turtles is 25°C to 35°C (77°F to 95°F).

Snake enclosures vary with the species of snake being housed: burrowing, arboreal, or semiaquatic. The minimum length of the enclosure should be as long as the snake. The enclosure should provide good ventilation yet should be able to retain heat. Most snakes are housed in glass aquariums or plexiglass tanks. Cages made of wood are not recommended. They are difficult to keep clean and provide the perfect environment for ectoparasites. Habitats need to have a heat source that will maintain the POTZ range for the species. Basking lamps or reptile heating pads placed under the cage are the usual and most reliable source. As with other reptiles, heat rocks should be avoided.

Many species of burrowing snakes are native to arid deserts where they burrow into the sand and wait for prey. Commercial reptile sand has been cleaned, is usually biodegradable, and is a suitable substrate for burrowing snakes. Newspaper, indoor/outdoor carpeting (reptile carpet), and reptile litter (similar to wood shavings) can be used as substrate for ground-dwelling snakes. Semiaquatic snakes should have a substrate similar to ground dwellers, but also a bowl or container large enough for them to enter the water and soak (**FIGURE 15-11**).

Every snake, regardless of species, should have a container of water large enough to allow it to be totally submerged. Many snakes hydrate in water by soaking rather than gulping water. Snakes that require a higher humidity, such as Rainbow boas and Rubber boas, need larger bowls of water and a bowl with damp sphagnum moss to help keep ambient humidity elevated.

Arboreal snakes, for example the Green tree python and the Emerald tree boa, require branches or sturdy perches. Both of these snakes fold and drape their bodies across branches in the highest part of an enclosure. Boas and pythons also require a higher humidity level and need to be misted daily. All snakes require a place to hide. Providing a hiding place decreases the chance of the snake having stress-related problems and may, depending on the species, reduce incidences of defensive or aggressive strikes.

FIGURE 15-11 All snakes should be provided with a water bowl for soaking. This is a juvenile Blood python submerged in a deep water bowl.

Diet

The correct diet for each species is essential to reptile health. As more information becomes available regarding the needs of individual species, diet recommendations are improved and modified. Reptiles are herbivores, omnivores, or insectivores (see Table 15-3 under Fast FACTS). There is no *one diet fits all* easy answer.

Herbivores include the Green iguana, Uromastyx, and Solomon Island prehensile-tailed skink. Omnivores include Bearded dragons, Blue-tongued skinks, tegus, certain geckos, and Rock iguanas. Old World chameleons, juvenile monitors, tegus, and anoles are insectivores. Juvenile diets are often different from adult diets. Researching each species is essential to understand and meet dietary requirements.

The staple herbivore diet consists of fresh dark leafy greens. This includes chard, kale, mustard greens, endive, bok choy, and collard greens. Spinach should be offered in limited quantities as spinach binds with calcium, blocking calcium absorption. Vegetables that can be added to the diet include grated carrots, broccoli, squash, peas, and green beans. Vegetables need to be chopped up in appropriately sized pieces, and should be *tossed* together and fed like a salad to avoid food preference. Iceberg lettuce should not be fed to any reptile. It has no nutritional value and may cause intestinal bloat. If fed a correct diet and provided with UVB lighting, reptiles should not require the addition of supplements to their food.

Herbivorous lizards take in large amounts of potassium salts in their diet. Because of this, they have nasal glands that excrete excess salts. It is normal to see these lizards with a white, crusty substance around their nares.

Lizards that are fed crickets should be fed daily. Crickets should be gut-loaded prior to feeding them to the lizard. Crickets are usually the main diet of omnivores in captivity but they should be alternated with mealworms and wax worms. Omnivorous lizards should be fed a pinky or fuzzy mouse once a week. A variety of canned diets are available for lizards, including *canned* (dead) crickets. Designed more for human convenience, they should not be considered a complete diet. The best diets are fresh and as close to a natural diet as possible.

Lizards are one of the few reptile species that use their tongues to lap water. Some snakes, turtles, and tortoises immerse their heads in water and gulp. Supplying an appropriate water source is essential in maintaining the health of the lizard.

> ## Chameleons Do Not Recognize or Drink from Standing Water.
>
> They need to be supplied with moving water, as they catch water droplets with their tongues. A drip system can be placed on top of the enclosure or a waterfall added to the habitat.

All tortoises require a diet that is high in calcium and low in protein and fat, and should be fed daily. Dark leafy greens, squashes, carrots, green beans, peas, and a small amount of fruit offer variety and meet nutritional needs. Tortoises also enjoy nibbling on grass hay, growing grass, and garden weeds like dandelions. It is very important that tortoises have grazing access only to areas of the lawn or garden that have not been treated with fertilizers, weed poisons, or insecticides.

Turtles are omnivorous. The bulk of the diet should be dark leafy greens with small amounts of animal protein. Animal protein includes crickets, mealworms, and earthworms. Aquatic turtles feed only in water. Commercial floating turtle sticks are available. If wild caught, the turtle may refuse to eat them, not recognizing the pellets as a food source. They should be added gradually and the consumption monitored. Uneaten pellets should be removed to prevent water fouling, bacteria, and mold growth.

Snakes feed on a variety of prey. They hunt and consume mammals, smaller reptiles, amphibians, eggs, fish, chicks, insects, and worms. Mice, rats, and rabbits are commonly offered as prey animals depending on the size of the snake. Semi-aquatic snakes also eat small fish.

Feeding on killed prey will prevent serious injuries to the snake. Live prey has the potential to cause serious bite injuries to the snake. Snakes that are chewed on by prey will try to escape but will not defend themselves. Some snakes will take only live prey. If feeding live prey, the snake should not be left unsupervised. If the snake does not strike within 10 minutes, the prey should be removed. Never leave live prey in the enclosure thinking the snake will eat it later. More often than not, the snake is the victim (**FIGURE 15-12**). Stunned or freshly killed prey is recommended to prevent the possibility of injury to the snake. Cervical dislocation is acceptable by the AVMA as a humane method to kill prey for the snake.

Some snakes are very picky about the color and type of prey they eat and the time of day they eat. Season also plays a part in the feeding habits of some species. During what would normally be a hibernation period in the wild, many

FIGURE 15-12 Snakes should never be left unsupervised with live prey. This Ball python has been severely chewed by a rat. (*Courtesy of Eric Klaphake, DVM*)

snakes stop eating in captivity. Ball pythons are the most difficult species to keep eating on a regular basis and they are usually the most timid about taking prey.

All snakes strike, holding the prey with their teeth. Smaller species often swallow their prey without constriction. Other species strike and constrict, tightly coiling their bodies around the prey and constricting, suffocating it to death before swallowing it whole. Prior to ingestion, a snake may release the dead prey item to reposition itself so the prey can be swallowed head-first. Digestion begins in the stomach. It may take up to five or more days to completely digest one prey item. Bones and hair of the prey are compacted in the feces and expelled. Adult snakes are usually fed one appropriately sized prey once a week or once every two to three weeks, depending on the feeding habits natural to the species. Smaller snakes and young, growing snakes should be fed weekly.

Some owners attempt to feed a large number of prey items in a short amount of time to increase the growth of their snake. This is called **slam feeding**, and it is not recommended. Slam feeding can cause the intestinal tract to become blocked, slowing down the metabolism and potentially causing regurgitation of the excess food. Victims of slam feeding techniques often require surgery to remove blocked fecal material and partially digested prey in the intestinal tract.

Force feeding (different to slam feeding) is a technique used to ensure that an anorexic snake has an adequate diet. It should not be attempted by inexperienced owners. Forceps and tongs used to push prey items into the mouth and down into the esophagus have caused such trauma that the snake has had to be euthanized. With some anorexic snakes, force feeding a mouse or rat head with a small pair of tweezers will stimulate their appetite. Snakes may also be fed with a red rubber catheter with an attached syringe of a prepared supplemental food. The catheter should be rinsed with warm water and the open tip placed behind the trachea to approach the esophagus. When correct placement is confirmed, the contents of the syringe are emptied into the catheter and deposited in the distal esophagus. With the catheter withdrawn, the patient's first third of the body should be held slightly elevated to allow the food to enter the stomach. Anorectic snakes should receive a comprehensive veterinary examination. There may be an underlying disease causing the anorexia, and *force feeding* may only complicate the issue (**FIGURE 15-13**).

FIGURE 15-13 Force feeding a snake may be necessary to provide nutritional support. A red rubber catheter is placed behind the trachea in the initial approach to the esophagus.

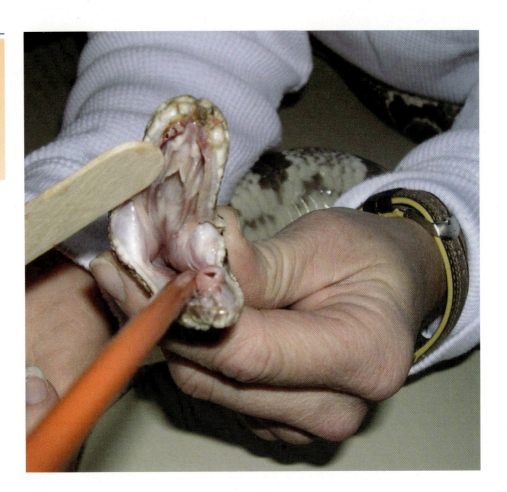

Restraint and Handling

Many lizards become stressed with close human contact. Some will flee or find a place to hide, while others become immobile. General restraint of lizards follows the rule of LEAST: the least amount of restraint required to get the job done, the better. Lizards do not struggle as much when they are held securely but not forcefully. Lizards are grasped behind the head and around the shoulders with one hand. The other hand should be placed around the pelvis holding the legs caudally and against the body (**FIGURE 15-14**).

Be cautious of a defensive/aggressive tail whip, especially from an iguana. The tail should be tucked under the same arm that holds the pelvis and rear legs. Once restrained, the lizard can be wrapped up in a towel for an examination or nail trim.

Another effective method of restraint, especially for radiograph and blood draws, is to place cotton balls over both eyes and loosely but securely wrap the head with a self-adhering bandaging material. When correctly applied, this also forms a muzzle and prevents the reptile from biting. This method puts moderate pressure on the vagal nerve and tends to keep even the most aggressive reptile in a subdued state.

FIGURE 15-14 A two-handed method of restraint for a very quick lizard, the basilisk. This holds the reptile securely and secures the tail, protecting it from injury.

Examining a chelonian can be difficult. To avoid internal injury, chelonians should not be held upside down or with the head lower than the body for long periods of time. This can cause respiratory difficulty and is very stressful for the patient. They should not be turned quickly from side to side as this might cause the intestines to twist. To examine the head of a small turtle, grasp the head on either side of the jaws and slowly pull the head forward (**FIGURE 15-15**).

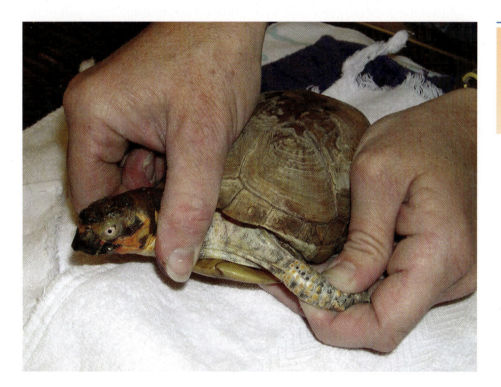

FIGURE 15-15 Restraint of a Box turtle. Once the foreleg is secured, the turtle will likely extend the neck, making it possible to quickly secure the head.

If the turtle retracts its head inside the shell, a gentle prod with blunt forceps near the back legs will usually encourage it to extend the head out from inside the shell. When the head starts to come out, it can be grasped quickly. Turtles are surprisingly quick in their responses and it may take several attempts to secure the head or a leg. If grasping a leg, be aware that the turtle can very quickly extend the head and full length of the neck to deliver a painful bite from the beak. Completely enclosed Box turtles can be tempted to open up if submersed in water, and if this approach fails, it may be necessary to use forceps. If using forceps to encourage the head to extend from the shell, they should *never* be applied to the mandible. The resistance of the turtle can be so strong that it will most likely cause a midline mandibular fracture. Forceps should be used with great care and applied only to the beak. Even then, there is a risk of breaking chips of keratin from the beak.

Large turtles and tortoises are very strong, and it is not possible to manually extend the head or legs without anesthesia. In some instances, large tortoises can be encouraged to stay out of their shells with the offer of food during the examination. Fresh strawberries work wonders.

Movement around snakes, especially near the head, should be careful and slow. Quick movements often elicit a strike and possible bite. Some of the more aggressive snakes, for example the Emerald tree boa and the Burmese python, are likely to strike and bite the instant visual contact is made. Before attempting restraint, it is important to understand exactly what behaviors are characteristic of the snake. In all snakes, the head should be secured first. Grasp the head gently but firmly from behind, caudal to the jaws. Once the head is secured, the rest of the body can be held loosely, allowing the snake some movement over the support of the handler's body and arms. For personal safety, any snake over five feet requires two restrainers: one for the head and first third of the body and the other to control the body to prevent the snake from wrapping around any part of the handler's body (**FIGURE 15-16**).

Snakes should never be placed around anyone's neck. Many people are seen carrying large snakes in this manner. Should *anything* frighten the snake, it will quickly tighten and constrict in self-defense or in preparation to strike.

Snakes that need to be transported can be placed in a pillow case or some type of cloth bag that can be tied securely at the top. A large snake can be placed in a sports bag or two-handled cooler with ventilation holes. The lid should always be secured.

To examine the oral cavity of a snake, a soft pliable speculum (rubber spatula) is used. Mucous membranes are normally pale to pink without a thick mucous discharge.

Medical Concerns

A physical examination should be performed on any reptile prior to purchase. Clients considering acquiring a new reptile can be advised in general terms what signs to look for in a potential purchase. The examination begins with the eyes, which should be clear and open. A sick reptile will typically have an ocular discharge with the eyes partially or completely closed. The eyes should be checked for any swelling that might suggest irritation or an eye infection. The aural openings

FIGURE 15-16 For safety sake, a snake that is greater than five feet in length requires two restrainers.

should also be checked for swelling that might suggest abscesses or inflammation. The aural opening is also a common place for mites to hide. The Leopard gecko in **FIGURE 15–17** was not fed the correct diet or housed properly. It was rescued from a pet store and slowly nursed back to health.

The rostrum should be examined for trauma or overgrowth of a beak on a chelonian. The nares should be clear from discharge. The oral cavity should be examined for trauma and areas that are abnormal in color. The normal color of the mucous membranes in most reptiles is pale pink to white. In Bearded Dragons it is normal to see yellow mucous membranes. Stomatitis (mouth rot), inflammation of the mouth, ulcers, open mouth breathing, and excessive salivation can indicate poor health. For example, upper respiratory infections in snakes will cause excessive salivation and open mouth breathing. Tongues in snakes and monitor lizards should be constantly flicking in and out of the mouth. Tongue flicking indicates an alert and healthy specimen.

Overall weight and appearance of the reptile needs to be assessed. Reptiles that are suffering from malnutrition feel *empty* when they are picked up. Their body weight will be substantially less than it should be for the species. In

FIGURE 15-17 A critically ill Leopard gecko; the eyes are swollen shut and encrusted with sand. It has stomatitis and is very emaciated. With veterinary care and several weeks of dedicated nursing care by the rescuers, the gecko is now reported to be in excellent health. (*Courtesy of Bev and Dan Ring*)

malnourished lizards, pelvic bones become more prominent and the tail appears shriveled or thin. Snakes have a decreased muscle mass with prominent dorsal processes and ribs. Obesity can also be a problem in reptiles, especially due to poor diets. Obese lizards have a distended abdominal cavity with palpable fat pads along the body. The tail base is also enlarged with fat pads. Obese snakes have fat pads along the lower third of the body.

Most reptile hearts are small and located more dorsally than those of mammals. Because of their scales and solid breast plates, it is more difficult to auscultate the reptilian heart. Wrapping a damp towel around the thorax of a lizard amplifies the sound and decreases interference from scale noise (**FIGURE 15-18**).

FIGURE 15-18 Wrapping a damp towel around the thorax of a lizard amplifies heart sounds and decreases interference from scale noise.

Reptiles are often diagnosed with bacterial infections in the joints and limbs. Penetrating wounds from enclosures or wounds caused by cage mates can be a source of infection. Infection begins locally, at the site of the injury, and then becomes systemic. If the infection is treated when localized, reptiles usually recover without complications. Stomatitis is common in lizards and snakes. Stomatitis can develop with immunosuppression, an inadequate diet, or an underlying disease. Signs include anorexia, gingivitis, an inability to use the tongue properly, bleeding, and oral discharge.

Rostral abrasions occur in lizards that are constantly trying to escape from the enclosure and bumping the walls causing injury to the rostrum. Treatment involves providing the correct enclosure and environment and applying ointments as recommended by the veterinarian to prevent a systemic infection. If the abrasion is severe, surgical debridement may be necessary (**FIGURE 15-19**).

FIGURE 15-19 A pair of basilisk lizards. The male (lower) has a rostral abrasion.

Abscesses are common in reptiles and involve a variety of bacteria. Abscess can form in the oral cavity, eyes, ear canal, and skin. There may also be internal abscesses on the liver and lungs. Abscesses in reptiles are different from those in mammals. The pus found in the abscess is caseous, *cheesy* and nonflowing. Subcutaneous abscesses are surgically opened, and the solid pus packet removed. The cavity may be left open or packed with sterile gauze. Internal abscesses are difficult to diagnose, and may be treated with systemic antibiotics (**FIGURE 15-20**).

Thermal burns are frequently seen by veterinarians. Faulty heat rocks and the close proximity of basking lights are common causes of these burns. Reptiles do not often remove themselves from a heat source that is too hot. Most often, these may be very ill reptiles simply too debilitated to move. Basking reptiles continue to lie on top of a hot item, often burning through the skin and underlying muscle tissue. Treatment depends on the severity of the burn. Sterile saline is used to irrigate the burned tissue. Debridement and possible surgery may be recommended

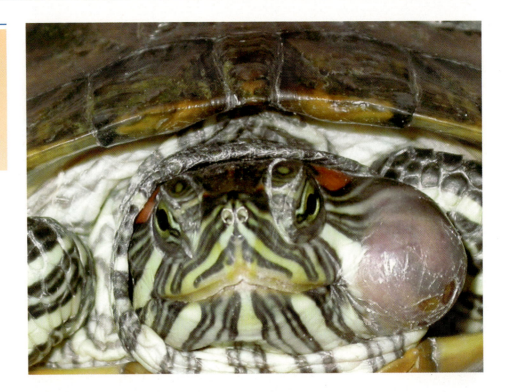

FIGURE I5-20 Aural (ear) abscesses are fairly common in captive turtles. The abscess on this red-eared slider is so large that it prevents the turtle from retracting its head. (*Courtesy of Eric Klaphake, DVM*)

by the veterinarian. Recovery from thermal burns may take months or years with several sheds that gradually replace the destroyed tissue and skin.

Respiratory Disorders

The most common cause of respiratory problems in reptiles is poor husbandry and nutrition. Signs of respiratory distress include open mouth breathing and bubbles at the sides of the mouth and from the nares. Depending on the degree of respiratory distress, there may be abnormal posturing, **gaping** (frequent opening of the mouth), and audible respiratory sounds. Anorexia and lethargy are also seen with respiratory problems (**FIGURE I5-2I**).

Stress contributes to illness or a delay in healing. One of the major stress factors in captive reptiles is the environment and the food they are offered. Reptile behaviors that may indicate stress include inflating the body with air, hissing sounds, and open mouth breathing. There may also be immobility, tenseness, a change in color, cautious movement, and evacuation of cloacal contents.

Gastrointestinal Disorders

Gastrointestinal disorders are associated with internal parasites, bacteria or viral infections, and foreign body ingestions. Mycobacteria are often diagnosed in reptiles. The invading bacteria cause deterioration and wasting of the whole body. White nodules develop on the viscera, slowly destroying the gastrointestinal tract. The prognosis is grave.

A viral disease of great concern in snakes is Inclusion Body Disease (IBD), a disease that affects boids. Research is ongoing, but recent studies have found that this disease is likely caused by a strain of arenavirus that had not previously

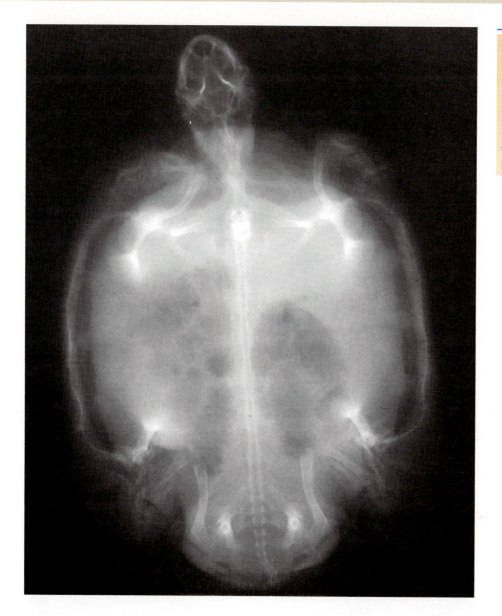

FIGURE 15-21
Respiratory infections in reptiles are often due to inadequate husbandry practices and not meeting POTZ requirements. This turtle has pneumonia in the left lung.

been known to affect snakes and is more commonly associated with wild rodents. There are several strains of IBD that may cause a variety of clinical signs. In young boas, the disease is acute, resulting in paralysis and death. Signs in adult boas include regurgitation one to two days after eating and upper respiratory tract infection. As the disease progresses, neurological signs may be present. In affected pythons, the disease progresses quickly and involves multiple body systems. There is frequently pneumonia, stomatitis, and neurological signs. The disease was previously referred to as *star gazing* as affected snakes lose the ability to right themselves, with much of the body twisted and upside down due to the loss of muscle control. There is no treatment and it is recommended that any snake testing positive for IBD be euthanized. IBD is contagious and recommendations are to quarantine all boids and cull affected animals in an attempt to stop the spread of the disease. Positive diagnosis is made with biopsies taken from several locations—liver, lungs, skin, and glottis—and sent to a specialty exotic animal laboratory for examination. Microscopically, cells reveal the invasion of clumps of viral protein: the *inclusion bodies*. PCR testing can be successful in providing an initial diagnosis.

Gastrointestinal foreign bodies are more commonly seen in terrestrial reptiles. Some reptiles ingest cage substrate, including sand and gravel. Chelonians sometimes ingest rocks and wood material. Snakes and lizards often ingest inanimate objects. It is not unusual for iguanas to swallow necklaces, earrings, and children's toys (**FIGURE 15-22**).

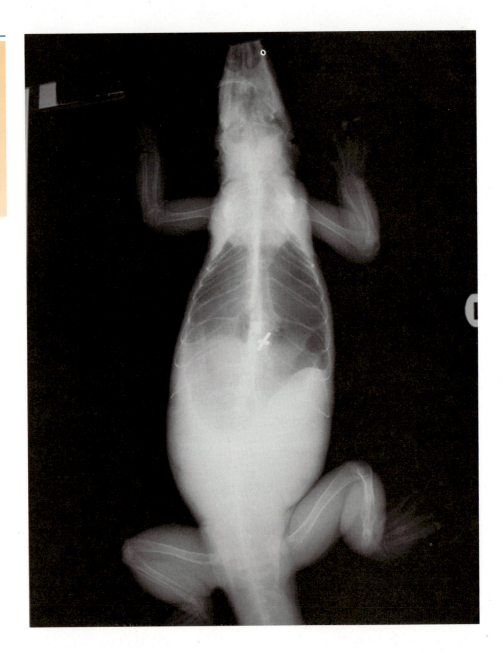

FIGURE 15-22 This iguana was presented with vague signs, lack of appetite, and lethargy. The radiographs revealed it had swallowed the owner's necklace, complete with the chain. (*Courtesy of Martin and Brandi Orr, Bird and Exotic All-Pets Hospital*)

Gastrointestinal neoplasia is a concern in reptile medicine. Tumors have been reported in all species of reptiles with no apparent age preference. Physical signs of tumors vary, depending on the type of neoplasia and location of the tumor. General signs may include poor growth, anorexia, constipation, melena, and palpable masses. In most cases, the patient does not survive this disease.

Constipation in combination with a decrease in appetite can be a result of ingestion of substrate, inadequate water consumption, poor hygiene, stress from a recent change, trauma, ovulation, or an insufficient diet. Treatment includes water soaks for at least 30 minutes daily, warm water enemas (20 to 30 ml/kg), and stimulation of the cloaca. This is accomplished with a lubricated cotton tip swab inserted into the cloacal opening.

Nondigestive Disorders

Nutritional secondary hyperparathyroidism (NSH) and renal secondary hyperparathyroidism (RSH) are metabolic diseases associated with metabolic bone disease (MBD). These diseases occur in reptiles that have not been fed an adequate diet or have not been provided with UBV lighting. These metabolic diseases are associated with a deficiency in dietary calcium, vitamin D deficiency, and excessive dietary phosphorus. A diet based on crickets or mealworms is too low in calcium to meet the needs of growing lizards. Full spectrum UVB lighting is essential for the absorption of calcium and homeostasis of the calcium/phosphorus ratio. Clinical signs of MBD, NSH, and RSH are similar and include muscle tremors, deformed bone structure, stunted growth, spontaneous fractures, and paralysis. Often, the diet has a calcium content that is too low and a phosphorous content that is too high.

Diagnosis is achieved through history, clinical signs, chemistry panels, and radiographs. Chemistry panels are the best diagnostic tests to determine the severity of the disease. Calcium levels will continue to decrease while the phosphorus levels will continue to rise due to the reabsorption of the bone. Evaluation of the patient for pathogenic fractures and bone density is performed through radiographs.

Treatment of MBD involves improving the diet, adding calcium supplements or giving calcium injections, and providing correct lighting. It is recommended to replace UVB lights every 6 to 12 months (or according to the manufacturer's recommendations) because the phosphorous produced by the lights is time limited. Fluorescent bulbs should not be more than 18 inches from the reptile. Mercury bulbs should not be more than 72 inches from the reptile. Depending on the severity of the disease, treatment may not be able to reverse some of the damage the kidney or bone received. Hypovitaminosis A is caused by an inadequate dietary intake of vitamin A. Signs include respiratory infections, neurological and ocular problems, and difficulty shedding. Supplemental vitamin A is often given by injection.

Oversupplementation of vitamins and minerals can be just as detrimental as a deficiency. Reptile owners need to be aware of the specific dietary needs of a species and provide the appropriate diet. Many owners are not aware that this may mean twice weekly visits to the grocery store to obtain fresh greens and time spent in food preparation. Supplements are not a substitute for a correct diet.

Gout is caused by the accumulation of excess urates in the intestinal tract and joints. Gout affects many species including snakes, lizards, and tortoises. Gout is associated with dehydration, poor nutrition, and chronic renal failure. Signs include swollen joints and painful movement. Treatment involves supportive therapy and the use of a diuretic, a drug that increases urination and elimination of excess body fluid. Treatment can slow the progression of the disease, but associated renal failure is fatal.

There have been reports that several species of reptiles and amphibians have died after being fed fireflies including Bearded Dragons, White's tree frogs, and other ground-dwelling species. One-half of a firefly can be lethal to a 100 g lizard. The chemical in fireflies is a type of cardiac glycoside. Signs of toxicity can appear within 15 minutes or up to two hours after ingestion. Sudden death occurs with ingestion of large numbers of fireflies. Signs of toxicity include head shaking, gaping, and dorsal color changes encompassing the head and extending to the tail. There is no known effective treatment. Other insects that have shown to be toxic include Asian ladybugs, bees, fire ants, Monarch butterflies, Queen butterflies, wasps, spiders, and all insects that feed on milkweed, which contains the same cardiac glycoside present in the sap of the plant.

Many reproductive problems can occur, especially in green iguanas and chameleons. Most lizards are oviparous. A common problem occurs when the lizard becomes egg bound. This may occur due to lack of a suitable site to lay the eggs, stress, poor nutrition, or hormonal problems. Reptile eggs do not have a hard shell like bird's eggs. The shell is soft, rubbery, or leathery, depending on the species. Medical management includes increasing the humidity and temperature, administering calcium gluconate injections, giving fluid therapy, and providing a suitable substrate for egg laying. The other option is surgical removal of the eggs. Due to the location of the ovaries, the mass of eggs during the preovulatory stage may completely block the GI tract due to their numbers and bulk (**FIGURE 15-23**, **FIGURE 15-24**, and **FIGURE 15-25**).

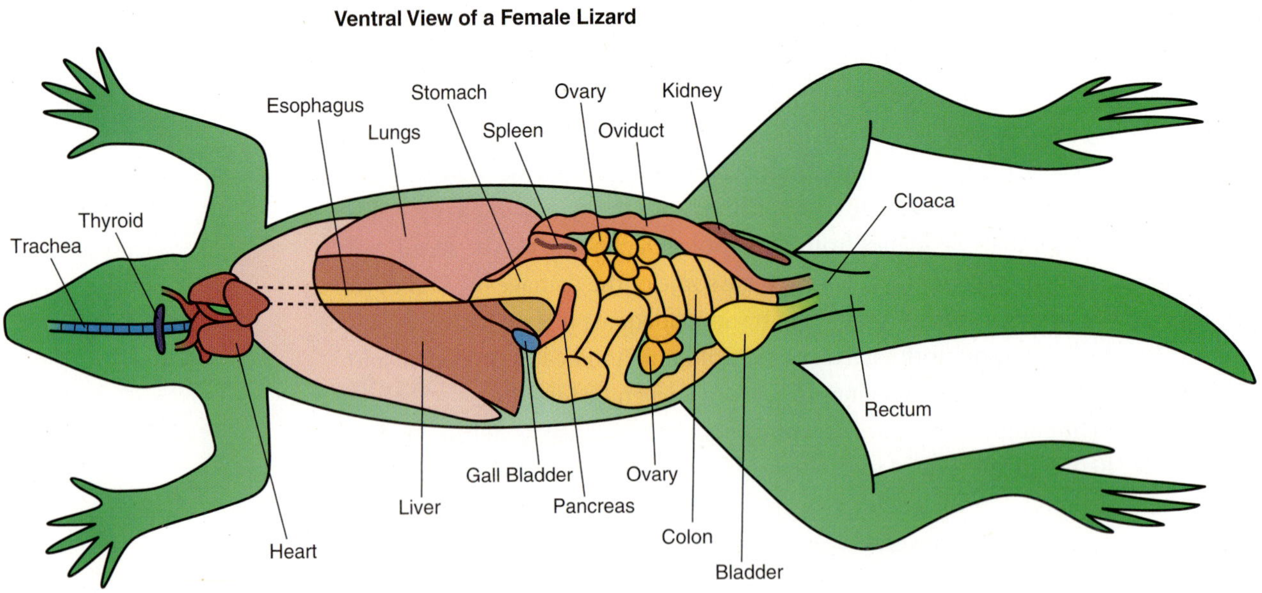

Ventral View of a Female Lizard

FIGURE 15-23 The anatomy of a female lizard showing the location of preovulatory eggs.

Shell Disorders

Shell repairs in chelonians may be necessary, often due to trauma. Frequent causes are dog bites or being hit by a car. The fractured area of the shell should

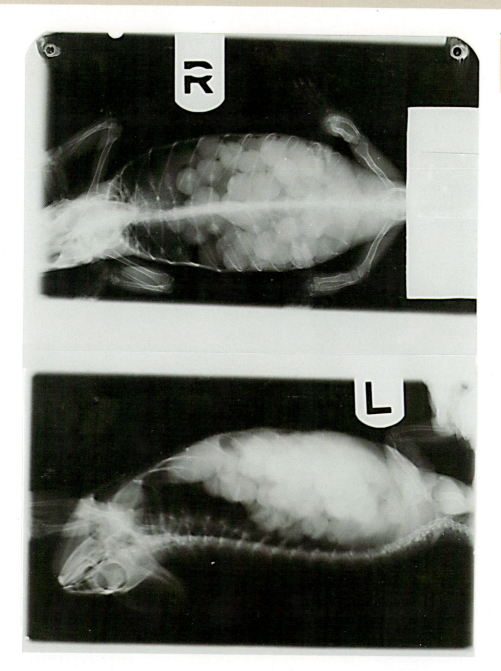

FIGURE 15-24 A radiograph of an egg-bound female Veiled chameleon.

be debrided to remove any fragments of shell, bone, or contaminants. The area should be lavaged with saline (PSS) to reduce the possibility of an infection, then cleaned with a surgical scrub and flushed again with PSS. Bleeding may be stopped with surgical glue or gelfoam. The damage can be repaired using fiber-glass patches, hoof acrylic, or dental material. The patch becomes a permanent addition to the shell. If the injury occurs in a juvenile chelonian, the fiberglass patch will need to be changed to allow for growth (**FIGURE 15-26**).

Septicemic cutaneous ulcerative disease (SCUD) is seen in aquatic turtles. This is a bacterial infection of the skin and shell. The condition is common with turtles kept in a dirty environment or not allowed a basking area out of the water. The bacteria cause ulcers of the skin and can cause loss of the scutes. This condition

FIGURE 15-25 Necropsy of a young Senegal chameleon that was found dead. Ten eggs were revealed from the necropsy.

FIGURE 15-26 A Box turtle that has a fiberglass patch to repair a shell injury.

can be life threatening because if the bacteria invade the blood stream they can cause septicemia. Treatment is focused on changing the environment and keeping the water quality high and well filtered. Affected areas of skin and shell should be cleaned and scrubbed with dilute iodine solution daily until healed.

Overgrown beaks and nails are common in chelonians because their captive environment does not allow for natural wear. Both beak and claws can be trimmed with the use of a dremel tool. Some beak deformities can be the result of an inadequate diet or trauma.

Pericloacal prolapse is common in reptiles. This may involve the penis, hemipenes, cloaca, intestines, oviduct, and bladder. The causes of a pericloacal prolapse include tumors in the cloaca, bladder stones, fecal impactions, MBD, dystocia, internal parasites, trauma, or microbial infection. Most prolapsed organs are cleaned with a surgical scrub, flushed, and reduced with a concentrated sugar solution. This allows the organs and tissues to be gently pushed back inside the body. Some veterinarians may partially suture the cloacal opening for several days to prevent a reprolapse (**FIGURE 15-27**).

FIGURE 15-27 A cloacal prolapse in a Bearded Dragon. The vent was partially closed with sutures to prevent its recurrence.

Parasites

Internal parasites are common in wild-caught lizards. Parasites may be protozoans in the gastrointestinal tract and circulatory system, and nematodes within the gastrointestinal tract.

In the wild, these parasites may not be problematic, but in captivity, when the lizard is exposed to stress, malnutrition, or inadequate husbandry, parasites can become a problem. Signs include anorexia, dehydration, regurgitation, and bloat. If the patient is already compromised, parasites may become overwhelming and cause death. Parasites can be diagnosed by fecal flotation or by obtaining a sample with an enema if there are no feces. A lubricated red rubber feeding tube is passed into the colon. Palpate the reptile prior to inserting the tube to determine how far to pass the tube to reach the feces in the colon. A warm saline solution (5 to 10 ml/kg) is syringed into the tube and then pulled back into the syringe to obtain the sample. A direct smear and fecal flotation is used to determine the presence of parasites and then treated with the appropriate anthelmintic as prescribed by the veterinarian.

Cryptosporidium invades the intestinal tract of reptiles and can cause regurgitation, weight loss, and death. Reptiles can be asymptomatic carriers of this protozoan or show signs associated with subclinical disease. Some reptiles can be infected by feeding wild-caught prey or drinking contaminated water. Acid fast stains or PCR testing is used to confirm the diagnosis.

External parasites, mites, and ticks are common in snakes and lizards. Red mites, *chiggers*, live under the scales and around the folds of the legs, eyes, and ear openings. Mites feed on blood and can cause anorexia, dehydration, and anemia. Injectable ivermectin is usually prescribed to remove mites from the patient; however, the enclosure needs to be completely disinfected. All natural wood furnishings should be removed and destroyed. Mites hide in wood, moss, and small crevices. Cage furnishings can be bagged and placed in a freezer for 24 to 48 hours, killing free-living mites.

Ticks are usually attached in *hard-to-reach* areas around the head and limbs. All visible ticks need to be carefully and completely removed. Ivermectin injections may also be used in the treatment of a major tick infestation in lizards or snakes, but it is toxic to chelonians and should not be administered. Permethrin has been safely and successfully used to treat tick infestations in chelonians.

Soaking the lizard in water is often recommended by pet stores; it will not *drown them* (the ticks), nor will a thick coating of petroleum jelly *suffocate them* and solve the problem. Professional veterinary treatment will save an owner much frustration and potential loss of the reptile.

The snake mite *Ophionyssus natricis* is a blood-feeding parasite. It can cause scales to lift and become necrotic, and cause dehydration, dermatitis, and anemia. It also transmits blood-borne diseases and is a likely vector of IBD. The mites burrow under scales and into crevices in the corners of the mouth and around the eyes.

Because their life cycle is unique, they are very difficult to eradicate. Mated females can lay both fertile and unfertile eggs. The fertile eggs develop as females and the unfertile eggs develop as males. Males are produced asexually, referred to as **parthenogenesis**.

Eggs hatch within 50 hours of being laid. This species of mites has a 40-day life span that is self-perpetuating. Snake mites have a five-stage life cycle: egg, larva, protonymph, deutonymph, and adult. Only mites in the last two stages of development feed on blood and they must have a blood meal before producing eggs.

The female mite is a voracious feeder, consuming more than 1000 percent of her body weight at each feeding. Mouth parts are similar to those of a tick with a penetrating blood-feeding tube.

There may be visual signs of infestation with scale disruption and dermatitis. Infested snakes become agitated and rub on any item in the enclosure. They will also soak in water more often. Over-the-counter products are available, but it is recommended that infected snakes be treated by a veterinarian.

Clinical Procedures

Intramuscular (IM) injections should be given in the front limbs of lizards and chelonians. If administered in the hind limbs, the renal portal system delivers the agent to the kidneys, where it is quickly filtered out and the effectiveness is lost. IM injections can have a prolonged effect due to a decrease in cardiac output and a lower metabolic rate (**FIGURE 15-28**).

Venipuncture sites in lizards include the ventral coccygeal vein and the ventral abdominal vein. The coccygeal vein is easily accessed in the lizard when restrained in dorsal recumbency. A tuberculin syringe with a 22- to 25-gauge needle is required. The needle is inserted between the scales at the midline of the tail. It should be inserted at least one-third of the distance from the vent to avoid penetration of the hemipenes (**FIGURE 15-29**).

The ventral abdominal vein is superficial and fragile. The lizard is restrained dorsally and a 22- to 25-gauge needle is used with a shallow approach just lateral to the vein.

Venipuncture sites in chelonians include the jugular vein, the dorsal and ventral coccygeal vein, the heart, the carotid artery, and the subcarapacial vein. The jugular vein is located laterally on the neck, beginning at the tympanic membrane. It follows the angle of the jaw along the length of the neck. This is a common site for blood collection in small turtles and tortoises when the head can be extended and restrained. The vein is held off at the base of the neck and a tuberculin syringe with a 22- to 25-gauge needle is used to collect the sample. Pressure

Renal Portal System of a Lizard

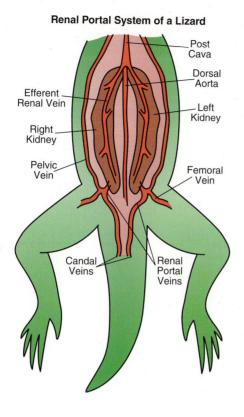

Post Cava

Dorsal Aorta

Efferent Renal Vein

Left Kidney

Right Kidney

Pelvic Vein

Femoral Vein

Candal Veins

Renal Portal Veins

on the vein should be applied for several minutes after blood collection to avoid a hematoma. The dorsal coccygeal vein is superficial and should be approached midline from the ventral surface of the tail. A 25-gauge needle is angled at 45 to 90 degrees and slowly advanced into the skin until blood enters the needle hub. The subcarapacial vein is found midline, directly under the carapace. The head should be retracted into shell. The approach should be with a 60-degree angle, just caudal to skin attachment to the carapace. In soft shell turtles, cardiocentesis can be performed for blood collection (**FIGURE 15-30**).

The two most commonly used blood collection sites in the snake are the ventral coccygeal vein and cardiac puncture. The ventral coccygeal vein is located caudal to the vent, along the midline of the tail. The snake should be restrained vertically. The appropriate-sized needle is inserted midline at a 90-degree angle until coming into contact with the vertebrae. The needle is then slowly backed out until blood enters the hub.

A cardiac puncture requires patience. The snake is restrained dorsally or held aloft by two people. The heart beat can be felt or seen at the distal end of the first third of the length of the snake's body. The heartbeat of a snake is substantially slower than that of mammals and it may take several seconds to locate the heart. Once a heartbeat is visualized, the site can be marked with a nontoxic felt-tipped pen. Using a 25-gauge needle and tuberculin syringe, enter straight into the heart, avoiding a lot of needle movement to prevent further damage or laceration of the heart.

Blood samples must be handled carefully and placed in the collection tube immediately after collecting the sample. Blood cells of reptiles are fragile and have a tendency to clot quicker than mammals. Because the blood volume is small,

FIGURE 15-30 To obtain a blood sample from the subcarapacial vein, the head should be retracted.

the use of microtainer blood tubes are recommended. The minimum amount of blood required in these tubes is 0.25 ml and the maximum is 0.5 ml. Purple-top tubes that contain EDTA are appropriate for most blood tests. Lithium heparin (green-top tube) can also be used but has shown to cause degradation of cells more quickly in the reptile.

Blood chemistries in snakes can be affected by a variety of conditions. Environmental conditions, reproductive status, nutrition, and time of year can affect the results of snake chemistries. Diet can affect the phosphorus and calcium levels, especially in female snakes. Electrolytes will vary considerably and can be affected by diet and other illnesses.

Medications for chelonians can be given orally or by injection. Intramuscular and subcutaneous routes are preferred. Intramuscular injections are given in the front legs due to the function of the renal portal system. Subcutaneous injections are given between the front leg and the neck. This site is the easiest to access. Be aware that even the smallest turtle or tortoise could attempt to bite.

Administering oral medication to a reptile can be difficult if the reptile does not open its mouth easily. To open a reptilian mouth without damaging it requires head restraint. Once the head is restrained, a rubber spatula, a plastic card (credit card), or a tongue depressor can be used to gently pry open the mouth. If metal forceps are used, the tissue around the mouth and in the oral cavity could be injured.

Nutritional support in chelonians can be given by placing a metal avian feeding tube into the esophagus with an attached syringe. To best avoid the trachea, the tube is advanced toward the right side of the oral cavity.

When administering fluids or medications, intramuscular injections should be considered as a last option. This is due to the sporadic uptake of medications when the patient is dehydrated or hypothermic. Several medications can be irritating to the reptile and complicated further by limited muscle mass. Fluid replacement in reptiles is administered by intravenous, intraosseous, subcutaneous, intracelomic, or per os (PO) routes. Chelonians can soak in and absorb a minimal amount of water through the cloaca if they are not severely dehydrated. If the reptile is slightly dehydrated, subcutaneous fluids or PO (per os, orally) can be administered. Snakes and lizards receive subcutaneous fluids between the lateral scales. Chelonians receive subcutaneous fluids in the inguinal fold or the ventral neck flap. PO fluids are administered by placing a red rubber feeding tube in the stomach.

ICe (intracelomic) administration can also be used to give small amounts of fluids to reptiles. In snakes, fluids are given toward the last fourth of the coelomic cavity, injecting between the scales. In the lizard and chelonians, the fluids are given in front of the hind leg. The fluids are directed toward the opposite shoulder. For maintenance, fluids are calculated at 5 to 10 ml/kg daily.

Blood volume for most species is approximately 5 to 8 percent of total body weight. Approximately 10 percent of this can be collected at one time. Blood should be collected in a lithium heparin (green top) tube. Reptile erythrocytes are nucleated and elliptical. Red blood cells are larger in size than in mammals and similar to birds. Packed cell volume (PCV) in most reptiles is between 20 percent and 40 percent. A PCV of less than 20 percent may suggest anemia.

Reptiles do not usually regurgitate under anesthesia with the exception of a recently fed snake. The reptile patient does not need to be fasted prior to anesthesia. Because reptiles are ectotherms, they can be slower to induce and more difficult to maintain under anesthesia. Recovery may be prolonged.

Injectable anesthetics lower cardiac output, causing slower blood distribution and resulting in a slower absorption rate. When using injectable anesthetics in reptiles, an increased time for drug effectiveness is expected. Administering a second volume will not shorten induction time and often leads to anesthetic overdose and death. Some injectable anesthetic agents can be very irritating if given ICe and there is a risk of lacerating internal organs. Most injectable anesthetics are administered intravenous or intraosseous in reptiles. This improves induction and recovery time and is less irritating to the reptile. Propofol is a commonly used injectable anesthetic that has shown to provide a rapid induction with a lower excitability stage and it is nonirritating to tissues. Ketamine and tiletamine, dissociative agents, are injectable anesthetics that are used in reptiles for surgical procedures. The major disadvantage with both of these agents is the prolonged recovery time. In some reptiles it might take 24 to 48 hours for the patient to fully recover.

Recommended inhalants for use in reptiles are isoflurane and Sevoflurane. Sevoflurane has a greater variance in reptiles and, occasionally, a surgical plane of anesthesia may be difficult to achieve.

The reptile heart is three chambered, with two atria and one ventricle. The ventricle has three subchambers with shunts that force the blood to the body and lungs. These valve-like shunts can function independently or in unison (FIGURE 15-31).

FIGURE 15-31 Reptiles have a three-chambered heart, with two atria and one ventricle. The diagram of a lizard heart illustrates cardiac blood flow.

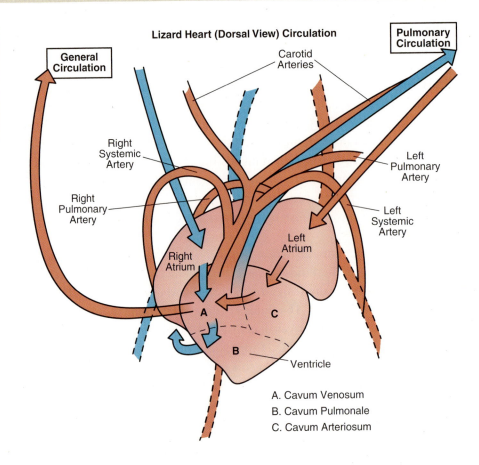

Lizard Heart (Dorsal View) Circulation

General Circulation

Pulmonary Circulation

Carotid Arteries

Right Systemic Artery

Left Pulmonary Artery

Right Pulmonary Artery

Left Systemic Artery

Left Atrium

Right Atrium

A

C

B

Ventricle

A. Cavum Venosum
B. Cavum Pulmonale
C. Cavum Arteriosum

The cardiac mechanism has a direct effect on oxygen saturation levels and the elimination of anesthetic gas during surgery and recovery.

Reptiles, like birds, do not have a diaphragm to assist with respiration. Abdominal and intercostal muscles move the air in and out of the lungs (in snakes, only the right lung is inflated). It is common for reptiles to become apneic when completely anesthetized. Reptiles are episodic breathers; they may take a couple of breaths and then stop breathing for a short period of time. When monitoring a reptile under gas anesthesia, the anesthetist should ventilate the patient manually several times in sequence. This inflates the lung(s) similarly to normal respiration and assists in maintaining the anesthesia level. Normal respiration for an anesthetized reptile is two to four breaths per minute.

It is fairly easy to intubate reptiles as the glottis is easily visualized. Uncuffed endotracheal tubes should be used. If the only appropriate-sized endotracheal tube is cuffed, it can be used, but not inflated. Inflation of the cuff can cause trauma to the trachea as many reptiles do not have complete tracheal rings to support the cuff. Pressure from an inflated cuff can cause tissue necrosis because it restricts blood flow. Chelonians have short, bifurcated tracheas. This can pose a problem when intubating for anesthesia. Endotracheal tubes should be placed at the top of the bifurcation to avoid inflating only one lung. When chelonians breathe, the abdominal and neck muscles assist with air flow. When a chelonian retracts its head into the shell, it stops muscle movement and halts respiration. An anesthetized chelonian should have its head and neck fully extended to prevent respiratory compromise.

A basic way of monitoring a patient under anesthesia is through body reflexes. During anesthesia induction, muscle relaxation starts at the midline of the body and moves cranially, then caudally. The tail is the last part of the body to relax. Loss of the righting reflex is considered an initial assessment to determine if the patient is anesthetize or sedated. The corneal reflex can determine the depth of anesthesia in a reptile. Take a cotton tip swab with a small amount of ophthalmic ointment on the tip and touch the cornea to assess the reflex. Tongue withdrawal in lizards and snakes is helpful to determine a light plane of anesthesia or recovery.

Cardiovascular function can be assessed with the use of a stethoscope, esophageal stethoscope, or a doppler. When placing an esophageal stethoscope, advance slowly to avoid entering into the stomach. The doppler probe can be placed over the heart, carotid artery, or coccygeal artery. Pulse oximeters are not the most reliable in assessing vital signs in reptiles. The primitive brain stem of reptiles continues to function, producing recognizable sounds of cardiac output when, in fact, the patient may be unrecoverable.

Hypothermia is a major problem with anesthetized reptiles. Without supplemental body heat, delayed recovery or death can occur. Heating pads, forced air warmers, and heat lamps are options to maintain body temperature in an anesthetized reptile.

Radiographs can be very helpful in diagnosing problems. Mammography film and cassettes or digital radiographs produce the best radiographic detail in reptiles. The use of digital radiography has eliminated the time it takes to develop radiographs and provides a higher quality diagnostic image for the veterinarian. Reptilian radiographs are useful in diagnosing gastrointestinal foreign bodies, reproductive problems, pneumonia, and liver disease (**FIGURE 15-32**). Most radiographs can be taken without anesthesia or chemical restraint, just with patience. Normal views for lizards and snakes are dorsal/ventral and lateral. Snakes should be uncoiled for radiographs. Depending on the length of the snake and the area of interest, films need to be taken in sequence, one length of the body at a time. If it is a large snake, restraint assistance is required. Snakes may also be tempted to enter a length of clear plastic tube, making it easier to obtain a fairly straight position (**FIGURE 15-33**).

Chelonians are the easiest to position for a radiograph. Dorsal/ventral is the most common view taken. If a lateral or cranio/caudal view is needed, a chelonian can be placed on top of an object with all four legs suspended. A cranio/caudal view is helpful in assessing the chelonian's lung cavity to determine problems associated with the respiratory system (**FIGURE 15-34**).

Branches can provide a climbing surface for the lizard to hold onto while positioning for a radiograph. Placing a reptile in a radiolucent container can reduce movement and help obtain a diagnostic radiograph.

Ultrasound can be used to locate the heart for blood collection in snakes, cardiac function, reproductive status, and organ function. It can also be utilized when performing fine needle aspirates and biopsies.

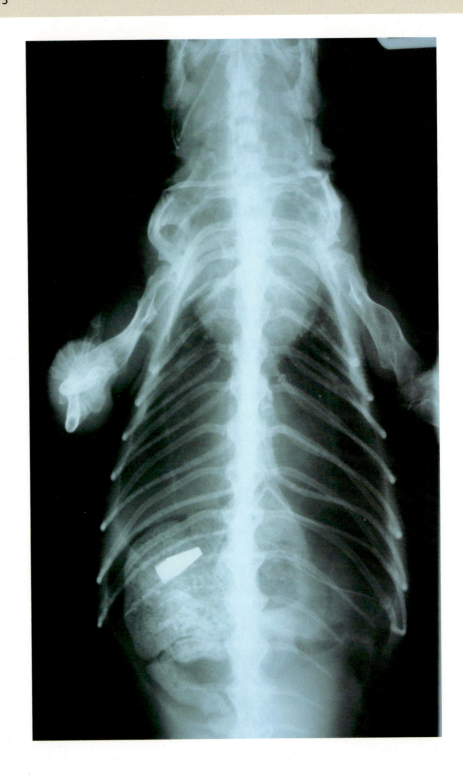

FIGURE 15-32 A D/V radiograph of an iguana with a foreign body in the GI.

FIGURE 15-33 A D/V radiograph of a snake with a vertebral fracture.

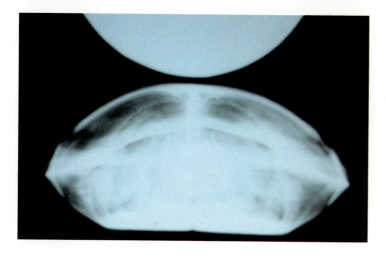

FIGURE 15-34 A craniocaudal radiograph of a Red-eared slider with a respiratory infection. There is decreased lung capacity in the turtle's left lung field.

Summary

Reptile medicine has become a skilled and specialized field. The most important aspects of care and preventive medicine are often species specific due to the great variety of species in captivity. Species can vary enormously in anatomy, physiology, habitat, and dietary requirements. The patient may be a lizard, a chelonian, or a snake, and all are native to specific ecosystems from subtropical to arid desert. Reptiles may be carnivores, herbivores, or omnivores and aquatic, semiaquatic, or land dwelling. The most diverse group of reptiles are the lizards, each with specific temperature and humidity requirements. They are divided between Old World species and New World species. Of the 2500 known species of snakes, more than half are colubrids and boas, both very popular snakes kept in captivity. Chelonians (turtles and tortoises) are facing many threats to their survival with approximately 257 species remaining. Some reptiles are egg layers, others give birth to live young, while others are ovoviviparous, meaning the eggs hatch with the female's body and are delivered as live young.

Restraint techniques for reptiles must be mastered to prevent injury not only to personnel but also to the patient. Clinical procedures for reptiles also vary greatly, depending on the species. Reptiles have a renal portal system that affects how injections are given; reptiles have a three-chambered heart with valve-like shunts, and, like birds, they do not have a diaphragm. Reptiles are episodic breathers and all of these anatomical differences need to be considered with treatment plans, radiographs, and anesthesia.

fastFACTS

REPTILES

ZOONOTIC POTENTIAL

- Cryptosporidium
- Salmonella
- Coccidia

Table 15-1

REPTILE TEMPERATURES AND HUMIDITY LEVELS

Lizards	POTZ	Humidity
African fat tail gecko	78 to 90°F	20 to 50%
Anole	73 to 84°F	70 to 80%
Basilisk	73 to 86°F	80 to 100%
Bearded dragon	84 to 120°F	30 to 40%
Blue-tongued skink	81 to 90°F	30 to 60%
Crested gecko	74 to 83°F	50 to 80%
Dwarf chameleon	70 to 85°F	70 to 80%
Giant day gecko	80 to 86°F	50 to 80%
Fischer's chameleon	72 to 85°F	70 to 80%
Gold dust gecko	80 to 86°F	50 to 70%
Green iguana	85 to 103°F	80 to 100%
Jackson's chameleon	70 to 80°F	50 to 75%
Leopard gecko	77 to 86°F	20 to 30%
Panther chameleon	85 to 90°F	70 to 100%
Savanna monitor	80 to 95°F	20 to 30%
Tegu	78 to 90°F	50 to 80%
Veiled chameleon	80 to 95°F	20 to 30%
Water dragon	77 to 93°F	50 to 80%
Uromastyx	85 to 110°F	20 to 30%
Snakes	**POTZ**	**Humidity**
Ball python	77 to 86°F	50 to 70%
Blood python	85 to 86°F	50 to 70%

Boa constrictor	82 to 93°F	70 to 90%
Red-tailed, Columbian, Burmese python	77 to 86°F	70 to 90%
Corn snake	78 to 88°F	20 to 50%
Garter snake	68 to 95°F	20 to 30%
Green tree python	78 to 85°F	80 to 100%
Hog Island boa	78 to 90°F	80 to 100%
Hog-nosed snake	80 to 90°F	20 to 30%
King snake	73 to 86°F	50 to 75%
Milk snake	78 to 90°F	20 to 40%
Rainbow boa	75 to 90°F	90 to 95%
Reticulated python	78 to 90°F	80 to 100%
Sand boa	77 to 86°F	20 to 30%
Chelonians		
Box turtle	75 to 84°F	50 to 80%
Leopard tortoise	68 to 86°F	40 to 50%
Painted turtle	73 to 82°F	80 to 100%
Red-eared slider	75 to 90°F	80 to 100%
Red-footed tortoise	80 to 85°F	70 to 80%
Russian tortoise	78 to 90°F	20 to 50%
Radiated tortoise (star)	80 to 100°F	20 to 50%
Sulcata tortoise	80 to 95°F	20 to 50%

Note: These are recommended average temperature ranges and humidity levels.

Table 15-2

LIFESPAN IN CAPTIVITY	
Colubrids	**Age (Years)**
Corn snake	32
Rat snake	22
Hog-nosed snake	19
King snake	19 to 44
Northwestern garter snake	15
Boids	
Boa constrictor	40
Solomon Island boa	16

(Continued)

Rubber boa	26
Emerald tree boa	31
Rainbow boa	31
Carpet python	26
Green tree python	19
Burmese python	28
Ball python	20 to 47
Reticulated python	29
Box turtle	20
Red-eared slider	20
Red-foot tortoise	20 to 30
Softshell tortoise	12 to 15

Table 15-3

REPTILE DIETS		
Species	**Diet**	**Recommended Feedings**
Pythons	Rodents, small mammals, chicks	2 to 3 weeks
Boas—larger species	Rodents, small mammals, chicks	2 to 3 weeks
Boas—smaller species	Rodents	1 to 2 weeks
Rat snakes	Rodents, chicks	2 to 3 weeks
Corn snakes	Rodents, checks	1 to 2 weeks
Gopher snakes	Rodents, chicks	1 to 2 weeks
Pine snakes	Rodents, chicks	1 to 2 weeks
King snakes	Amphibians, rodents, fish, small lizards	1 to 2 weeks
Water snakes	Amphibians, rodents, fish, small lizards	1 to 2 weeks
Garter snakes	Amphibians, rodents, fish, small lizards	1 to 2 weeks
Hog-nosed snakes	Amphibians, rodents, fish, small lizards	1 to 2 weeks
Anoles	Insects	Daily
Chameleons	Insects, limited greens, small lizards	Daily
Geckos	Insects	Daily
Water dragons	Insects, pinkies	Daily
Most skinks	Insects	Daily
Swifts	Insects	Daily
Ameivas	Insects	Daily

Species	Diet	Recommended Feedings
Lacertas	Insects	Daily
Small monitors	Insects, pinkies	Daily
Tegus	Insects, pinkies	Daily
Day geckos	Invertebrates, insects	Daily
Large monitors	Vertebrates	Weekly
Large tegus	Vertebrates, fruit	Weekly
Bearded dragons	Greens, invertebrates	Daily
Blue-tongued skinks	Greens, invertebrates	Daily
Uromastyx	Greens, invertebrates, seed	Daily
Green iguanas	Greens	Daily
Prehensile-tail skinks	Greens	Daily
Softshell turtles	Carnivores	Daily
Box turtles	Earthworms, fruits, veggies, crickets	Daily
Wood turtles	Earthworms, fruits, veggies, crickets	Daily
Sulcata tortoise	Grasses, greens, fruits	Daily
Leopard tortoise	Grasses, greens	Daily
Radiated tortoise (star)	Grasses, greens	Daily

Table 15-4

REPRODUCTIVE (STATUS) OF REPTILES		
Species	Oviparous	Viviparous
Chelonians	×	
Monitors	×	
Iguanas	×	
Water dragons	×	
Geckos	×	
Veiled chameleons	×	
Panther chameleons	×	
Jackson's chameleons		×
Python snakes	×	
King snakes	×	
Milk snakes	×	
Rat snakes	×	

(Continued)

Species	Oviparous	Viviparous
Corn snakes	×	
Most boas		×
Most vipers (rattlesnakes)		×
Colubrids—some		×
Garter snakes		×
American water snakes		×

Review Questions

1. What does POTZ mean and why is this important in housing reptiles?

2. Why is it important to provide UV lighting for reptiles?

3. Gout is a common disease in reptiles. What are the signs often associated with this disease?

4. What are the common signs of a reptile with respiratory difficulties?

5. What is an episodic breather?

6. What type of habitat is required for chameleons?

7. What are the causes of metabolic bone diseases?

8. What are common blood collection sites for the following species?
   a. lizards
   b. snakes
   c. chelonians

9. What are the different ways of determining the sex of the following reptiles?
   a. snakes
   b. chelonians
   c. lizards

10. Explain the renal portal system in reptiles.

Case Study I

History: A juvenile green iguana has been presented by the owner because it is not eating and has difficulty moving. The owner purchased the iguana two weeks earlier from a local pet store and houses it in a 10-gallon aquarium. She was told to feed it lettuce and crickets. Questioned further, the owner states that she keeps the aquarium by the window so there is light, and when the sun comes in, it seemed "warm enough."

Physical Examination: The left rear leg is swollen and the veterinarian suspects a spontaneous fracture of the femur. The iguana rests on the table with its eyes closed. The veterinarian also notes that the pelvic bones are pronounced and there is little muscle mass in either hind limb or tail.

a. What disease could have caused the fracture?

b. How is this condition treated?

c. In assisting this client, what advice would you give regarding diet, lighting, and POTZ?

Case Study II

History: A seven-year-old Ball python is presented by the owner, concerned that the snake is not eating. He has had the snake for two years and reports that it seemed healthy and has been eating regularly. He tried feeding it three mice last week, but the prey was refused. He thinks it may be "trying to hibernate because it yawns a lot."

Physical Examination: The veterinarian hears audible respiratory sounds and she makes a note, "dyspnea." In her assessment, the patient is not "yawning" but gaping.

a. What medical problem do these signs indicate to the veterinarian?

b. What further questions would the veterinarian ask of the owner with regard to the habitat?

c. What are the veterinarian's likely treatment recommendations?

Case Study III

History: A medium-sized Red-eared slider is presented by a rescue organization. It has an unknown history, except that the previous owners bought it when it was "very small." The volunteer is concerned about the turtle's health in general and has also noticed that some of the scutes are missing.

Physical Examination: There are several scutes missing and the plastron has areas of redness. The skin of all four legs is inflamed and exhibit edema.

a. What is the veterinarian's likely diagnosis?

b. What conditions would predispose a turtle to this condition?

c. Would the veterinarian recommend environmental changes as a part of the treatment?

For Further Reference

Donoghue, S. (2002). Assist-Feeding Sick Lizards. *Exotic DVM, 4*(2), 38–39.

Garner, M. M. (January 2005). *Common and Emerging Diseases of Reptiles: Bacterial.* North American Veterinary Conference, Orlando, FL.

Girling, S. (2003). *Veterinary Nursing of Exotic Pets, 2nd Edition.* London: Blackwell Publishing.

Heard, D., Fleming, G., Lock, B., & Jacobson, E. (2002). Lizards. In *BSAVA Manual of Exotic Pets, 5th Edition* (pp. 223–240). London: British Small Animal Veterinary Association.

Hernandez-Divers, S. J., & Cooper, J. E. (2001). Reptile Hepatic Lipidosis. *Proceedings Association of Reptilian and Amphibian Veterinarians, 8,* 193–199.

http://www.anapsid.org (accessed May 30, 2014).

http://www.aza.org (accessed June 5, 2014).

http://www.labs.vetmed.ufl.edu (accessed June 5, 2014).

http://www.ncbi.nlm.nih.gov (accessed June 5, 2014).

http://www.ucst.edu (accessed June 6, 2014).

http://www.veterinarypracticenews.com (accessed June 6, 2014).

Johnson-Delaney, C. (Ed.). (2000). *Exotic Companion Medicine Handbook.* Florida: Zoological Education Network.

Mader, D. R. (2006). *Reptile Medicine and Surgery, 2nd Edition.* St. Louis, MO: Saunders/Elsevier.

Mayer, J., & Bays, T. B. (2006). *Exotic Pet Behavior.* St. Louis, MO: Elsevier.

McArthur, S. D. J., Wilkinson, R. J., & Barrows, M. G. (2002). Tortoises and Turtles. In *BSAVA Manual of Exotic Pets, 5th Edition* (pp. 208–222). London: British Small Animal Veterinary Association.

Means, C. (November 2006). *Color Change in a Bearded Dragon.* Tulsa, OK: NAVC Clinicians Brief, pp. 27–28.

Mitchell, M. A. (2010). Managing the Reptile Patient in the Veterinary Hospital: Establishing a Standards of Care Model for Nontraditional Species. *Journal of Exotic Pet Medicine, 19*(1): 56–72.

Mosley, C. A. E. (2005). ACVA. Anesthesia and Analgesia in Reptiles. *Seminars in Avian and Exotic Pet Medicine, 14*(4), 243–262.

Raiti, P. (2002). Snakes. In *BSAVA Manual of Exotic Pets, 5th Edition* (pp. 241–256). London: British Small Animal Veterinary Association.

Stahl, S. J. (January 1999). *Medical Management of Bearded Dragons.* North American Veterinary Conference, Orlando, FL.

Wright, K. (2005). Beyond POTZ: Environmental Influences on Reptile Healing. *Exotic DVM, 7*(4), 11–15.

Wright, K. (2008). Two Common Disorders of Captive Bearded Dragons and Nutritional Secondary Hyperparathyroidism and Constipation. *Journal of Exotic Pet Medicine, 17*(4), 267–272.

Wyneken, J. (2001). Respiratory Anatomy—Form and Function in Reptiles. *Exotic DVM, 3*(2), 17–22.

UNIT V

AMPHIBIANS

OBJECTIVES

After completing the chapter, the student should be able to

- Describe the correct habitat for different amphibian species.
- Provide appropriate client education to new amphibian owners.
- Deliver basic nursing care to common amphibian species.
- Demonstrate appropriate restraint techniques with each species of amphibians.
- Describe diets for each different amphibian species.
- Provide the correct temperature and humidity (POTZ) ideal for each common amphibian species.
- Identify potential problems with inappropriate housing, restraint, and diet of amphibians.
- Understand common medical disorders in amphibians.

Introduction

Throughout time, amphibians have been associated with myths and folklore. In ancient Egypt, the frog goddess Heket represented fertility and was believed to have been responsible for regenerating the limbs of Osiris. Salamanders and newts have traditionally been connected to fire and the mysterious healing brews of witchcraft. Not much was understood about amphibians centuries ago; frogs rained from the sky, touching toads caused warts, and salamanders drank the milk of cows. With present-day knowledge, one can easily understand how some of these stories came to be. Amphibians have the ability to regenerate limbs; frogs are fertile, some producing as many as 35,000 eggs at one time (*Bufo marinus*); and salamanders live in moist woodlands and crawl out of firewood when it is burned.

Immature amphibians have no resemblance to adults. All must go through developmental stages of metamorphosis of changing from jelly-like eggs to aquatic larva with gills. As they become adults, they will crawl onto land, breathe air, and live on land as if appearing by magic. The word *amphibian* is derived from Greek and reflects a greater awareness of what must have seemed, then, a very strange group of creatures. *Amphi* means both, and *bios* mean life: life both in water and on land.

Today, there are more than 4000 different species of amphibians, ranging from the tropics to the subarctic regions of the world. Their numbers are decreasing at an alarming rate. Many have become extinct just in the past few decades, others have suffered massive die-offs, and still others are exhibiting deformities that cannot be explained. It is believed by many scientists that amphibian die-offs and these mutations are a direct result of environmental pollution both locally and globally. The alarming rate of amphibian extinction was first noted in the late 1970s. Theories include climate change, the use of insecticides that disrupt the food chain, and a recently discovered chytrid fungus found in water.

Toxins Produced by Amphibians Produce a Range of Effects. Bufo species contain toxins potent enough to kill a small dog, should it be unfortunate enough to pick it up in its mouth. Other effects of amphibian toxins include neurotoxins and the vasoconstrictive effects of poison darts. Many amphibians produce potent toxins, some of which are powerful **hallucinogens** (drugs that cause hallucination).

The increase in human population and urban development and the destruction of wetlands and old growth forests have undoubtedly played a major role in the decline of amphibians, as the food sources have depleted because of the use of insecticides and contamination of habitats. Many scientists are concerned that air and water quality have become so polluted that, coupled with global warming, the demise of amphibians is a probable indicator of a dying planet. The number of declining amphibian species worldwide is accelerating. The World Conservation Union lists nearly 4000 amphibian species for which the numbers are declining; their status is unknown (Red Data Book, 2013).

Herpetology, the study of reptiles and amphibians, has long been a recognized science. This science is constantly changing as new reptiles and amphibians are discovered and studied. However, amateur collectors and *hobbyists* have created a new demand and a new threat. Many species have been exploited beyond recovery to supply the market. Many species do not breed in captivity. Supplying the pet market with wild-caught specimens of already rare amphibians has placed many species on CITES endangered list. For example, the tiny (under 1 inch) golden mantella frog was frequently found in pet stores as recently as 2002. It is now critically endangered in its native habitat of Madagascar. This tiny frog has been successfully bred in captivity and is being screened for disease prior to being introduced into the wild population in an attempt to increase their numbers.

Amphibians can be aquatic, terrestrial, or arboreal. Generally they are grouped as *Anura*, those without tails (frogs and toads); *Caudata*, those with tails (salamanders, newts, and sirens); and *Gymnophiona*, the caecilians, or the limbless amphibians.

The family Anura has the greatest diversity with approximately 3500 species. Most frogs are semiaquatic, with the exception of the Surinam toad and the African clawed frog, which are entirely aquatic. There is no biological difference

FIGURE 16-1 The Cuban tree frog is one of the few species of frog with nodulated skin, similar to that of toads.

FIGURE 16-1 The Cuban tree frog is one of the few species of frog with nodulated skin, similar to that of toads.

between frogs and toads, but the terms are used to separate terrestrial, aquatic, and semiaquatic species. Toads have drier, thicker, and nodulated skin. Frogs, with the exception of some tree frogs, have thinner, smooth, mucoid skin (**FIGURE 16-1**).

Frogs are used in the study of developmental biology and cloning research. They have been raised as food and used as classroom dissection specimens. The African clawed frog (Xenopus) is the most commonly used frog in biomedical research. The American bullfrog (Rana) is harvested for human consumption (*frog legs*) and is also preserved for student study.

Caudata, those with tails, contains approximately 375 different species. Salamanders are used in studies of limb regeneration as they have the ability to completely regrow a limb that has been lost. This phenomenon is not completely understood.

Salamanders and newts are usually separated by habitat. Salamanders prefer moist woodlands and riverbanks, while the smaller newts are mainly aquatic. Some salamanders, for example the axolotl and European olm, retain a state of neoteny, in that they remain in an immature larval state throughout their lives.

Gymnophiona are limbless, worm-like, and do not have pelvic girdles or pectoral muscles. This is the smallest group, with approximately 160 species. They are rare in captivity.

Argentine horned frogs are very popular terrarium specimens. They have bony, tooth-like projections on the lower jaw that can cause a painful bite. The common color variations of this species are green, cream, and reddish brown (**FIGURE 16-2**).

Poison dart frogs have become popular additions to terrariums. These tiny frogs are brightly colored, a natural warning to predators that they are poisonous. Most poison dart frogs are not toxic in captivity. The toxins are derived from feeding on ants that eat toxic plants in the wild (**FIGURE 16-3**).

The White's tree frog is another popular pet frog and one of the largest of subfamilies, Litoria. These frogs are big and sturdy and become relatively tame in captivity. The frog's appearance of being *overweight*, coupled with its folds of skin, is appealing to a large number of amphibian owners. White's tree frogs are frequently called *dumpy tree frogs* (**FIGURE 16-4**).

FIGURE 16-2 Argentine horned frog.

FIGURE 16-3 *Bumblebee* poison dart frog.

FIGURE 16-4 White's tree frog, often called *dumpy tree frog.*

Red-eyed tree frogs have striking colors with crimson-red eyes. They are delicate arboreal frogs that require a more experienced herpetologist (**FIGURE 16-5**).

FIGURE 16-5 Red-eyed tree frog.

The tiger salamander is one of the world's largest land-dwelling salamander species. This nocturnal salamander adjusts well to captivity and tends to burrow under logs (**FIGURE 16-6**).

Correct identification of many species is clouded and confused. Suppliers may not know the genus and species. Shipments arrive in the retail market with

FIGURE 16-6 Tiger salamander.

a variety of common names that vary from location to location and wholesaler to wholesaler. Pet stores also assign incorrect nomenclature to species, based on nothing more than appearance. Many **tadpoles** (immature specimens) are sold as *fish tank scavengers* and develop not into frogs but into salamanders.

In One Such Instance, what was purchased as a frog tadpole was dropped off in a classroom in Utah. It was left in a 30-gallon aquarium with about four inches of foul-smelling water, rotting vegetation, and dead, bloated earthworms. This *tadpole* became a fully grown, air-breathing *terrestrial* salamander that measured nine inches long. When identified, it turned out to be a rare species, native only to the Kentucky River Palisades. It was not determined how the larva came to be sold in a pet store in Salt Lake City.

Common to all amphibians is their life cycle. Most have four stages: egg, aquatic larva with gills, aquatic larva with limb buds, and a complete metamorphosis into air-breathing adults. Frogs and toads in the larval stage are called tadpoles, while salamanders and newts are referred to simply as **larval stage**. Depending on the species and environmental factors, it may take two or more years to complete metamorphosis (**FIGURE 16-7**).

Amphibians usually return to the same water in which they were **spawned** (hatched). This may entail mass migrations, crossing freeways or using specially designed under-road tunnels constructed specifically for toad and frog migration safety. Migration sites are carefully monitored by wildlife authorities. Road signs are posted either by governmental agencies or by the caring public to help prevent the thousands of deaths that were occurring every spring when amphibians crossed the roads to reach the ponds and waterways where they were hatched.

FIGURE 16-7 The life cycle of amphibians.

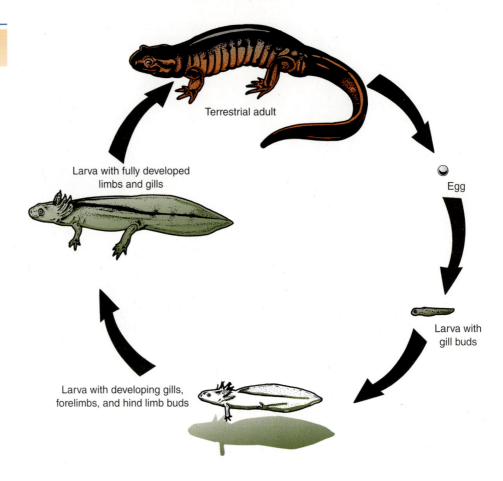

Terrestrial adult

Egg

Larva with fully developed limbs and gills

Larva with gill buds

Larva with developing gills, forelimbs, and hind limb buds

Pond keepers are well aware of this strange event and regularly patrol their ornamental ponds to remove gelatinous masses of spawn, transporting them to other, more desirable locations. It would not take long for frogs to completely overrun an average-sized fish pond. While the songs of frogs may be pleasant to some, the vocalizations of spring can quickly become a nuisance. In the frenzy of **amplexus** (the male clasping the female's forelegs for breeding purposes), it is not unusual for male frogs to clasp onto anything that swims, including ornamental fish, resulting in scaring and loss of prized specimens.

Most amphibians are oviparous, similar to fish, and lay gelatinous strings or masses of eggs that are fertilized outside of the body. The eggs, or spawn, do not have shells and are deposited directly in water, very wet reed beds, or cups formed by plant leaves. Without water, the eggs dry up and the tadpoles die. There are exceptions, and some caecilians are viviparous, giving birth to live young and producing nutrients in the oviduct to feed the embryos.

Sexing amphibians can be difficult. Males have internal testes and the sexes may be distinguishable only during breeding season. Some male frogs and toads have **nuptial pads** near their toes, which assist in amplexus. Male frogs and toads are more vocal and are, in many species, approximately half the size of females. Some species of salamanders and newts are sexually dimorphic. In many species of newts, the males have crests just behind the head and along the dorsal surface. There may also be color differences. Some males, especially newts, become very brightly colored during breeding season.

Behavior

Amphibians do well in captivity if provided with the proper care, diet, and environment. Some species are more adaptable to captivity than others. The ornate horned frog and the tiger salamander are very adaptable and can survive under a variety of conditions. Careful selection must be made if housing different species together. Some amphibians produce toxins that could kill other species, and adult species are carnivores, which could lead them to feed on one another. This is especially true with frogs. Most amphibians hunt at night. During the day, amphibians hide under moss, logs, and rocks. They become active at dusk, catching insects and other food items that are moving about.

Housing

All amphibians are poikilotherms; that is, they are totally dependent on their environment to regulate body temperature and metabolic activity. Many captive amphibians die because they have specific temperature requirements that are either unknown or are not met.

Amphibians are unable to thrive and adapt to temperature ranges outside of their species-specific POTZ requirements. Amphibians kept above their normal range (too hot) become anorectic and agitated, have changes in skin color, and can easily develop secondary medical problems. Amphibians housed below their normal range (too cold) have decreased appetites and lower metabolic activity. They exhibit lethargy and bloating and have slower growth rates.

Before acquiring any amphibian, the habitat should be set up and allowed to stabilize as a microenvironment before the addition of any animal. A vivarium can be created that is correct for the species, as well as an enjoyable and attractive focal point.

Basic requirements for amphibians include a glass aquarium or plastic tank of an appropriate size. It needs to be able to accept and safely accommodate electrical cords for heating and lighting requirements, remembering that there will be the addition of a water supply. It should provide good ventilation and have a secure, ventilated lid. Habitats can be attractively furnished with the addition of living plants and waterfall features, misters and foggers, natural rocks, ledges, and logs (**FIGURE 16-8**).

Diurnal amphibians require full spectrum lighting. Nocturnal species need *night bulbs* that produce black or red light. These lights do not produce UVB light but do provide heat. Light bulbs and UV tubes are sold under a variety of names, but they all clearly state the spectrum of light that is produced and that they are manufactured specifically for use in reptile/amphibian habitats. All of these lights have a time-limited ability to produce the full spectrum. While the habit may be lighted, chances are that if the light is past the recommended number of hours for use, it is not producing the full spectrum light. Bulb packages should be carefully read and replaced when recommended by the manufacturer if they are to be fully effective.

There are several ways of maintaining environmental temperature. If the habitat is small, the UVB light may provide enough heat, but most will require an additional heat source that will warm the environment as a whole, rather than warming

just the animal. Padded, under-the-tank heating units are available in a variety of sizes to adequately cover enough of the habitat bottom floor. As heat rises, it will also warm the substrate. Reptile basking lamps should be used with caution. Many become very hot and, if placed too close to the cage, can cause death from dehydration and, in some instances, complete desiccation. Temperature ranges are generally higher in the daytime and should be lower at night. The POTZ range for the chosen species should be considered when setting up the habitat.

All amphibians need a specific level of humidity. The range of humidity is usually 75 percent to 90 percent. The use of natural sphagnum moss and tropical plants, such as **bromeliads**, can help to maintain humidity. The water needs to be easily accessible, be it an ornamental waterfall or a plain shallow bowl. The water supply will also increase the humidity. Hygrometers and thermometers should be placed so they can easily be viewed to monitor environmental conditions.

The floor of the habitat needs several inches of a suitable substrate. Depending on the species, it could be natural moss, combinations of peat and sterilized organic soil, shredded coconut, and bark fibers. All of these make suitable substrates, as they tend to hold moisture. Wood chips and shavings should not be used, as they pull moisture away, quickly drying out the flooring and, consequently, the amphibian.

Many amphibians die because they are confined within contaminated environments. Water quality should always be monitored, as amphibians breathe through their skin and lungs and take in water through skin membranes by **osmosis**. Amphibians do not drink water, but absorb water through the skin from moisture in the environment or by soaking in water bowls. If the water quality is poor, bacterial and fungal infections are likely to result. The same applies to substrates; if skin permeability is compromised by injury or fungal or bacterial diseases, it inhibits the animal's ability to breathe.

The water in the habitat should be treated with dechlorinating chemicals. There are several products for water treatment and they are found with tropical fish supplies. Distilled water is not recommended, as all the minerals have been removed, including those that are beneficial. Water should be changed frequently and the container cleaned regularly to prevent bacterial growth. Frogs should also be misted twice daily with chlorine-free water.

For aquatic species, the use of submersible pumps and aquarium filters helps keep the water clean and oxygenated. The tank may be left bare or contain a layer of aquarium gravel or pebbles. Bags of gravel or smooth stone should be thoroughly rinsed to remove dust before being added to the tank and be large enough that the amphibian cannot ingest pieces.

It is best to obtain captive-bred specimens to avoid the introduction of parasites or diseases brought into the collection from wild-caught specimens. Although there is no way to really know if a specimen is wild caught or captive bred, the best approach is to establish a relationship with a reputable supplier who has the same concerns. Display habitats offering captive-bred species are usually marked with the letters CB (captive bred).

Diet

All adult amphibians are carnivorous and eat only live food. Tadpoles and larvae feed on a variety of food stuffs, including the newly hatched fry of fish, algae, and one another. Some species also secrete a substance that inhibits the growth of other tadpoles, effectively providing a ready food source for siblings. Adults consume a variety of insects, moths, worms, invertebrates, and small fish. Large species of frogs also prey upon rat and mouse pups, immature birds, and other amphibians. The smaller species of frogs, mantellas and poison darts, require much smaller food items such as fruit flies or pin-head crickets. Salamanders eat insects and other invertebrates, including slugs, earthworms, and snails. Aquatic frogs eat tubifex worms, bloodworms, and mosquito larva. Small fishing worms known as *red wrigglers* may also be taken. Before acquiring any species, one needs to know the species' natural diet, how often it should be fed, and, most importantly, that suitable prey items can be readily obtained.

Dietary inadequacies are a frequent concern in keeping amphibians healthy. A contributing factor is that owners receive incorrect information regarding the feeding of an individual species, or substitute convenient prey items that are refused.

Most Species Sold in Retail Stores would never encounter a cricket in their natural habitat. Often, when crickets are dumped into a habitat, the crickets feed on the amphibian. Different species of insects, beetle larva (mealworms), wax worms, flies, roaches, and other prey items, have different nutritional values. Many amphibians suffer not only from starvation but also from the stress, harassment, and trauma from crickets feeding on them.

Handling and Restraint

Amphibians should be handled only when necessary and never with bare hands. They absorb bacteria, soap residue, and lotions from bare hands directly through their skin. Additionally, the comparative dryness of human skin can damage amphibian skin and its protective layer of mucous. Non-powdered and wet disposable latex or nitrile gloves should always be worn when handling any amphibian. Many amphibians produce potent toxins from specialized glands in their skin or, in the case of the marine (cane) toad, from highly developed parotid glands. This giant toad is able to squirt its toxin several feet. Handlers of this species, *B. marinus*, should also wear protective eye glasses or goggles. Toxins may enter through open skin wounds on the hands and the conjunctiva and mucous membranes of the mouth and nose. Always wear gloves to protect the skin of the amphibian and to protect the handler. Remember: *toxins in—toxins out*. Even when gloves are worn, hands should always be thoroughly washed after handling any amphibian.

The least amount of restraint possible should always be applied so as not to increase the stress to the patient or cause trauma. A towel soaked in dechlorinated water should be placed on the examination table to help prevent dehydration during the examination. Frogs and toads can be restrained by cupping one hand around the chest, just behind the front legs. Salamanders, with their elongated bodies, should be held with both hands: one hand behind the neck with the fingers encircling the chest, and the other hand cupped around the pelvis.

Care must be taken with all species of horned frogs. They can and will deliver a painful bite. They are also reluctant to *let go* and attempt to swallow whatever is in their mouths.

Totally aquatic species should be examined in water. Small, clear, plastic tubs should be on hand and disposed of after each use. Clinics that regularly see amphibians could benefit from having prepared, dechlorinated water available in glass jugs. Routinely changing and refreshing the supply ensures that a safe water supply is always readily available.

Medical Concerns

Healthy amphibian skin is shed regularly. Many owners do not even realize that shedding has occurred. The fine layer of loosened skin is consumed, with many species pulling it away from their bodies and using their forelegs to push it directly into their mouths. It is not cast off and left in the habitat, as with reptiles.

An incomplete or patchy shed is usually an indication that the humidity is too low. If this is the case, it is also an indication that the amphibian is dehydrated. Dehydrated skin darkens and the mucous layer will feel tacky. The eyes may appear sunken and dull. Severely dehydrated amphibians are unable to right themselves if turned over. At this stage the prognosis is grave, as the condition has progressed from dehydration to near desiccation, a condition where cellular fluid and internal organs may have irreversible damage or may already be destroyed.

Digestive Disorders

Every now and again, the story surfaces that a live mealworm chewed its way out of the gut of the amphibian that swallowed it. There are various stories circulating and these are often repeated as absolute fact. While it is true that mealworms have mouthparts, there is no substantiated evidence that this occurs. What can be the cause of death in an amphibian is that too many mealworms are fed at one time. This causes **gastric overload**, or impaction. Gastric overload occurs when an amphibian is fed a large quantity of prey items in a short amount of time. This can happen not just with mealworms but with any prey fed in excess.

Foreign bodies are often ingested because of the way amphibians feed. They swallow prey whole, capturing it with sticky tongue pads or with a *snap and gulp*. Pieces of substrate and small pebbles are frequently found in the stomach. Surgical removal may be required, as they are unlikely to be able to eliminate foreign body blockages.

Nondigestive Disorders

Chytridiomycosis is a fungal skin disease that has been the cause of recent amphibian deaths and population declines in several countries. The disease originated in Africa and has spread all over the world, affecting the amphibian population. Signs of this disease include loss of muscle coordination and hyperkeratosis of the outer layer of the skin. If detected early, this condition can be treated topically by placing the affected animal in a tank with antifungal medication.

A strict diet of only crickets, coupled with inadequate UV lighting, can cause nutritional secondary hyperthyroidism and metabolic bone disease. This disease is characterized by spine and limb deformities. Mandibular bone degeneration (MBD) results in a condition commonly called *rubber jaw*. This condition can also be caused by an imbalance in the ratio of calcium to phosphorus, decreased calcium and increased phosphorus. MBD may be treated with several weeks of calcium gluconate, either by bath, injection, or oral medication. This is often supplemented with vitamin D3 to help with calcium assimilation. Gut-loading, feeding an enriched diet to the prey prior to feeding it to the amphibian, may help to avoid MBD. Amphibians and reptiles are both susceptible to rubber jaw. It may be stabilized with dietary correction, but the skeletal changes are not reversible. Many amphibians and reptiles die due to this preventable disorder, as they become unable to capture prey and feed.

Most bacterial infections are a result of poor husbandry. **Red leg** is caused by Aeromonas, bacteria that thrive in moist conditions. Signs of red leg include ulcers on the toes, edema, and skin necrosis. There is erythema on the ventral side, causing the diseased tissue to become red. The bacteria can become systemically invasive, eventually causing sepsis, seizures, and death. Treatment is rarely successful, as patients are usually not seen until advanced signs of the disease are noticed.

There is sometimes a generalized edema that can result from septicemia. Bacteria invade the lymph system and capillaries, causing body fluid to pool around internal organs. If the condition is recognized early on, daily soaks in *distilled* water may help to pull the fluid out and restore osmotic balance. The causative bacteria should be identified and treated with appropriate antibiotic therapy.

Corneal lipidosis has become an increasing problem in captive amphibians. It is seen more frequently in White's tree frogs. The cause is thought to be an incorrect diet or imbalance of fatty acids and fats in prey items fed in captivity. White, cloudy lesions form over the eye and cornea. These lesions start out superficially and can penetrate deeper into the cornea.

Rostral abrasions and lacerations can be caused by trauma, tank mates, prey bites, rough handling, and inappropriate substrate. The injured area can be rinsed with 0.9 percent NaCl, and silvadene cream may be prescribed to help prevent a bacterial or fungal infection. Dermatitis can be caused by poor water quality. High ammonia levels and extreme pH levels can predispose the amphibian to dermatitis.

Cloacal prolapse is commonly associated with stress, poor water quality, shipping, and overcrowding. The cloacal tissue needs to be kept moist with a water-soluble lubricant prior to seeking veterinary care.

Parasites

Amphibians can be infected with internal parasites, including trematodes (flukes), cestodes (tapeworms), and many species of nematodes (roundworms). They may also harbor intestinal protozoans, including coccidia and giardia. Many White's tree frogs have presented with small nodules on the rear limbs. With careful dissection of the nodule, each may be found to contain a small living tapeworm just under the skin. These may be carefully removed with the tip of a 25-gauge needle.

Parasitism varies from species to species. For example, a protozoan population may not be problematic in one species but devastating to another. Introducing wild-caught specimens into a collection could also introduce parasites to captive-bred specimens that have little or no ability to cope.

Clinical Procedures

Drawing blood from amphibians is complicated by restraint concerns and their extremely small, fragile blood vessels. Cardiac puncture can be performed with an insulin syringe if diagnostic blood work is required. This procedure is not without the risk of lacerating the skin with the needle and the possibility of causing severe cardiac damage. The lateral tail vein and ventral abdominal vein may be more easily accessed in salamanders. A light source can help visualize the abdominal vein and heart.

Subcutaneous and intramuscular injections both cause tissue trauma and possible necrosis. When injections are required, they are best delivered by intraperitoneal or intracoelomic (ICe) routes. Fluid therapy for an amphibian is provided by soaking the patient in dechlorinated water or dilute 0.9 percent NaCl with 50 percent water. In critical instances, isotonic fluids may be delivered ICe.

A medicated bath is an effective method for antibiotic delivery. Enrofloxacin (Baytril®) is an often used antibiotic for bacterial infections in amphibians. The drug is combined with 0.9 percent NaCl for therapeutic soaks. If water baths are not feasible, soaking a paper towel with the diluted antibiotic and placing the amphibian on top of the paper towel or wrapping it around the patient can also be effective. Topical medications are not effective in species that have a waxy coating on their epidermis, for example, the Waxy tree frog.

Offering nutritional support to an amphibian may involve removing the head of the cricket prior to placing it in the mouth. Some amphibians will need to have the mouth opened by a speculum. The following items can be used: business or credit card, a small piece of developed radiographs, or a rubber spatula. Some amphibians will open their mouths at a light touch of the food item.

In some cases, it may be necessary to tube feed the amphibian. A polypropylene urinary catheter (tom cat catheter), red rubber feeding tube, or small avian gavage tube can be used to tube feed. This method can be less stressful on the amphibian patient and the semisoft or liquid food is easier for them to digest.

The Amphibian Heart Is Similar to the Reptile Heart.
It is three chambered, with one ventricle and two atria. Blood flows through the right atrium and then directly to the lungs before returning, oxygenated, to the heart.

Anesthesia in amphibians is similar to fish anesthesia. The most common agent used is MS-222 (tricaine methanesulfonate), a powder that is dissolved in water. Concentration depends on the volume of water used. The patient is placed in the bath in a clear plastic container. A main component of clove oil is eugenol, which contains antiseptic and analgesic properties. Mixed in dechlorinated water, clove oil has been documented to be an effective anesthetic in leopard frogs and salamanders. The anesthetic agent is absorbed through the skin. Anesthesia is not immediate and it may take 20 minutes to induce a surgical plane of anesthesia. The patient, especially a nongilled amphibian, needs to be carefully observed to prevent drowning. A lid needs to be placed on the container as some patients will have an excitement phase and attempt to jump out. During induction, observable respirations will drop and may cease. Even if respiration ceases temporarily, there is still cardiac function. During induction, the abdomen reddens and many patients exhibit a *floating tilt*. During this phase, they should be supported and held horizontally with a double-gloved hand to prevent total submersion. When an appropriate plane of anesthesia is reached, the patient will be unable to right itself and the corneal reflex will be lost. Anesthesia can be maintained by trickling the induction solution over the body.

Anesthetized patients should be placed on towels presoaked with dechlorinated water. The amount of wetness needs to be carefully monitored to prevent the heat of the surgical lights from drying out the towels. They may be rewetted by using a 60 cc syringe, filled with warmed, dechlorinated water. The patient and the wetted towels can be placed on a large, rubber tub lid or on plastic sheeting to help contain the moisture. Anesthesia can be reversed by placing the patient in *distilled*, oxygenated water. The water may be allowed to flood over the skin in a recovery tray.

Local anesthesia, for example lidocaine, can be used on small lesions. Lidocaine can be injected or put on topically around the lesion to be surgically removed.

Righting reflex, palpebral reflex, and superficial pain response (pinch the skin over the foot) are ways to monitor the amphibian under anesthesia. It is recommended that prior to anesthetizing the patient, these reflexes should be evaluated to use as baseline reflexes.

Radiographs are easier to take if the patient is placed on a wet towel and lightly covered with another wet paper towel. The paper towel covering should be quickly removed before the taking the radiograph. Positioning for amphibian radiographs is usually dorsal/ventral, *as they sit*, and most will stay still for that short *snap* in time.

Summary

Immature amphibians have no resemblance to adults. All amphibians mature through a five-stage life cycle of metamorphosis, beginning as a jelly-like egg to aquatic larva with gills. Almost all mature to crawl from the water onto land and breathe air. Amphibians include frogs and toad, salamanders and newts, and the lesser known gymnophiona, the limbless amphibians. All amphibians must be handled with care, only when necessary, and never with bare hands. Many amphibians produce toxins and have a protective layer of mucous. Adults are carnivorous and eat only live food. Blood sampling and other clinical procedures is complicated because of restraint concerns and extremely small and fragile blood vessels. Anesthesia in amphibians is similar to fish in that the anesthetic agent is dissolved in water and the patient absorbs the solution through the skin.

*fast*FACTS

AMPHIBIANS

REPRODUCTION
- **Sexual maturity:** varies; some anurans are capable of reproduction at one year of age; salamanders species (spp.) are capable of reproduction at two years; some amphibians are not capable of reproducing until six years of age.
- Male amphibians mature faster than females.
- **Caecilians:** viviparous
- **Anurans:** oviparous (external fertilization in the water)
- **Salamanders:** oviparous

VITAL STATISTICS
- **Environmental temperatures:**
Tropical lowland species: 24–30°C (75–85°F)
Subtropical species: 21–27°C (70–80°F)
Tropical mountain species: 18–24°C (65–75°F)
- **Humidity:** 70–100%

ZOONOTIC POTENTIAL
Bacterial
- Salmonella
- Campylobacter
- Mycobacteria

Review Questions

1. How do frogs differ from toads?

2. Diagram the stages of amphibian metamorphosis.

3. Discuss the environmental concerns that arose with the discovery of a number of frogs that exhibited developmental abnormalities.

4. List some of the reasons that amphibians are disappearing.

5. Define POTZ and its importance in keeping an amphibian healthy.

6. Why should only dechlorinated water be provided for amphibians?

7. What are the two main causes of MBD?

8. What are the reasons for wearing latex gloves when handling an amphibian?

9. What causes red leg and what are the signs that indicate an amphibian has contracted this disease?

10. What could be the cause of small nodules on the legs of a White's tree frog?

Case Study I

History: A mature leopard frog of undetermined age has been brought in by the owner. The frog usually has a healthy appetite but has stopped eating. She also reports that the frog has sores on the hind feet, between the webbing.

Physical Examination: The veterinarian is able to examine the patient through the clear container and small amount of water the owner has brought the frog in, and he is able to see the that there are ulcers on the legs and toes with inflammation and edema of the hind limbs.

   a. What is the likely cause of this frog's condition?

   b. What is the recommended treatment for this condition?

Case Study II

History: An amphibian is presented to the clinic staff for identification. It was bought some months earlier as a "giant tadpole" but now it has developed an elongated body, four limbs, and a tail. It is approximately four inches long. It is definitely not the frog she was expecting, but she is still very interested in keeping it and learning more about what she has. The staff refer to field reference guides and determine that it is small semiaquatic salamander.

   a. What general information could be given to the client regarding the habitat she should provide for her salamander?

   b. What type of diet should be fed?

For Further Reference

Amphibian Folklore. http://www.livingunderworld.org/ (accessed May 2006).

Amphibian Medicine. (2009). *Journal of Exotic Pet Medicine* 18(1).

Girling, S. (2005). *Veterinary Nursing of Exotic Pets.* Oxford, England: Blackwell Publishing.

Grayson, K. (2006). *Regeneration in Urodeles.* http://www.biodavison.edu/ (accessed July 6, 2006).

http://www.worldwildlife.net (accessed March 17, 2014).

The IUCN Red List of Threatened Species. (2006). IUCN World Conservation Union Species Survival Commission. http://www.iucn.org (accessed June 20, 2006).

Johnson-Delany, K.A. (2005). *Exotic Companion Medicine Handbook for Veterinarians.* Lake Worth, TX: Zoological Education Network.

Lafortune, M., Mitchel, M., & Smith, J. (2014). Evaluation of Medetomidine, Clove Oil and Propofol for Anesthesia of Leopard Frogs. *Journal of Herpetological Medicine and Surgery* 11(4):13–18.

Lutz, J. (2002). *Survey Shows Few U.P. Frogs with Deformities.* http://www.wupcenter.mtu.edu/ (accessed February 12, 2002).

Mader, D. (2006). *Reptile Medicine and Surgery, 2nd Edition.* Philadelphia, PA: Saunders/Elsevier.

O'Malley, B. (2005). *Clinical Anatomy and Physiology of Exotic Pets.* Edinburgh, Scotland: Elsevier/Saunders.

Toads on Roads. http://www.froglife.org/ (accessed July 7, 2006).

Wright, K. (2006). *Common Bacterial and Fungal Diseases of Captive Amphibians.* Conference Notes, North America Veterinary Conference, Orlando, FL, January 7–11, 2006.

Wright, K. (January 7–11, 2006). *Fluid Therapy for Amphibians.* Conference Notes. Presented at the North America Veterinary Conference, Orlando, FL.

UNIT VI

SCORPIONS

OBJECTIVES

After completing the chapter, the student should be able to

- Describe the correct habitat for a scorpion.
- Provide appropriate client education to new scorpion owners.
- Identify anatomical organs of a scorpion.
- Demonstrate appropriate safe handling techniques when working with a scorpion.
- Describe the correct diet for a scorpion.

Introduction

Invertebrate animals do not have a spinal column, but rather an exoskeleton that gives the body form and support. They account for over 95 percent of the earth's animal species. There are many classes of invertebrates and they inhabit every area of the planet, including the seas, with the exception of the arctic and subarctic regions. Their species include some of the most ancient life forms. Many species living today have remained unchanged for more than 400 million years.

The largest phylum (division) in the animal kingdom are the arthropods, animals with hard, segmented bodies and jointed legs. They have an exoskeleton, which is molted and replaced throughout its lifetime. The exoskeleton is made of chitin, the outer shell that gives the body support and protection. There are over half a million recognized species, divided into 12 classes. Included within this group are crustaceans, arachnids, and insects.

Crustaceans are aquatic species and include crabs, lobsters, and shrimp. Arachnids include spiders (tarantulas), scorpions, ticks, and mites. All arachnids have four pairs of legs. The greatest numbers of species are the insects. Insects are different from arachnids in that they have only three pairs of legs.

People have either a fascination for or a fear of members of this group. So much so, for example, that *arachnophobia*, the *fear of spiders*, is a term familiar to many. With some understanding of what they are, and what they are not, arthropods can be interesting specimens to keep. While they are not usually considered *pets*, some species of invertebrates have become very popular and are readily available from private breeders and pet stores. Those commonly available are the Emperor scorpion, tarantula species, and Hermit crabs.

Scorpions inhabit many ecosystems. They live in arid deserts and the tropics, grasslands, and forests. The emperor scorpion is native to West Africa, a semi-tropical region where it digs burrows in the forest floor. Emperors are large and impressive, reaching a potential adult length of eight inches. They are the largest living species of scorpions (**FIGURE 17-1**).

FIGURE 17-1 An Emperor scorpion.

Common to all scorpions is venom that is delivered by a tail sting. The tail part ends at the **telson**, where the stinger is located. The stinger, or **aculeus**, is sharp, hollow, and connected to the venom gland. All scorpions are carnivores and use the stinger to stun or paralyze their prey. They are visual hunters and capable of moving very quickly. They have four pairs of running legs and capture prey with the paired claws (pincers). Once secured, the scorpion arcs the tail and injects the prey with venom from the telson. Once subdued, the claws move the prey item to the mouthparts where it is macerated into tiny pieces, liquefied, and ingested. Female scorpions also capture and macerate food items for the young traveling on her **cephalothorax**, where the head is joined to the thorax.

Emperor scorpions are not known for being aggressive or for the potency of their venom, which has probably contributed to their popularity as captive specimens. Regardless, all scorpions must be handled with care. There is always the potential of an anaphylactic reaction to the venom of any scorpion (**FIGURE 17-2**).

If housing more than one, there may be a chance that the scorpions will reproduce. Sex determination is not the easiest purely for reasons of restraint in order to examine them closely. If the scorpion is placed in a clear-bottomed container, it may be possible to see the **pectines** on the underside of the thorax. Pectines are feathery-looking appendages just behind each pair of legs. Comparatively,

FIGURE 17-2 Anatomical
drawing of a scorpion.

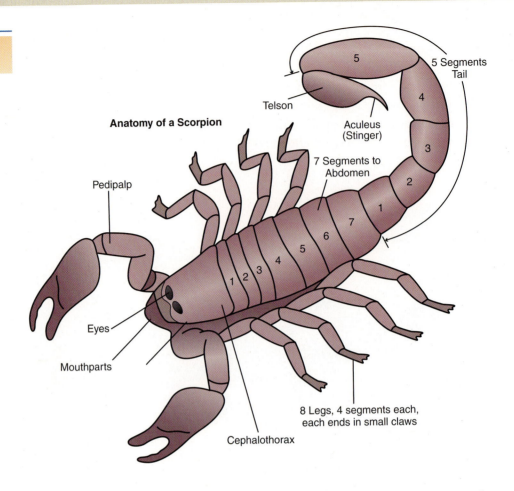

FIGURE 17-2 Anatomical drawing of a scorpion.

male pectines are larger. In mature emperor scorpions, those that no longer molt, males are approximately one-third to half the size of females.

Scorpion reproduction occurs when the male deposits a **spermatophore** (sperm packet) on the ground. The male manipulates the female so that she is directly over the spermatophore. The female places the packet into her genital opening, located between the fourth pair of legs.

All scorpions give birth to live young. The gestation period is long and can vary from *several months* to a year and a half. There may be as many as 50 to 60 in each clutch. The young are born white and have small black eye dots (**FIGURE 17-3**). They crawl up onto the mother's back, where they remain at least until their first molt. The degree of maternal care varies. An incidental report of observed behavior suggested that a female scorpion appeared to rip up and predigest food to provide for her young. It may also be that the female was not attempting to provide food for the young, but just did not consume her own prey and abandoned it for some reason. This same female, a few days later, cannibalized the entire clutch. Cannibalism in captivity is known to occur frequently, but the cause of this behavior is not clear. It may be that the female is disturbed too often or that there is insufficient food.

Release of nonindigenous species is illegal. Before any deliberate attempt is made to breed scorpions in captivity, there must be a clearly defined goal and market for the potential multitude of young scorpions. Permits may be required to establish a captive breeding program.

FIGURE 17-3 Female scorpion and her brood. This female Bark scorpion and her brood were found under a tarp in Baja, Mexico. Upon capture, the mother was fiercely protective of her young and repeatedly attempted to sting the container. She and her young were released although it is doubtful that all of the young survived because of the stress of captivity, however brief.

Behavior

Scorpions are nocturnal and hunt at night. Most scorpions are also prey for other animals, such as birds, lizards, and mammals. Because of this, they hide under rocks and bushes while they lie in wait to ambush their prey.

Housing

A 10-gallon aquarium with a secure screen lid is adequate, as scorpions do not need a great amount of space. Much of their time is spent in lairs or burrows where they wait for prey. There should be at least four inches of substrate that can retain some moisture. Potting soil, mulch, and orchid bark are all suitable. Half-round cork bark, small clay flower pots, and PVC pipe sections can be used as places for a lair. Shallow water dishes should be provided. If the habitat is large enough and multiple lairs are provided, scorpions may be kept with others of their same species.

Lighting is not required, as scorpions are nocturnal. The seasonal photoperiods of 8 to 12 hours provide enough natural light. Many people use a black light to observe their scorpions at night. Under black light, scorpions glow green.

Emperor scorpions require a relatively high humidity that mimics their natural environment. The habitat should be kept at around 65 to 75 percent humidity. Humidity can be assisted by the water bowls provided for the scorpion. Misting the enclosure several times daily will also increase the level of humidity. If condensation forms on the sides of the tank, it is an indication that the humidity level is too high. A humidity gauge, hygrometer, is a useful addition in order to monitor humidity levels. Scorpions also need supplemental heating, provided by under-the-tank heating pads. POTZ, for most species, is 80°F to 85°F.

Diet

Although scorpions are carnivores, they are liquid feeders. Once the prey has been captured and subdued, it is transferred to the mouthparts by the pedipalps, large claws that resemble those of a lobster. The chelicerae, or mouthparts, rip the food to pieces, turning it to pulp while the digestive system secretes enzymes that liquefy the unrecognizable mass. The liquid is then sucked up into the gut.

Captive scorpions are usually fed crickets. If this is the only food source, the crickets should be dusted with a mineral supplement or gut loaded. Large scorpions should also be given a pinkie mouse, alternating the pinkie and the crickets. Adult scorpions should be fed twice a week and juveniles every other day.

Handling and Restraint

Even though emperor scorpions are reportedly hard to provoke, it is not recommended to pick them up or have them sit in a hand. When the scorpion needs to be moved for habitat cleaning, it should be herded into a cup with a secure lid and set safely aside until being returned to the habit and released from the cup. Most scorpion stings are defensive and the amount of venom delivered can be controlled. Should a handler be stung, one thing that has helped reduce the pain is soaking the site in very hot water and adding a powdered astringent such as Domboro®. The astringent is thought to reduce the pain by drawing out the venom. Other than the sharp pain of the sting, it is difficult to locate the exact point of venom injection as the puncture mark is very small. As with all animal bites or stings, the area should be cleaned thoroughly with hot, soapy water.

Medical Concerns

There is very little information available regarding disease or health problems with scorpions. Providing that the habit is stable and there is sufficient food and water, scorpions have an average life span of seven to eight years although some have reportedly lived for 10 to 15 years in captivity. A healthy, immature scorpion will completely molt several times until it reaches maturity, which can vary, depending on food source, from one to six years,

Summary

Scorpions are invertebrates and, as such, do not have a spinal column but instead an exoskeleton that gives the body form and support. All scorpions hunt at night, capturing their prey and injecting venom that is delivered by a tail sting. All species of scorpions give birth to live young. The neonates crawl up onto the mother's back where they will remain until at least the first molt or period of shedding and replacing the exoskeleton as they mature. All scorpions should be handled carefully and should be herded into a cup with a secure lid. It is not recommended to pick them up or expect them to sit quietly in a hand.

*fast*FACTS

SCORPIONS

LIFE SPAN
- 7–8 years

REPRODUCTION
- Viviparous and cannibalistic
- Clutches of 50–60 common

VITAL STATISTICS
- **POTZ:** 80–85°F
- **Humidity:** 65–75%

ZOONOTIC POTENTIAL
- None reported

Review Questions

1. How is venom delivered by a scorpion?
2. What type of reaction can occur after a sting from a scorpion?
3. What is the purpose of the following mouthparts of a scorpion?
   a. Pedipalps
   b. Chelicerae
4. How is sex determined in a scorpion?
5. What does a newly born scorpion look like?
6. What are common causes for a female scorpion to cannibalize her young?
7. What is the best way to move a scorpion?
8. What is the common diet for scorpions?
9. Where do most scorpions spend their time during the day?
10. What is the name for the part of the tail that delivers the sting?

Case Study I

History: A client has brought in a large black scorpion in a clear plastic container. His complaint is that the scorpion does not do anything all day except hide out in a log. He wonders if it is sick.

Physical Examination: The scorpion appears to be active and healthy; it moves around the container with the tail curled up and over the back.

a. What advice could be given to the owner about scorpion behavior and the best way to observe the scorpion?

b. What does the tail position indicate regarding the scorpion's attitude?

Case Study II

History (Situation): A client telephones the clinic and needs advice. Her 12-year-old son wants to "get a bunch of scorpions." She has several doubts, one being that they are highly poisonous and very aggressive.

Physical Examination: None.

    a. What helpful advice would you offer this client and why?

    b. What specifically is a concern regarding "a bunch of scorpions"?

For Further Reference

Calzoo's Reference Sheet. (2006). *Emperor Scorpions*. http://www.calzoo.com/html/emperorscorpions.html (accessed February 5, 2006).

http://arizonesis.org (accessed January 25, 2014).

http://desertmuseum.org (accessed January 25, 2014).

http://www.enchantedlearning.com (accessed February 5, 2006).

Hunziker, R. (n.d.). The Emperor Scorpion, an Animal That Time Forgot. *Reptile Hobbyist*.

Wake, G., & Judah, V. (2006). *Observation and Discussion of Abandoned Emperor Scorpion in Rescue Facility*. Jordan Applied Technology Center, Utah.

TARANTULAS

OBJECTIVES

After completing the chapter, the student should be able to

- Describe the correct habitat for a tarantula.
- Provide appropriate client education to a new tarantula owner.
- Identify anatomical features of a tarantula.
- Demonstrate appropriate handling techniques when working with a tarantula.
- Describe the appropriate diet for a tarantula.

KEY TERMS

cephalothorax
chelicerae
booklungs
setae
urticating
spiderlings
epigastric furrow
spermathecae
dry bite

Introduction

Tarantulas are arachnids and belong to the same family as spiders. They inhabit all temperate regions of the world, but the greatest number of species is found in Central and South America and in the desert southwest of the United States. There are more than 800 recognized species. All have fangs and venom, are covered in *hair*, and possess spinnerets that produce silk. Tarantulas are the largest of all arachnids: the largest is the Goliath bird-eating tarantula (*Theraphosa blondi*), which has a leg span that can exceed 12 inches (**FIGURE 18-1**).

> **Goliath Tarantulas Are Native to South America.** They live in deep burrows in the ground and do not feed on birds. Common prey items include adult rats and mice, lizards, snakes, and any other prey item smaller than itself. Wild-caught specimens are offered as pets and are not recommended due to their large size and aggressive nature.

Tarantulas have two anatomical body divisions: the cephalothorax, where the head is joined to the thorax, and the abdomen. They have eight eyes clumped together at the front of the head. Other sensory organs of the cephalothorax are the pedipalps, which are used in feeding and reproduction. Two fangs fold back, under the front of the cephalothorax. The tip of each fang is hollow and used to inject the venom that either kills or paralyzes the prey. Tarantula venom, depending

on the species, is either a neurotoxin that affects nerves or a cytotoxin that attacks cells. Tarantulas can also give a *dry-bite* where no venom is released. The **chelicerae**, or teeth, are in a row above the fangs.

Tarantulas have eight legs with two small claws on the end of each leg. Tarantulas have the ability to release their legs in an attempt to escape or if a leg becomes trapped. This is called autonomy, and is similar to tail release in reptiles. Missing legs are replaced with the next molt.

Tarantulas have an *open* circulatory system. Instead of having a blood and lymph system, they have hemolymph that fills the entire body cavity. Hemolymph is transported back to the heart by seeping through small internal channels. The heart is more like an elongated tube that lies along the length of the abdomen. Tarantula hemolymph is pale blue.

Breathing, or gas exchange, takes place in folded, *accordion-pleated* tissue called **booklungs**. Air enters and exits through the ventral side of the abdomen directly into the booklungs, where respiration occurs.

Tarantulas have internal organs, including silk-producing spinnerets located in the dorsal caudal area of the abdomen. Depending on the species, there may be four to six spinnerets. Silk is used to line their burrows and ensnare prey. Female tarantulas lay their eggs in a protective, dense silk sack, and males use a silk pad to transfer sperm to the pedipalps prior to fertilization.

Tarantulas are referred to by many as *big, hairy spiders*. Their bodies and legs are completely covered in **setae** (bristle-like structures), and many species have **urticating** hairs on the dorsum. These hairs are sharp and have barbed tips. Used for defense, tarantulas flick these hairs off with their rear legs. When they penetrate skin, they cause an irritating condition, urticaria, which resembles hives. A definite bald patch on the dorsal surface of a tarantula is an indication that these hairs have been flicked off. They are not replaced until the next molt.

Tarantulas stop eating just prior to a molt. Adults normally molt once a year, and growing **spiderlings** (immature tarantulas) molt an average of four times a year. The entire exoskeleton is cast off. During this time, the tarantula is very delicate and should not be disturbed. If they are disturbed, it can result in death. Molting tarantulas lie on their backs with their legs up. The exoskeleton splits on the dorsum, allowing the tarantula to withdraw the body first, followed by the legs. This process may take several hours. When a molt is complete, it will appear as if there are two tarantulas. The castoff exoskeleton is an exact cast of the living tarantula. It may take three or four days for the new exoskeleton to harden completely (**FIGURE 18-2**).

Many enthusiasts have been successful in breeding tarantulas. Sexual maturity varies with the species, and some females may not be mature until they are five years old. Males generally mature much earlier. Sexing immature tarantulas is very difficult, and young pairs are often acquired by chance.

Mature specimens can be sexed from castoff molts. Careful examination of the ventral abdomen shows the **epigastric furrow**, the female genital opening. Mature tarantulas have subtle visual differences: males develop spurs on the first pair of legs and the pedipalps are more swollen in appearance than those of females. Females have much larger abdomens. Even if breeding is not a consideration, purchasing a male could be a very short-term investment. Males have a life span of only one to two years. Females of many species may live as long as 30 years.

Male tarantulas deposit sperm on a silk pad and transfer the packet to their pedipalps, where it is stored. From there, a needle-like injector transfers the sperm to the epigastric furrow of the female. Sperm is stored by the female in pouches called **spermathecae** until eggs are produced. If the female does not produce eggs from a mating prior to molting, the sperm will be lost, as the spermathecae are discarded with the exoskeleton.

Commonly kept species include the Mexican red-knee (also known as orange-knee), the Chilean rose-hair, and various species from the genus *Avicularia*, commonly called *avics*, all of which may be sold as *pink toes*. By the time *Avicularia* arrive in the retail market, they are not identified by species. *Avicularia* species are arboreal, while most of the other species are ground dwelling.

Almost every pet store offers two or three varieties, but it is rare to have them identified correctly in a nonspecialist store. Before purchasing a tarantula, the prospective owner should do a considerable amount of research, recognize the more common species, and be able to differentiate males from females. It would be of great value to join a club or society with experienced members who are more than willing to share their knowledge.

The venom of a tarantula (in most cases) is nontoxic to humans. Both males and females are capable of biting and injecting venom. The bite can be painful and the surrounding tissue may itch and have a rash due to the urticating hairs flicked off and imbedded in the skin. If the venom is injected into a human, there may be slight swelling with some numbness and itching. A **dry bite** is one in which the tarantula does not inject venom.

FIGURE 18-2 The molted exoskeleton of a healthy tarantula.

Behavior

Tarantulas spend most of their time hidden in burrows or under bushes, rock ledges, and logs. They are most active in the late afternoon to early evening. They are crepuscular predators, killing the majority of their prey at dusk.

> **No One Should *Ever* Place a Tarantula on Another Person as a Prank.** Many people have a deep-seated fear of spiders. Tarantulas have been killed when the *victim* instinctively jumps away and the tarantula crashes to the floor. Unless startled or threatened, tarantulas rarely bite. When they do, it can be very painful and there is always the risk of a severe allergic reaction to the venom.

Housing

Tarantulas are usually kept in glass aquaria with secured screen lids. Whether ground dwelling or arboreal, they are adept climbers. All enclosures should be longer than they are tall to prevent a fall that could cause autonomy or abdominal rupture. Even tree-dwelling species need only an assortment of branches on the floor to use as anchor sites for webs.

The substrate can be a mixture of bark, peat moss, and potting soil. Adding vermiculite helps to retain moisture. All tarantulas should be provided with a den area, a place to retreat. Half-round cork logs, small flower pots, or one of the many suitable *huts* available in pet stores will all be used by tarantulas.

Humidity and temperature requirements vary with different species. Increased humidity will assist with the molt, but humidity that is too high will encourage mold and fungal growth. For most species, POTZ is between 72°F and 82°F. Tarantulas should be housed singly or cannibalism will occur. A shallow water dish should be provided. This can be as simple as a jar lid turned upside down. There are *tarantula sponges* (small pieces of natural sea sponge) that are soaked in dechlorinated water and placed in the water dish. The use of these is controversial and many feel that the sponges encourage bacterial growth. Sponges are used to increase humidity, but also so the owner will not have to change the water as often. All water dishes, with or without sponges, should be cleaned and replenished daily.

Extra lighting is not required, as tarantulas are crepuscular/nocturnal. Indirect exposure to natural sunlight is sufficient. Nighttime activities can be observed by using a black or red light.

Diet

All tarantulas are carnivores and should be offered live prey. Most captive specimens are fed a staple diet of crickets. Meal worms, wax worms, and grasshoppers are also readily taken. Depending on the size of the tarantula, pinkie mice should also be fed occasionally. Adults should be offered food every other day.

Spiderlings are fed daily. Tarantulas are liquid feeders. Much like the scorpions, tarantulas rip prey into shreds and inject it with digestive enzymes. The liquid is then sucked into the stomach. Live insects should not be left in the habitat, as they will chew on the legs and abdomen of a resting tarantula.

Handling

As a general rule, tarantulas are not receptive to being handled and are best observed in the habitat. When it is necessary to move a tarantula, a large, soft aquarium net (one large enough to encompass fully extended legs) can be used. Protective eyewear is strongly recommended to prevent urticarial hairs from penetrating the eyes if they are flicked from the back of an agitated tarantula.

The tarantula should be allowed to crawl out of the net and not be shaken away from it. Alternatively, it can be herded into another, smaller container while still in the enclosure. Many experienced keepers allow their tarantulas to crawl into a scooped hand. They should not be poked or prodded with any hard object. To move a tarantula from an area within the habit, a long-handled, soft-bristled paint brush gently stroked down the back will encourage it to move away. Whatever method is used, the safety of the tarantula should be uppermost. Tarantulas are easily frightened and are very quick to react. The abdomen can split and rupture if the tarantula falls even a short distance. Because of the open circulatory system, a tarantula may quickly bleed to death.

Medical Concerns

Sick and dying tarantulas sit hunched upright with their legs tucked underneath them. They are reluctant to move and become anorectic. Many losses are due to incorrect housing, lack of sufficient food and water, or excessive attempts to handle.

Nondigestive Disorders

The most common problems that affect tarantulas relate to trauma. Normal, willful loss (autonomy) of a leg includes the entire limb. If a tarantula becomes caught in a habitat and loses only a part of the leg, the remainder should be removed. If the partial leg is left, the tarantula could bleed out from the remaining stump. The stump should be grasped firmly with a pair of forceps and pulled straight up in one quick movement. Surgical glue or any common superglue should be applied to the stump of a forcibly removed leg. Abdominal splits, if addressed before there is too much hemolymph loss, may also be sealed in this manner.

Parasites

Tarantulas could be infested with mites, nematodes, and roundworms, all of which may be likely in wild-caught specimens. Particularly insidious is the fly larva of *Acroceridae* species. They are commonly called *small-headed flies*. Female flies lay their eggs near resting spiders. When the larva hatches they crawl up the tarantula's legs and enter the booklungs. The larvae may stay dormant for many years before beginning to feed on the tarantula's internal organs.

Nematodes have become a fairly common parasite in captive-bred tarantulas. Enclosures with high humidity can predispose the tarantula to this very small species of roundworm. Anorexia, change in posture (*standing on the tips of their toes*), and behavior changes occur, which may be indicative of a parasite infestation. In the later stages, when the infestation has become overwhelming, there may be a thick white oral discharge. Death follows the onset of clinical signs. To date there is no recommended treatment.

Summary

Tarantulas are arachnids, related to spiders. All have fangs and venom and are covered in hair. Like other spiders, tarantulas have spinnerets and produce silk. They have eight eyes clumped together on the front of the head. The tip of each fang is hollow and is used to inject the venom that either kills or paralyzes the prey. Tarantulas may also deliver a defensive *dry bite* where no venom is released. Tarantulas have an open circulatory system. Instead of blood and lymph systems, they have hemolymph that fills the entire body cavity. As a general rule, tarantulas are not receptive to being handled. They are easily frightened and very quick to react. The abdomen can split and rupture if the tarantula falls even a short distance. The most common problems faced by tarantulas are trauma related.

*fast*FACTS

TARANTULAS

LIFE SPAN
- **Females:** 15–30 years, depending on species
- **Males:** 1–2 years or less, depending on species

REPRODUCTION
- **Sexual maturity:** females: average five years; males: one year

VITAL STATISTICS
- Molting tarantulas lay on their backs, legs up; sick tarantulas remain upright, legs folded underneath.

- Adults molt yearly, juveniles may molt four times a year

WEIGHT
- **Average:** 15–16 g

ZOONOTIC POTENTIAL
- None reported

Review Questions

1. What two types of venom do tarantulas possess?

2. What is an *open* circulatory system?

3. Tarantulas do not have blood, they have ___.

4. What are the purposes of web production?

5. What are urticating hairs?

6. What is a *dry bite?*

7. What parasites commonly infest tarantulas?

8. Why is it important to know the sexual differences?

9. What happens to the spermathecae if a female does not produce eggs?

10. What is meant by *liquid feeder?*

Case Study I

History: A client brings his tarantula into the office for an examination. His chief concern is that the tarantula seems to be agitated and stands up on its toes and it has not been eating.

Physical Examination: The patient has backed itself into a corner of the container and is reluctant to move. However, the veterinarian can see that there is a thick white discharge near the mouth parts.

a. Based on the description given by the owner regarding the posture of the tarantula and the visible white discharge, what will be the likely diagnosis?

b. What is the prognosis for this patient?

Case Study II

History: A client telephones concerned that her tarantula is dead. She says it is up-side down on its back and won't move. She refuses to bring in a "dead tarantula" but demands some information.

Physical Examination: none

a. What advice can be offered to the owner regarding the status of the tarantula?

b. Describe the differences in appearance between a tarantula that is ill/dying and one that is molting.

For Further Reference

ATS Tarantula Care Sheet. American Tarantula Society. http://www.atshq.org (accessed July 11, 2006).

Breene, R.G., III. *Spider Digestion and Food Storage.* American Tarantula Society. http://www.atshq.org (accessed December 21, 2005).

Flank, L., Jr. (1998). *The Tarantula.* New York: Howell Book House.

http://www.enchantedlearning.com (accessed January 7, 2006).

Pizzi, R. (2009). Parasites of Tarantulas. *Journal of Exotic Pet Medicine* 18(4):283–288.

Walls, J.G. (1995). Red Knees, Legs and Rumps: An Introduction to Brachypelma Tarantulas. *Reptile Hobbyist* 1(4):48–55.

HERMIT CRABS

OBJECTIVES

After completing the chapter, the student should be able to

- Describe the correct habitat for a hermit crab.
- Provide appropriate client education to new hermit crab owners.
- Identify anatomical features of a hermit crab.
- Demonstrate appropriate handling techniques when working with a hermit crab.
- Provide an appropriate diet for a hermit crab.

KEY TERMS

decapod
cheliped
asymmetrical
stridulation

Introduction

Hermit crabs have been kept for many years and are not new to the pet market, but now there is much more knowledge about them and their care in captivity.

Hermit crabs are **decapod** (10-legged) crustaceans and belong to the genus *Agoura*, or *Anomala*. Because hermits are decapods and do not live within their own shells, they are not considered to be *true crabs*.

All species of hermit crabs begin life in the ocean and must return to the ocean to breed and lay their eggs. The larval crabs mature in the shallow waters of the tide line. When they complete metamorphosis, they crawl out on to land in search of a shell.

Most species of hermit crabs remain on the ocean floor, but land hermits may be found along all the shore lines of the temperate seas of the world. There are many species of land hermits (also called *tree crabs*) that are harvested for the pet trade. There are approximately 500 different species of hermits worldwide. There may be several different species in the same shipment, but all are commonly called hermit crabs.

Hermit crabs are invertebrates and have an exoskeleton, but it does not provide protection for their soft bodies. They do not produce their own shell but occupy abandoned shells of other marine invertebrates, with a preference for whelk (snail) shells. As the hermit grows, it needs to find a larger shell to allow for the larger body size (**FIGURE 19-1**).

When a hermit crab is tucked into a shell, it is fully protected by folding a **cheliped** completely across the opening. Chelipeds are clawed front legs. They

FIGURE 19-1 A large but unidentified species of Hermit crab.

are **asymmetrical**, with one always being much larger than the other. These are used for defense, blocking the entrance to the shell, and transferring food to their mouths. These large claws are attached to the cephalothorax (the carapace), which is the hard external part of the crab. The carapace and claws are made of chitin. Behind the chelipeds are two pairs of walking legs. The mouthparts and gills are dorsal to the walking legs. Hermit crabs have two body divisions, the cephalothorax (joined head and thorax) and the softer abdomen, which is coiled within the shell. The abdomen contains all the internal and reproductive organs. Within the shell are small pairs of fourth and fifth legs. These legs help hold onto the inside of the shell and are also used to clean the gills (**FIGURE 19-2**).

Gills are used for respiration even in specimens found inland. In addition, the ventral surface of the abdomen contains numerous sinus pockets that function similar to gills but *breathe* atmospheric air. Gills extract oxygen from water. This adaptation allows them to travel away from the ocean but not away from a damp environment.

On the head are two pairs of different-sized antennae. In a healthy crab these sensory organs are constantly in motion. The antennae, or *feelers*, are sensitive to motion, vibration, and potential food sources. Antennae also have specially developed filaments that detect smell. Hermit crabs have two eyes on the ends of mobile stalks, but their vision is poor. They rely more on the antennae for sensory input.

It is very difficult to determine the exact species of hermit crabs offered for sale. Size can be confusing and does not indicate age. A small crab is not necessarily a younger one. Some species never exceed three or four inches, while others, sold as *giants* or *jumbos*, may be five or six inches and are probably a different species that is larger by nature. They are generally sorted and priced by size.

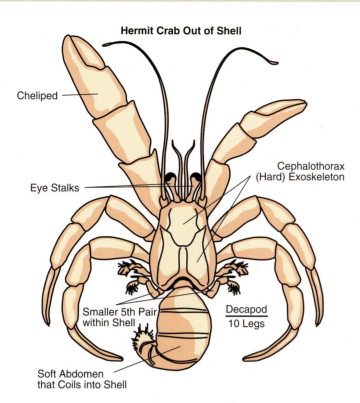

Hermit Crab Out of Shell

Cheliped

Eye Stalks

Cephalothorax
(Hard) Exoskeleton

Smaller 5th Pair
within Shell

Decapod
10 Legs

Soft Abdomen
that Coils into Shell

FIGURE 19-2 The anatomical features of a Hermit crab.

Because of their special reproductive requirements, all hermit crabs are wild caught. It is very difficult or impossible to determine the age of wild-caught hermit crabs. The natural life span of hermit crabs has not been determined. Some are known to have lived in captivity for over 30 years, while others die within a year or two. This may be due to species differentiation, the age when caught, stress and handling during shipment, and inadequate care, especially a lack of moisture.

Hermit crabs not only exchange and trade shells with each other but also molt their exoskeleton. Molting allows the crab to grow. When the crab emerges from a molt, it will seek out a larger shell. Alternatively, some crabs will move into larger shells just prior to molting. Most crabs burrow into the substrate for protection while molting. At this time, they are very susceptible to predation and trauma. Hermit crabs are compatible, often traveling and migrating in large numbers, but cage compatibility can be a challenge when molting due to the potential of cannibalism. Many experienced keepers recommend that molting crabs should be kept in separate habitats for their own protection.

When a hermit crab is preparing to molt, there are behavioral changes. They consume more food and water and digging activity increases. Once in-bedded, the crab should be left undisturbed. The molt may take from several days to a month or more to complete. The molted exoskeleton is usually eaten by the crab.

Behavior

Hermit crabs can be very noisy at night during their exploring and foraging activities. They also make stridulation noises that may sound like the chirps of a cricket. Stridulation is the creation of sound caused by body parts that are rubbed together.

Hermit crabs are gentle by nature, but they can use their one large claw to pinch predators or humans. If a handler is pinched, immersing the hermit crab under water for a brief moment will usually cause the crab to release its grip. Alternately, place the crab on a solid surface and remain still. In all probability, the crab will let go and move away.

Housing

Hermit crabs are often sold with the recommendation to keep them in a tall screened enclosure called a *crabbitat*. All hermit crabs are adept climbers, and the screen or wire provides an excellent gripping surface. However, this type of housing is not well suited to the safety and health of the crab. Many crabs have fallen from the top, losing legs and damaging shells. In an open, screened environment, it is very difficult to maintain adequate levels of humidity and surface moisture.

A ten-gallon or larger aquarium with a secured screen top provides enough space and adequate ventilation yet still has the ability to retain moisture. A substrate of sand, gravel, or forest floor mulch that is kept moist, but not soaked, will keep the humidity higher and the environment damp enough to hydrate the gills, Deep sand is a good, natural substrate because it allows the hermit crab to bury itself and dig tunnels. If gravel is used, the pebbles should be smooth to prevent abrasions to the exoskeleton. Colored sand or gravel is safe as long as it has not been treated with a water-soluble dye. Ground nutshell should not be used as bedding because it clumps when wet and can quickly become moldy. A shallow water bowl should also be provided. This should be filled with dechlorinated water. Natural sea sponges are available that can be soaked in dechlorinated water and placed in the dish to help retain moisture and humidity. Water that is used to keep the substrate moist should also be dechlorinated (**FIGURE 19-3**).

FIGURE 19-3 Example of a Hermit crab enclosure.

Many attractive tanks are set up to resemble a shore line, with multiple levels of substrates that slope down to a small beach area. A current trend in keeping hermit crabs healthy is to provide both freshwater and sea water. Bottled sea water is available in many pet stores specifically for this purpose. Salt water is not the same as *salted* water. Many hermit crabs have been lost when table salt has been added to the water supply in a well-intended attempt to mimic the salinity of the sea. Products used for maintaining salt water aquaria are also used in hermit crab habitats. Specific gravity needs to be measured with a hydrometer, to determine the concentration of synthetic sea water.

Providing multiple levels with branches, smooth beach pebbles, and other natural furnishings will give the crabs exercise. Hermit crabs are normally very active and will explore every nook and cranny of their habitat. They also seek out places to hide, and that should be considered when setting up the tank (**FIGURE 19-4**).

FIGURE 19-4 A variety of empty shells needs to be provided for Hermit crabs.

Humidity should be maintained between 70 and 90 percent. Additional heating is not required, as hermit crabs will thrive in a temperature range between 75°F and 85°F. Special lighting is not required as hermit crabs are nocturnal. The habitat should be exposed to indirect light, but not placed directly in front of a window. Direct sunlight can dramatically increase the temperature inside the aquarium.

Diet

Hermit crabs are omnivores and scavengers. They will eat anything, both plants and animals. Commercial hermit crab diets are readily available. These are nutritionally balanced and easy to feed. Pellets should be supplemented with fresh, dark leafy greens and vegetables. Pellets that become damp should be removed to prevent the growth of mold, and any uneaten fresh foods should be removed the next morning. If fed a variety of foods and a commercial diet, hermit crabs do not need vitamin or mineral supplements. What could be added is a cuttlebone. Cuttlebones are the internal calcified structure of cuttlefish (related to squid). They provide a natural source of calcium and trace amounts of iodine. Cuttlebones are readily available from pet stores and may be found with the bird supplies.

Handling and Restraint

Hermit crabs should be held securely from the back of the shell to avoid a painful pinch from the claws. This is not a sign of aggression but simply a way of hanging on. This pinch can be very strong and the crab will be reluctant to let go. If placed back into the habitat or on a solid surface it is familiar with, it will usually release its hold and move off.

Medical Concerns

The cause of most hermit crab death is neglect and poor husbandry. An inappropriate cage setup without adequate water and substrate moisture will quickly dehydrate the crab. People often underestimate the amount of food hermit crabs consume. Carefully monitoring the amounts given and the amounts consumed will ensure that the crabs are getting enough nourishment.

A sick hermit crab may show signs of lethargy, color change, behavior change, and decreased appetite, and there may be a discharge from the leg joints.

Parasites

Some hermit crabs can be infected with mites. These mites do not seem to cause problems to the hermit crab and are left untreated. Mites can sometimes be seen crawling all over the hermit crab.

Clinical Procedures

Common veterinary procedures performed on larger hermit crabs include amputation of injured limbs, cleaning of injuries inflicted from other hermit crabs, and stopping any bleeding. Severely dehydrated crabs may be sprayed with sterile saline to help rehydration.

Summary

Hermit crabs are decapod (10-legged) crustaceans. They do not produce their own shells but occupy abandoned shells of other marine invertebrates and must be provided with a variety of different-sized empty shells. Hermit crabs have two very large asymmetrical clawed front legs (chelipeds) that are used for defense, blocking the entrance to the shell, and transferring food to the mouth parts. Hermit crabs are gentle by nature but can be very noisy at night as they forage for food. The also make stridulation noises, caused by body parts being rubbed together. Hermit crabs are omnivores and scavengers. Captive losses are often attributed to an insufficient diet both in variety and in the amount of foodstuff provided. Because of their reproductive requirements, all hermit crabs available in the pet trade are wild caught.

fastFACTS

HERMIT CRABS

LIFE SPAN
- Unknown, but may live in captivity for 10–30 years

REPRODUCTION
- Hermit crabs are unable to reproduce on land or in freshwater. Hermit crabs must lay their eggs in the ocean.

VITAL STATISTICS
- Environment temperature: 75–85°F
- Humidity: 70–100%

ZOONOTIC POTENTIAL
- None reported

Review Questions

1. What does decapod mean?
2. Explain how hermit crabs are able to survive away from the shore line.
3. List three functions of the antennae.
4. List three functions of the cheliped.
5. How does the behavior of a hermit crab change prior to molting?
6. What is a hydrometer?
7. Why is it important to supply an assortment of empty shells in the habitat?
8. What is stridulation?
9. With regard to temperature and humidity, what is the POTZ range for hermit crabs?
10. List two advantages to adding a cuttlebone to the habitat.

Case Study I

History: A client telephones with a few questions regarding the three hermit crabs that she won in a carnival game. She reports that they are very inactive and won't eat the pellets or crickets she purchased at the pet store.

Physical Examination: The client is reluctant to bring them in for an examination and a diagnosis should never be offered over the telephone; however, with this history:

a. What further questions would you ask regarding their habitat and diet to help guide the new owner in caring for her hermit crabs?

b. Why would advice regarding the diet be especially important?

Questioned further, the client reports that she has made a "crab cave" from a cardboard box.

Case Study II

History: A clients brings a small plastic habitat into the clinic. It contains two hermit crabs. His complaint is that they will not breed although they have been together for six months and he has tried everything from placing them in a shallow fish tank to adding boiled eggs and dead goldfish to supplement the diet. He feeds them an assortment of fresh greens daily and reports that they eat everything offered.

Physical Examination: The crabs have been well handled, and when placed on the surface of a padded table, both appeared interested in exploring their surroundings. The veterinarian examines each one carefully and can detect no obvious signs of illness.

   a. What is the veterinarian going to advise this client regarding the captive breeding of hermit crabs?

For Further Reference

de Vosjoli, P. (2005). *Land Hermit Crabs*. Los Angeles, CA: BowTie Publications, Inc.

Don Crablover 2000 & Wilkin, C. *Molting Demystified*. http://www.hermit-crabs.com (accessed July 2, 2005).

http://www.enchantedlearning.com (accessed January 2, 2014).

Pavia, A. (2006). Species Profile 'The Hermit Crab' Critters. In *USA Annual*. Los Angeles, CA: BowTie Publications Inc.

UNIT VII

ALPACAS AND LLAMAS

OBJECTIVES

After completing the chapter, the student should be able to

- Describe the correct housing for alpacas and llamas.
- Provide client education to new alpaca and llama owners.
- Interpret the body language of alpacas and llamas.
- Be familiar with the common vaccines available and the recommended vaccination schedule for alpacas and llamas.
- Provide basic nursing care to alpacas and llamas.
- Demonstrate restraint techniques for common procedures.
- Understand common medical disorders in alpacas and llamas.

Introduction

Alpacas and llamas are members of the camelid family. Camelids are divided into two groups: Old World species and New World species. Old World camelids are of two types: the Bactrian camel from Asia, with two humps and long hair, and the Dromedary from Africa, with one large hump and shorter hair. The distinctive humps are used to store fat, not water. Both have been domesticated and used for centuries as beasts of burden, a method of moving goods and services across the vast deserts of North Africa and the Middle East. Nomads live on camel's milk and meat, make rugs and blankets from camel hair, and burn camel dung for fuel. Even today, camel trains can be seen traveling alongside busy highways in thriving modern cities of North Africa and the Middle East, creating a very stark contrast between ancient civilization and modern life.

New World camelids are distantly related. All four species are from South America. They are much smaller and without humps. These are the vicuna, the guanaco, the more familiar llama, and the alpaca. Guanacos and vicunas are wild species. Llamas and alpacas are not wild but domesticated, agricultural animals (**FIGURE 20-1** and **FIGURE 20-2**). Although a part of the Incan civilization and culture for more than 5000 years, llamas and alpacas only recently have become popular in the United States. The alpaca is greatly appreciated for its many qualities, not least of which is the value of alpaca wool, known as fiber. It is soft, strong, and warm (warmer than most sheep wool),

FIGURE 20-1 An alpaca dam with her cria.

FIGURE 20-2 A shorn llama in the kush position.

making it very valuable and much sought after. Llama wool is coarse and not as valuable. Because of their larger size, llamas are used as pack animals or guard animals, patrolling and protecting sheep, goats, and the smaller and more timid alpacas.

Unlike other exotic species, alpacas and llamas produce an income for their owners. They are an investment, with alpacas sometimes costing thousands of dollars each. Alpacas are not only investments but are also gentle and loveable companions. The National Registry for Alpacas requires that all animals be DNA tested prior to acceptance in the registry. Today, there are more than 250,000 alpacas registered in the United States.

There are two types of alpaca: the **huacaya** and the **suri**. The difference between the two is the type of fibers in the coat. Huacaya fiber is short and crimped. Suri have long fibers without any crimping. Recently, selective breeding of llamas has improved their hair fiber, and it is sometimes difficult to tell the difference between llama and alpaca hair.

Llamas have two types of hair: coarse guard hairs and a fine undercoat. Alpacas have only dense, fine hair. Alpaca wool is measured in microns to evaluate the density and quality. Hair grows an average of four inches a year. Alpacas and llamas are shorn yearly like sheep.

Hand-spinning is a growing hobby and a profitable pastime. Angora rabbit hair is sometimes blended with alpaca and llama wool to create different fibers for spinning. The yarn is used for making jackets, hats, socks, blankets, rugs, and a variety of art objects. It may also be compressed into heavy duty felt (**FIGURE 20-3**).

Aside from hair coat, there are visual differences between llamas and alpacas in their size, ear shape, and tail set. Llamas have banana-shaped ears with a tail set high on the rump. Alpacas have straight ears that are set more forward on the head than those of the llama. The tail of the alpaca sits lower on the rump. Llamas are larger than alpacas.

FIGURE 20-3 The fibers of alpaca wool are spun and dyed to make beautiful yarn.

The feet of camelids are unique. They have two padded digits on each foot. The underpad is soft, similar to the pads on a dog's foot. There is also a protective horny growth made of keratin that protects the top of the toes. The nail- or hoof-like growth may need to be trimmed periodically if the animals are kept on soft ground and not worn down naturally.

A balanced flock usually has one dominant male with four to five females and their **crias** (young) living as a family unit. Away from the unit are the young bachelors that have been driven away by the dominant male. Many farms separate males and females until breeding. Males in mixed groups fight with their canine teeth to establish dominance. Once dominance is established, sexual behavior is less aggressive. The male chases the female and pushes her into sternal recumbency, allowing the male to mount. Alpacas and llamas are induced ovulators. When ovulation occurs after several days of breeding, females refuse the male and further attempts to breed are met with **spitting**, a well-aimed and foul-smelling stream of gastric contents.

Gestation is approximately 335 to 360 days. Most crias are born between 8 am and 12 pm, during the warming hours of the day. Alpacas and llamas give birth standing up. Dystocia is uncommon in alpacas and llamas. All camelids have four teats. Crias rotate nipples nursing from teat to teat. Twins are extremely rare. Two weeks after the birth of the cria, the female may be rebred.

Occasionally, crias will need to be bottle fed because the dam is unable to produce enough milk or has died. Whole cow's milk is considered an adequate milk replacer for a cria. **Aberrant Behavior Syndrome** (ABS) is an aggressive behavior that has become a problem with bottle-fed crias. This aggressive behavior is a result of inappropriate human bonding during bottle feeding. A young animal may jump or rear up at the caregiver, become overfriendly, or wrap its own neck around a person's neck. This is most often seen in male crias, but can also be seen in females. As with any animal being bottle fed, crias need to be treated and taught in the ways of their species, not as a human baby. The recommendation to correct this behavior is to discipline the animal immediately following the onset of the behavior. This can be done by placing the cria in a kush position. Teach the cria to respect your space and set boundaries of what is acceptable behavior. If possible, try and socialize the cria with other alpacas or llamas while bottle feeding (**FIGURE 20-4**).

FIGURE 20-4 Crias are sometimes given bottle feeds more for social bonding than for nutritional need.

Behavior

Alpacas and llamas are social animals and flock oriented. They produce muted but distinctive **humming** sounds to communicate with one another. Humming is heard frequently with a dam and her cria. They produce a clicking sound as a warning to alert others to potential danger, and *grumble* at one another if personal space is invaded. Alpacas and llamas can also produce high-pitched screams if they are extremely frightened or during breeding.

Alpacas and llamas demonstrate annoyance by spitting or regurgitation. This starts as a gurgle, bringing up stomach content that is then sprayed from the mouth. The spray, or spit, can reach as far as six feet and is very accurate in striking a target. The fluid has a strong, obnoxious odor that cannot be easily removed, especially from clothing.

Ear and tail positions are good indicators of how an alpaca or a llama is likely to behave. In an unconcerned and relaxed animal, the tail lays flat against the rump. The tail rises higher with the degree of intensity of aggression or alarm. When an animal is aggressive or threatening, the ears are pinned back, flat against the head. The ears of a submissive alpaca will be slightly forward, with the tail curved over the back. The head and neck are lowered, and there may be a slight crouch (**FIGURE 20-5**).

FIGURE 20-5 It is important to understand the body language of alpacas and llamas. Attitudes can be evaluated by (a) ear and (b) tail positions.

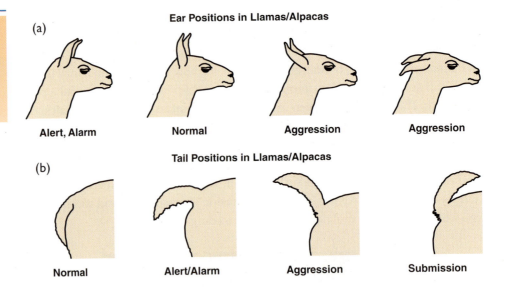

Ear Positions in Llamas/Alpacas

(a)

Alert, Alarm   Normal   Aggression   Aggression

Tail Positions in Llamas/Alpacas

(b)

Normal   Alert/Alarm   Aggression   Submission

Housing

Alpacas and llamas have few demands or special accommodation requirements for housing. They need a shelter from the weather, a small area to roam and graze, and a strong, safe fence that will keep predators out. Alpacas and llamas do not jump or challenge fences unless they feel threatened.

Barbed wired should never be used because of the danger of hair becoming entangled and caught in the fencing. A struggling animal is easily injured from the sharp, cutting barbs of the fence. Animals have become so entangled in barbed wire that they have been **disemboweled** (removal of the entrails) and have bled to death in their panic to escape. There should be no more than a two-inch gap between a fence post and a gate. A gap that is too wide can trap the head, causing trauma or cervical injuries in the struggle to become free.

Alpacas need a fresh water supply, a small hay rack, and a feeding tray. Cleaning the pen is simple. One common dung pile is used by all animals to urinate and defecate, creating little mounds. The dung pile also makes a great garden fertilizer. Alpacas and llamas have a spiral-shaped colon, so they produce smaller fecal pellets similar to goats and sheep (**FIGURE 20-6**).

FIGURE 20-6 Alpacas and llamas create communal fecal piles, keeping much of their paddock clean.

Diet

Llamas and alpacas are grazers, used to sparse, rough forage. One of the biggest problems is that if they are given access to rich grass, they will overeat. Alpacas and llamas should be fed grass hay with only 10 to 12 percent protein. This type of grass hay is not considered a high enough quality for most other large animals and so costs less. If given the opportunity, alpacas and llamas will eat an equivalent of nearly two percent of their total bodyweight in foodstuff daily. Access to hay and lush grazing areas needs to be restricted. Depending on the type of forage available, the diet can be supplemented with commercial alfalfa pellets formulated specifically for alpacas and llamas. Pellets provide vitamins and minerals that may be deficient in the soil and hay. Oats should not be fed because of the potential for causing stomach ulcers from the indigestible, barbed oat husk. A mineral block can also provide minerals that may be missing in the diet.

Alpacas and llamas have lower incisor teeth and a tough dental pad instead of upper incisors. Similar to goats and sheep, they shear vegetation close to the ground. The lower incisors are open rooted and grow continually. The teeth may need to be trimmed if grazing land does not provide for natural wear. The top lips are divided, allowing them to manipulate food items independently from each side of the mouth. They do not use their tongues to gather food while eating, nor do they use them to lick or groom each other (**FIGURE 20-7**).

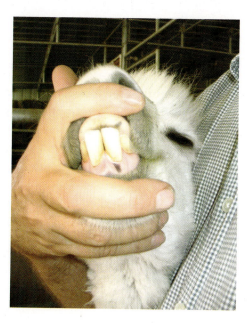

FIGURE 20-7 Camelids do not have upper incisors but instead have a hard dental pad.

Camelids have a three-chambered stomach. True ruminants, such as cows, have four-chambered stomachs. Bacteria in the first two chambers provide for fermentation of food. Like ruminants, camelids regurgitate and continue to chew preingested forage. When the **cud** (wad of preingested forage) is swallowed, it enters the third compartment, where it is further digested, and then moved to the small intestine for absorption. Unlike other ruminants, camelids are efficient feeders, able to extract protein and carbohydrates from poor-quality food. Because there are only three compartments and different bacterial flora in the stomach, they are also less prone to bloat.

Handing and Restraint

Before attempting to restrain an alpaca or llama, a visual assessment of ear and tail position will help determine the animal's behavior (refer to Figure 20-5). To restrain an alpaca or llama, herd the animal into a small area and approach slowly with arms outstretched. With one arm around the chest and neck, use the other hand to push lightly downward on the back. If the alpaca or llama is likely to be a spitter, place a towel over the nose, tucking it into the nose band of the halter. An alternative method is to place a loosely woven sock over the nose and mouth. Alpacas and llamas are obligate nasal breathers, and care must be taken not to close off their nostrils during restraint. The safest place to stand during restraint is close to the shoulder. This helps to prevent being kicked with a hind foot or being covered in spit. Alpacas and llamas can *cow-kick*, a forward striking movement with a hind leg that comes out to the side. They rarely bite or strike with a foreleg. Alpacas and llamas can be held with an ear restraint. Enclose the base of the ear with the palm of the hand and gently squeeze. *Do not twist the ear*. Ear restraint is acceptable, and if performed correctly, it will not damage the ear or make the animal head shy. It is common for alpacas and llamas to quickly lift the head up and forward, hitting the restrainer in the face. To prevent this, place one hand around the neck and hold the head close to the restrainer's body. Place the other hand at the shoulders, pressing downward on the patient. Take care not to put excessive pressure on the bridge of the nose, as the nasal cavity is short and easily blocked or constricted with pressure, blocking the airway. Alpacas and llamas may also be restrained in the **kushed** (recumbent) position. Kush is achieved by initially restraining a hind limb. A soft cotton rope loop with approximately six inches of slack is placed around the abdomen, just in front of the pelvis. Reach across the back of the alpaca and grasp the opposite hind limb and place it in the loop. The near hind leg is then lifted into the loop, causing the animal to sit down (**FIGURE 20-8**). When trying to open the mouth or calm the alpaca, the restrainer can place a finger inside the alpaca's mouth and rub the dental pad (roof of the mouth). This can be accomplished by entering the mouth from the side, just behind the lower incisors.

Restraining a cria is similar to restraining a large standing dog. A cria under 20 lbs can be picked up with one arm around the chest and the other arm around the rear legs.

An Alpaca (Kushed)

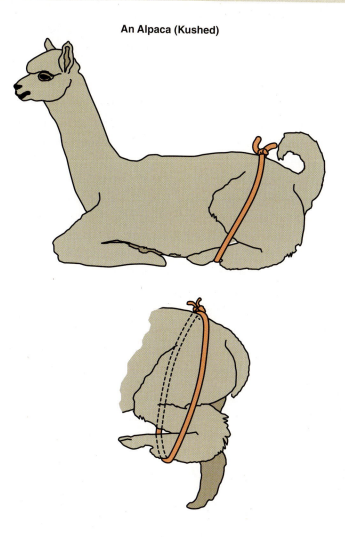

FIGURE 20-8 For ease of handling, alpacas may be placed in the kushed position.

Medical Concerns

The normal body temperature of alpacas and llamas fluctuates more than that of other species. They do well in cooler temperatures because their hair fiber is so dense and retains heat. Core body temperature will vary two to three degrees during the day. To determine if an individual alpaca or llama's temperature is normal, it is measured against the average temperature of the flock, which is taken at the same time. At night, crias begin to cavort and run around. While it certainly appears playful, the real purpose is to increase body temperature before nightfall. Normal resting body temperature of adults ranges from 99.5°F to 102°F. During hot months it may increase to 104°F and still be considered normal.

Alpacas and llamas are very hardy animals. Other than obesity, there are few health concerns. Viral diseases are not fully determined in alpacas and llamas. They are susceptible to some bovine diseases, but not all of them.

Because alpacas and llamas are frequently transported to many states for exhibitions and fiber festivals, interstate regulations must be followed. All animals need a certificate of veterinary inspection, *a health certificate*. Depending on the destination and the various states traveled through, there may be additional requirements, including proof of a negative test for tuberculosis and brucellosis. A current rabies vaccination certificate may also be required.

The recommended vaccination schedule for crias is 8 to 12 weeks of age, followed by a booster three to four weeks later. Adults are revaccinated yearly.

There is only one routinely given vaccine. It is a **toxoid** for *Clostridium perfringens*, types C and D. This vaccine also includes tetanus toxoid but has recently been questioned on its effectiveness in alpacas and llamas. Most vaccines are *off-label* for use in camelids, which means the manufacturers do not guarantee effectiveness and safety of the vaccine. The recommendation is to consult with the local veterinarian on the use of vaccinations and vaccine protocol. Alpacas and llamas are not susceptible to all eight clostridium diseases known in cattle. Clostridium bacteria are found in the soil and become pathogenic when they invade the body. *C. perfringens* targets the intestinal tract. This bacterial disease is often a result of overfeeding on pasture that is too lush, or it may be secondary to other intestinal problems. If a vaccine is recommended, tetanus toxoid is the most common one administered.

Respiratory Disorders

Respiratory infections are also seen in llamas and alpacas. Stressful conditions can predispose them to a Streptococcus infection known as alpaca fever. Incidents of chronic obstructive pulmonary disease (COPD) have increased due to frequent changes in feeding and poor-quality hay. Clinical signs include shortness of breath, coughing, and dyspnea. Treatment involves the use of a bronchodilator through a nebulizer or face mask.

Digestive Disorders

Stress is a significant factor in ulcer development in camelids. Changes in environment affecting social structure, severe injuries, and other illnesses can contribute to the formation of gastric ulcers in these animals. Treatment is based on history and clinical signs. Oral medications have not been effective. Supportive therapy including pain relief and antibiotics are recommended. The onset of clinical signs of gastric ulcers can be rapid and lead to sudden death.

Bovine viral diarrhea (BVD) is a viral disease that has shown to affect alpacas and llamas. Clinical signs include diarrhea, anorexia, fever, oral ulcers, abortion, and overall poor condition. Currently, there is no vaccine available for camelids. It is unclear how this virus is spread, but in cattle it is spread through ingestion and inhalation. The prognosis is poor with low survival rate in crias. Research on BVD needs to continue to fully understand the impact this virus can have on camelids.

Nondigestive Disorders

Hyperthermia can be a concern when alpacas and llamas are kept in hotter climates. It is not uncommon to see an adult maintain a body temperature of 104°F, which is close to being at risk for clinical hyperthermia in other species. Core body temperatures as high as 113°F to 116°F can cause organ failure and tissue death. Soaking the fleece is not an effective cooling method for alpacas or llamas. Over-the-fleece soaking does not penetrate the wool. The ventral abdomen where there is very little abdominal hair needs to be bathed with cool water. In severe cases, cool water enemas should also be administered. Alpacas and llamas need to be shorn of their wool and have access to shaded areas during the hottest months.

Many infectious diseases in alpacas and llamas do not always produce clinical signs similar to disease in other species. One example is the West Nile virus. Clinical signs in the alpaca or llama may be body tremors, head shaking, ataxia, and death. Research indicates that alpacas and llamas develop protective antibodies that may offer lifelong immunity against West Nile disease. It is recommended to check with the local veterinarian regarding current protocol for vaccinating against West Nile. Currently two vaccines are in use to prevent camelids from contracting West Nile: Fort Dodge West Nile-Innovator® from Zoetis and Recombitek Equine WNV®, from Merial.

Eastern equine encephalitis (EEE) is not common in alpacas and llamas; however, recently there have been a few isolated cases of camelids contracting this mosquito-borne disease. The virus causes inflammation of the brain and the surrounding tissues. Clinical signs include star gazing (staring off into space with the head tilted upward), ataxia, and difficulty in standing from a sternal position.

Equine herpes virus has been incidentally diagnosed in alpacas and llamas. This disease may not show any clinical signs in the alpaca or llama. It has been reported that the number of confirmed cases has started to increase slightly, but it is not known if this could be correlated to the population increase.

Parasites

Parasites can become a problem in alpacas and llamas because they affect not only the health of the animal but also the fiber of the coat. Common internal parasites also include protozoa and tapeworms. These parasites can cause poor growth and productivity. Alpacas and llamas should be dewormed twice yearly, rotating anthelmintic products. There is some question on the effectiveness of oral wormers and the way they are absorbed in the gastrointestinal system compared to injectable dewormers. Valbazen, Ivermectin Plus, Dectomax, and Panacur are common products used.

The most devastating parasites for alpacas and llamas are nematodes. Crias do not have a well enough developed immune system to cope with a nematode invasion. They fail to thrive, rapidly lose weight, and can develop bone deformities.

A parasite of growing concern, especially on the eastern coast of the United States, is the meningeal worm. The natural host is the white-tailed deer. The parasite can be transmitted to alpacas and llamas when they ingest a snail or slug from the same grazing land and pastures that deer graze on.

Meningeal worms are hair-like roundworms that invade the spaces between the brain and the surrounding tissue (the meninges). The meningeal worm is commonly called the *brain worm*. The adult stage of the worm causes paralysis in the alpaca and llama. To date, there is no treatment, and affected animals are humanely euthanized.

Clinical Procedures

Venipuncture of the jugular vein, the usual site for blood collection, can be a little more difficult in alpacas and llamas than in some other large animals because of the location of the jugular vein and the thick skin on the neck. Two sites can be accessed on the neck. One is low, near the thoracic inlet, and the other is more cranial, near the first two cervical vertebrae. Attempts in other areas of the neck are more difficult, because the jugular vein is deeper. For smaller amounts, the ventral tail vein may be used (**FIGURE 20-9**).

FIGURE 20-9 Obtaining a blood sample from the jugular vein.

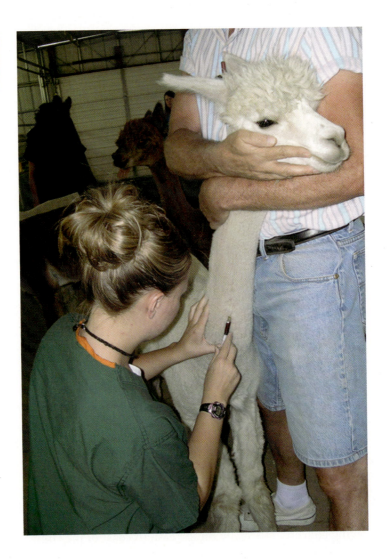

Camelids Have Flat, elliptical red blood cells without a nucleus. The leukocyte count in camelids is normally higher than that of other mammals.

Administration of vaccinations is not difficult with correct restraint. The vaccine is given subcutaneously in the axillary area. There is a little area of looser skin that can be tented to place the needle under the skin. Use of the axillary area helps reduce the possibility of adverse vaccine reactions and causes less pain for the animal (**FIGURE 20-10** and **FIGURE 20-11**).

FIGURE 20-10 Vaccinations should be given in the axilla, or just behind the elbow.

FIGURE 20-11 Optional position for vaccination.

Administering oral antibiotics and anthelmintics is a frequent procedure in alpacas and llamas. The syringe tip is placed in the oral cavity from the side of the mouth. Attempts to place the syringe from the front of the mouth will result in the medication being sprayed all over the animal, the restrainer, and the person administering the medication. The alpaca's mouth can be opened by placing a thumb in the side of the mouth and gently pulling down on the lower jaw (**FIGURE 20-12**). Care should be taken to make sure the medication is not deposited in the buccal aspect of the jaw because medication-induced ulceration can occur.

FIGURE 20-12
Administering oral medication to an alpaca.

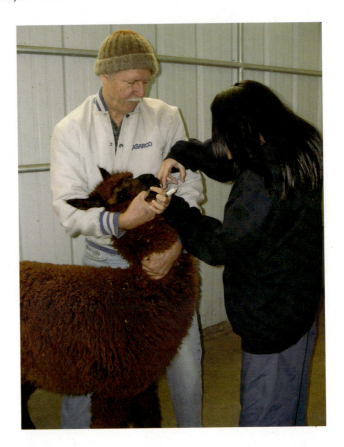

Many procedures can be performed with the use of local anesthesia. A local anesthetic is injected to infiltrate the procedure site. Local anesthesia or regional blocks are used to treat abscesses and lacerations and to perform C-sections and laparoscopic procedures. Examples of local anesthetics include lidocaine, bupivacaine, and mepivacaine. Veterinarians often use a local anesthetic in combination with sedation to perform a standing castration. Injectable anesthetics can also be used in combination with inhalant anesthesia. Care needs to be taken when administering IV anesthetics such as ketamine, xylazine, and butorphanol. The carotid artery lies directly underneath the jugular vein and the induction agent can be easily injected into the carotid artery if the needle is placed too deep. For general anesthesia, alpacas and llamas should be fasted 12 to 18 hours. They are prone to regurgitation during anesthesia and recovery. Dorsal recumbency should be avoided if possible, as it increases the chances of regurgitation during surgery and also makes it more difficult to adequately inflate the lungs. Most camelids do not require premedication unless they are fractious when administering the anesthetics or when placing a jugular catheter.

Camelids have very little space in the oropharynx and a long mandible that prevents easy visualization of the epiglottis. Endotracheal tube placement is the same *blind technique* used in horses. The anesthetized alpaca is placed laterally with its head and neck fully extended and a mouth speculum is required. If intubation is difficult, a nasotracheal tube can be placed instead.

Isoflurane is used successfully in both alpacas and llamas. Patients are monitored with the use of a doppler and ECG for heart rate and blood pressure. A pulse may be felt over the auricular, femoral, or dorsal pedal artery. A pulse oximeter can also be used to monitor the patient's oxygen and heart rate. Normal heart rate is 60 to 90 beats/min in the adult camelid. In general, most ruminants usually maintain good blood pressure (75 mm Hg or above) under anesthesia. The average respiratory rate under anesthesia is 10 to 30 breaths/min, and manual ventilations should be performed when necessary.

Hypothermia can be a complication during general anesthesia. Core body temperature should be monitored carefully. If the patient becomes hypothermic, administering warmed IV fluids or placing the patient on a heating pad is recommended.

Camelids are considered *sensible* animals during recovery. They rarely attempt to stand until they are capable of standing without falling over. During recovery, they should be placed in sternal recumbency and not extubated until they are actively trying to cough or spit out the endotracheal tube. As with many species, a smooth induction often produces a smooth recovery.

Radiographs are useful in diagnosing fractures, bone deformities, dental problems, and respiratory and gastrointestinal disease (**FIGURE 20-13** and **FIGURE 20-14**). To radiograph the cheek teeth and check for abnormalities, the

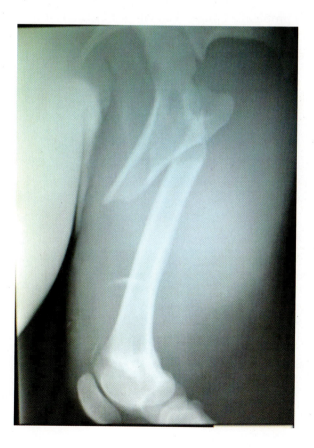

FIGURE 20-13 Fractured femur on a two-month cria.

FIGURE 20-14 Fractured femur on a young cria. Severe tissue trauma and edema were associated with the fracture.

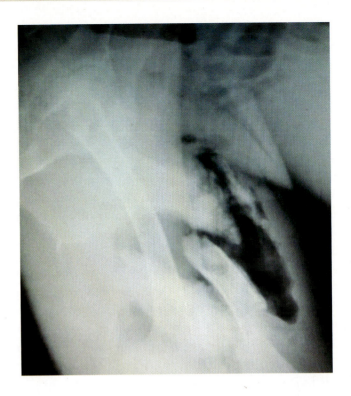

best views are lateral/oblique and ventro-dorsal of the head with the mouth open slightly to separate the mandibular and maxillary teeth. Root problems, malocclusion, periodontal abscesses, oral neoplasia, and cheek teeth infections are common dental problems. Congenital defects are also detected on radiographs (**FIGURE 20-15**).

FIGURE 20-15 Congenital defect of the pelvis. The alpaca was unable to walk normally or kush.

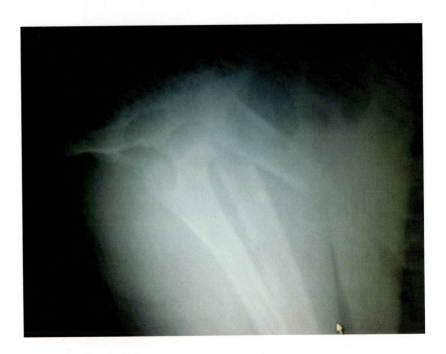

Contrast studies using iodine and barium are common in camelids and provide a better diagnostic picture than endoscopy or ultrasound. The most common study is the esophagram. Choke, esophagitis, stricture, and megaesophagus are abnormalities easily detected using radiographs. Gastric studies are not commonly performed due to the delayed gastric retention.

Summary

Alpacas and llamas are members of the camelid family, distantly related to the Bactrian and Dromedary camels of the Old World. Alpacas and llamas are smaller, New World camelids, and all four species are native to South America. Unlike other exotic species, alpacas and llamas not only represent a major investment but also produce an income for their owners. Alpacas are raised for their wool or fiber while the larger llama is often kept as a guard or pack animal. Alpacas and llamas are social animals and flock oriented. Their feet are unique in that they have two padded digits on each foot with a protective horny growth that covers the top of the toes. Alpacas and llamas are considered *easy keepers* with few special demands regarding husbandry. Both are grazing animals used to sparse, rough forage. They have lower incisors and a tough dental pad instead of upper incisors. Methods of restraint need to be mastered to safely handle an alpaca or a llama for the animals' safety and that of the handler. Both spit when annoyed and are capable of delivering a swift *cow kick*. They may be easily placed in sternal recumbency (kushed) for ease of handling. They are unique from other mammals in that their body temperature fluctuates through the day. Parasites have become a growing problem, and owners are recommended to consult with a local veterinarian regarding recommended vaccinations and parasite control.

*fast*FACTS

ALPACAS AND LLAMAS

ADULT WEIGHT
- Llama: 113.6–215.9 kg (250–475 lbs)
- Alpaca: 55–90.9 kg (121–200 lbs)

LIFE SPAN
- 20 years

REPRODUCTION
- Sexual maturity: male: 2–3 years; female: 1 year (over 100 lbs bodyweight)
- Gestation: 11 months (335–360 days)
- Birth size: llama cria 8–20 kg (18–45 lbs); alpaca cria 3.6–10.4 kg (8–23 lbs)
- Weaning age: 6 months

VITAL STATISTICS
- Temperature: 36.6–40°C (97.5–104°F)
- Heart rate: 60–90 beats/min
- Respiratory rate: 10–30 breaths/min

DENTAL
- Dental formula 2 (I 1/3, C 1/1, P 1-2/1-2, M 3/3)

ZOONOTIC POTENTIAL
- None reported

Review Questions

1. Explain how alpacas and llamas are placed in kush.

2. Why are body temperatures so variable in alpacas and llamas?

3. How is digestion different from monogastric species?

4. What could cause limb deformities in crias?

5. List the interstate documentation required to transport alpacas and llamas.

6. What are the recommended vaccinations for alpacas and llamas?

7. List the visual differences between alpacas and llamas.

8. What are some of the reasons alpacas and llamas have become so popular?

9. What differences are seen in camelid blood cells?

10. How and why do camelids spit?

Case Study I

History: In August, a client requests a farm call regarding her pet alpaca. All the owner reports is that it will not get up.

Physical Examination: The veterinarian examines the fleece, which is very long and matted. It has not been shorn since the owner bought the alpaca last year as she prefers "the shaggy look." A black tarp has been tied between two trees to provide shade but at 4 pm, the tarp offers no protection from the sun. While obtaining a rectal temperature, the veterinarian comments on the heat and humidity of the summer and guesses that it must still be "more than 90°F." The veterinarian records a rectal temperature of 105.8°F. He then asks the client if there is a garden hose nearby.

  a. What is the diagnosis?

  b. Why has the veterinarian asked for the hose?

  c. What recommendations will be given to the owner?

Case Study II

History: A client brings in a 20-week-old cria. His concern is that it may have a fractured leg.

Physical Examination: The veterinarian palpates the leg but cannot detect any sign of a fracture. He also examines the other three legs carefully and notes their comparative length and density, flexes each joint, then palpates the length of the spine. The owner is greatly surprised when the veterinarian asks about the owner's deworming schedule. The owner replies that he hadn't done it yet, but intended to later in the year.

a. What problem does the veterinarian suspect with this cria?

b. What is the connection between the deworming program and the lameness of the cria?

For Further Reference

Anderson, D.E. http://www.vet.ohio-state.edu (accessed March 2006).

Fowler, M.E. (1998). *Medicine and Surgery of South American Camelids, 2nd Edition*. Ames, IA: Iowa State Press.

Hoffman, E., & Fowler, M.E. (1997). *The Alpaca Book: Management, Medicine, Biology, and Fiber*. Cincinnati, OH: Clay Press, Inc.

https://www.alpacaregistry.com/public/about (accessed January 7, 2014).

http://www.merckmanuals.com (accessed January 7, 2014).

http://www.vetmed.usu.edu (accessed June 1, 2014).

Johnson, L.W. (2005). *Taking a Closer Look at Alpacas*. Fort Collins, CO: Colorado State University, CSU Research (seminar papers).

Otterness, C. & Otterness, J. Alta Mist Alpacas, Herriman, Utah 2003–2006. Personal conversation, advice, and practical experience with the flock of Alta Mist.

Schoenian, S. (2005, December). *Meningeal Worm, Brain Worm*. Western Maryland Research and Education Center, Maryland Cooperative Extension. www.sheepandgoat.com (accessed 2004).

UNIT VIII

Linn Currie/Shutterstock

21 MINIATURE PIGS

OBJECTIVES

After completing the chapter, the student should be able to

- Describe the correct housing requirements for miniature pigs.
- Provide client education to new miniature pig owners.
- Be familiar with the common vaccines available and when and how vaccines are administered.
- Evaluate the diets available to meet the nutritional needs of a miniature pig.
- Provide basic nursing care to a miniature pig.
- Demonstrate appropriate restraint techniques for common procedures without causing too much stress and anxiety in the pig.
- Understand common medical concerns in miniature pigs.

Introduction

In the late 1980s the *ultimate new pet* was introduced to the United States through Canada. Despite a cost of several thousand dollars each, the Vietnamese potbellied pig became America's newest fad pet. A Canadian, Keith Connell, maybe never thought (nor was it his intention) that this miniature pig would become a house pet, sought after as a companion animal and family member. Originally, the pigs were imported to zoo collections as a rare and different breed of pigs. They came from Vietnamese stock and had very distinctive characteristics. The biggest appeal was their small size. People did not understand that smallness alone did not make the animal any less of a pig. They were soon confronted with the reality of trying to cope with and understand exactly what they had brought into their homes (**FIGURE 21-1**).

In the Past Several Years, Micro-minis (super-micros, teacups, juliana, and micro-mini juliana crosses) have increased in popularity, yet none of these pigs are *mini*, *micro*, or *teacup* sized when fully grown. These *micro-mini pigs* have been selectively bred from smaller-than-average, or stunted, potbellied pigs. While these newly named types of the potbellied pig may not get as large as the standard pot belly, they certainly will not fit in a teacup. Most of the pictures advertising these pigs are photographs of piglets that are only one or two days old. *Mini pigs* grow until they are three years old and can weigh up to 200 lbs at maturity.

FIGURE 21-1 A healthy mixed breed miniature pig.

Although the popularity of the Vietnamese potbellied pig has declined and they are certainly more affordable, many who bought the *adorable little piglet* soon discovered it grew up to be *a pig*. There are countless pig rescue organizations and shelters overflowing with pigs that have been abandoned. Reasons often cited for not keeping them are: *unmanageable, aggressive—attacks my friends, it bites, it's too big, the landlord said it had to go, can't find/afford a vet,* and *he won't get off the couch*!

The original imported stock was small and averaged between 75 and 150 lbs. They were approximately 15 to 18 inches tall. *Miniature* was often a selling point. It was not long before they were out-crossed and the *mini-pig* matured much larger than promised, often at 350 lbs or more, but compared to agricultural pigs that can weigh up to 1000 lbs or more, they were still *miniature*. Despite educational organizations and multiple resources available, owners held on to their own expectations and the *fun* of having a pig. Perceptions changed dramatically when confronted with reality.

Many people are dedicated to their pet pigs. In the right circumstances, pigs are devoted to their human companions, can be easily trained, and become enjoyable and compatible companion animals. Many owners understand exactly what a pig is, what it can never possibly be, and how to cope with behavioral issues and emotional problems and accept the responsibility of providing veterinary care, hoof trims, ear cleaning, tooth trimming, and yearly vaccinations.

Piglets are precocial and born with very sharp deciduous teeth. The neonatal canines are referred to as **needle teeth**, and are clipped off when the piglets are one day old. Needle teeth lacerate the sow's udders and injure littermates. Permanent canine teeth erupt between five and seven months. If these teeth are not kept trimmed, they develop into large **tusks** that are used for fighting. These teeth are open rooted and grow continually. Due to the lateral placement in the mouth, tusks are sharpened by jaw movement and soon develop into formidable weapons. Both sexes have four permanent canine teeth.

Pigs have four toes on each foot. Two are walking toes and two are dew claws. Hoof trims should be done routinely to prevent overgrowth and lameness. One of the most common lameness problems found in pigs is directly caused by overgrown hooves.

Female miniature pigs reach sexual maturity between three and four months. Males are sexually mature at three months. Males are often neutered as soon as the testicles descend, as early as 10 days. Females can be spayed between 6 and 12 weeks. If left unspayed, females cycle approximately every 20 days for a period that lasts for two to three days. Gestation is approximately 114 days, also calculated as *three months, three weeks, three days*. Litter sizes in miniature pigs vary from four to eight. Sows have been known to cannibalize their young, especially weaker, smaller piglets or piglets that have been unable to establish a place in the litter hierarchy.

An adult male pig is a **boar**. When neutered, they are called **barrows**. Females are called **gilts** unless they produce a litter, after which they are called **sows**.

The eyesight of pigs is poor, but they have an excellent sense of smell. They can locate food underground, with rooting and sniffing, and can also scent and follow a trail in the air. Pigs are known for their keen sense of smell and their ability to locate truffles, a hard-to-find and much sought-after edible fungus that grows underground.

The snout of a pig has developed to dig. It has a distinctive disc of cartilage that strengthens it and enables the pig to **root**. They are able to root up fences, bushes, and small trees and work their way through wooden barriers. Rooting behavior is a natural instinct that cannot be changed. Outdoor pigs will root up areas of lawn and gardens. They are so efficient at this that many owners deliberately pen them over areas of a garden that need to be tilled. Household pigs root up carpet and flooring. Pigs root to explore, find food, and create cool wallows to lie in. Metal *pig rings* that pierce the end of the snout will not deter a pig from rooting. This is a common agricultural practice, and a nose ring is also used as a method of controlling an animal in conjunction with a **pig hook**—a long pole with a hook at the end that is intended to lead the pig *around by the nose*. More often than not, there is no control and the pig becomes irate and aggressive.

Miniature pigs are not *odorless*, as many breeders have claimed. Their odor resembles iodine or something metallic, not particularly unpleasant, but evident. Papaya fruit is often fed to house pigs to help remove the *piggy aroma*.

Pigs, regardless of their status as pets, are agricultural animals. Zoning ordinances usually do not permit the keeping of livestock in towns and cities. Always check with the zoning office before obtaining a pet pig.

Behavior

Pigs of any breed are noted for their intelligence. It is only recently that any discussion of animal emotion was considered valid. **Ethologists**, scientists who study natural animal behavior, previously insisted that *emotions* were attributed to animals only because of human feelings and interpretation—that it was subjective and anthropomorphic. Researchers are now openly discussing emotional intelligence in animals.

Pigs (as well as many other species) have demonstrated very clear, appropriate *emotional* reactions. Owners and rescuers speak of *piggy temper tantrums*; pigs can be embarrassed when caught causing mischief, cry tears when distressed or with the loss of a beloved companion, exhibit joy, and have a sense of humor. Pigs can have their feelings easily hurt with harsh words.

When Elephants Weep: *The Emotional Lives of Animals* by Jeffrey Moussaieff Masson and Susan McCarthy (1996) is groundbreaking in human understanding and acceptance that animals have emotions.

Miniature pigs have natural behaviors that are no different from those of other breeds. Pigs are herd animals and establish a hierarchy from the moment they are born. They will fight and scramble for a teat, push, shove, and vocalize with each other until the order of dominance is established. As they mature, dominance issues continue, both with other pigs and with the *new herd* of humans. Pigs will fight, charge, swing their heads, and bite to establish dominance. This can be very intimidating to humans, and it is meant to be. Most people confronted with a pig vying for dominance will back off in a reasonable fear of being injured, and the behavior then becomes reinforced. It is not the owner who is at the top of the herd, but the pig. Dominant pigs have, and demand, certain privileges: They have the choicest food and the best wallow. Subordinate pigs are presented with the exposed abdomen of the dominant pig in the expectation of a massage. In the kitchen, pet pigs nip at their owner's ankles, refuse to vacate furniture, and flop down in front of family members, exposing their bellies. Humans respond by saying, "Isn't that cute. Molly wants a tummy tickle," and proceed to do it, reinforcing the dominance that the pig has established. This is probably the most common misinterpretation of pig behavior and it quickly leads to other behavioral issues. Exposure of a pig's belly is not to be confused with the submissive posture of a dog.

Pigs should be trained with positive reinforcement, not physical punishments. Food treats used as training aids usually only teach the pig to beg, and begging quickly becomes a nuisance behavior. Toys make better rewards for good behaviors.

Housing

Pigs are very clean. They are easily housebroken because it is natural for them to urinate and defecate away from bedding and food. Household pigs use a litter box if they have no access to the outdoors. Large *doggie doors* are installed when pigs are allowed access to an outdoor area. When outside, pigs use one designated area of the yard to urinate and defecate.

Keeping pigs as indoor pets is not recommended. They should be housed outside and provided with a shelter to protect them from the extremes of weather, both hot and cold. Pigs do not have sweat glands nor do they pant.

In addition to shade, they should be given an area in to which to dig a wallow or be provided with a shallow child's pool. The skin of pigs is very similar to human skin and can be easily sunburned, and protective sunscreen should be applied daily.

Pigs do not have much hair. They do not have enough of a coat to provide any warmth. They should have a deep straw bed to keep them off the cold ground and an ample amount of bedding material to root under and be able to cover themselves. Sleeping bags, quilts, and blankets provide comfort and warmth. In very cold areas, supplemental heating should be provided for them. Heat lamps should be suspended over the sleeping area. Pigs are very susceptible to pneumonia, and the biggest contributing cause is an inability to stay warm.

Large, shallow basins should be provided for ample fresh drinking water. The supply should be checked regularly, as pigs will pick up and carry water and food bowls. A good choice would be the type of heavy rubber tubs sold by agricultural suppliers. These are tough, designed for heavy-duty use, and they are easy to scrub out and clean.

Diet

By far the greatest problem seen in companion pigs is obesity followed closely by constipation. Pigs are omnivores and will eat anything. Commercial pig diets are not suitable for pet pigs as they are formulated for maximum growth in the shortest amount of time. Pig chow should be specific to miniature pigs where nutritional goals are for average growth and longevity. Feed should be measured, rationed, and fed according to manufacturer's directions. Feed producers have developed specifically formulated diets for companion pigs.

Treats, which are as varied as dog biscuits and strawberry shortcake, need to be severely restricted. Companion pigs are often fed treats as behavioral distractions and *bribes*. Many favorite treats can cause constipation. One easy remedy to constipation is to feed canned pumpkin. Pigs enjoy the taste, and it has laxative effects. Fresh greens, grass hay, and root vegetables can also be added to the diet. These should be purchased fresh and be of human food quality. It is illegal to feed raw garbage to pigs. *Pig slop*, the combination of whatever might be left over from human consumption, is prohibited by federal and state laws that were enacted to control swine diseases. This law also applies to companion pigs.

Restraint and Handling

Pigs are not receptive to being held. Even the smallest of piglets will struggle and scream loudly in an attempt to escape. Pigs can be trained to wear a harness and walk with a leash, but the route is usually determined by the pig. They are certainly intelligent enough to learn obedience and leash behaviors, but they are also strong willed enough to ignore them. The weight of an adult pig usually prohibits any attempt to lift it for restraint or examination.

Pigs are reluctant to move in bright light and are more easily persuaded to move in subdued lighting. A **pig board**, a large solid barrier held from behind the pig, can be used to maneuver a pig in a certain direction. Many routine procedures can be performed with the pig restrained vertically on its hindquarters or in dorsal recumbency. A pig on its back needs to have the chest and head propped up to avoid respiratory distress from weight of the body putting pressure on the diaphragm. For most procedures performed on pigs, chemical restraint is required to reduce stress-related problems.

Chemical restraint refers to the use of drugs for sedation and anesthesia. The ideal chemical restraint is one that is short acting and easily induced and the effects of which can be reversed. Azaperone (Stresnil) is the only FDA-approved tranquilizer used in pigs. Dose dependent, it can be used to modify aggression or as an anesthetic. Used as a restraint for short procedures or radiographs, azaperone can last 20 minutes and there is usually a smooth recovery. Other combinations for minor procedures or restraint involve dissociative agents with phenothiazines or benzodiazepines. Dissociative agents can cause violent recoveries with any type of stimulus and should not be used alone. Combinations frequently used are ketamine/diazepam, ketamine/acepromazine, and telazol/xylazine. Recently, midazolam (Versed®) has been used as a sedative in pigs. It is an injectable benzodiazepine that can be administered through injections, orally, or rectally. If used as an injectable, it takes 10 minutes before the patient becomes affected, while the oral route takes approximately one hour. If administered orally, the drug is put on a piece of bread for the pig to consume. If given rectally, the drug is administered with a syringe and it takes 30 minutes to take effect.

Medical Concerns

When people obtain a pet pig, they usually do not consider veterinary services. Veterinarians and their staff should become experienced and knowledgeable with pet pigs and their behavioral issues, become skilled in performing hoof and teeth trims, and provide for vaccinations and health-care concerns of companion pigs. Just as importantly, the veterinarian and staff should recognize that the human–animal bond is just as strong with a companion pig as it is with a dog or a cat. Companion pigs are frequently presented to small animal practices because many large-animal practitioners are involved in herd management and production, with quite different approaches to pig or swine health.

There are many diseases of swine, most of which have a very low incidence or are not reported in companion pigs. This is not to say that they are immune, but the disease is more likely attributed to their being isolated from other pigs due to their companion animal status (**FIGURE 21-2**).

Only vaccines approved for use in swine should be given to miniature pigs. Vaccination protocols are variable and dependent on the disease incidence and the density of swine. In an area where swine production has a major agricultural impact, vaccination protocols would be quite different than that of a companion pig in an urban environment.

FIGURE 21-2 A rescued miniature pig that had been neglected and fed only small handfuls of dry dog food in an ill-advised attempt to keep it *small*. Multiple chronic health concerns have developed as a result.

Gastrointestinal Disorders

Constipation is commonly seen with pigs that are suffering from other illnesses and during winter months when they are not outside or very active. It is not uncommon to add a laxative to the pig's diet when constipation is suspected. This may include olive oil, canned prunes, canned pumpkin, and Metamucil®, which contains psyllium seed husks for added fiber.

Nondigestive Disorders

All pigs should be vaccinated against erysipelas, an infectious bacterial disease that can cause swollen joints, lameness, systemic infection, and skin lesions. It is also known as *red diamond disease* because of the distinctive diamond-shaped areas of affected skin. Piglets should be vaccinated at 8 to 12 weeks and revaccinated three weeks later, with booster vaccinations every six months. Erysipelas is a zoonotic disease.

Leptospirosis vaccine is given at the same time as erysipelas and should be boosted annually. Pigs should also be protected against mycoplasma pneumonia and tetanus. Recommendations for vaccinations by the veterinarian should be followed carefully to ensure the pig's health and immune status.

It has often been stated (by breeders of miniature pet pigs) that pigs are immune to rabies. This is incorrect, and specific concerns regarding companion pigs and rabies should be discussed with a veterinarian or directed to the CDC. The incidence of rabies in the United States fluctuates by region. The chance of rabies in pigs is extremely low, but not impossible. There is currently no approved vaccine for use in pigs.

Pseudorabies is caused by infection with a strain of herpes virus. It is a reportable disease of pigs and should not be confused with rabies. It is called *pseudo* because the affected animals exhibit erratic behavior (similar to rabies), the most obvious being intense itching and seemingly *mad* behavior in an attempt to scratch. It is also called Aujeszky's disease and *mad itch*. Mature pigs usually survive pseudorabies and become carriers that can infect other species coming into contact with carrier pigs. It is contagious to cattle, horses, sheep and goats, dogs, and cats. Dog and cats do not survive and usually die within a few days of showing clinical signs. It does *not* affect humans. Currently, all 50 states are free of this disease due to the strict laws regarding herd testing, transportation of swine, and eradication programs.

Rectal prolapse can occur in pigs that are stressed or constipated or it may be secondary to other illnesses. Veterinary care is required to shrink the prolapsed tissue and gently push it back into the rectum. Temporary sutures may be required to keep the tissue in place and allow the area to heal (**FIGURE 21-3**).

Dippity Pig syndrome (erythema multiforme) is becoming more prevalent in potbellied pigs during the spring months. The onset of Dippity Pig syndrome is sudden and can vary in severity from pig to pig. Pigs become sensitive to being touched on the rear quarters and may demonstrate a *hunkered-down* posture with the tail tucked between the rear legs. The pig may have weakness in the back legs and may be unable to walk. Moist, red areas may develop into lesions that appear along the back, oozing serum or blood. The primary cause of this syndrome is stress related; environmental changes including diet and weather changes. The average duration of this syndrome is 24 to 72 hours but may be prolonged. Treatment includes applying a topical cream to treat the lesions and providing rest and an ideal environment to allow the pig to recover.

Parasites

Potbellied pigs are susceptible to a variety of internal and external parasites. Coccidia can be found in piglets and causes diarrhea. Pigs may suffer from lice or mite infestations. The pig louse, *Haematopinus suis*, is a large blood-sucking parasite commonly seen around the neck and ears. *Mange*, caused by mites, is usually seen around the ears and face of the pig. Fecal examination for parasite ova should be included in each wellness examination.

Clinical Procedures

In addition to regular hoof trims, (**FIGURE 21-4**, **FIGURE 21-5**, and **FIGURE 21-6**) companion pigs should have routine veterinary examinations. They need to have their teeth trimmed, ears and eyes cleaned, vaccinated, and dewormed (**FIGURE 21-7**). Not all veterinarians accept pig patients or are able to provide the services needed. In addition, there is the problem of transporting the pig to the office. In most instances, for anything other than a small piglet, veterinary procedures are carried out during house call visits.

Injections can be given in the rear leg above the hock, in the lumbar area, behind the ears, and in the loin of the shoulder. Injections should not be given in the thick gluteal muscles of the ham to avoid the subcutaneous fat layer. Regardless of the site, there will be some blood as a result of the needle (hence the expression *bleeding like a stuck pig*). Vaccinations are usually given at the same time as hoof trims and other routine procedures, which are all performed under sedation.

Blood collection sites can be difficult to access because of the pig's dense layer of fat. The easiest sites to access are the lateral saphenous vein and the medial and

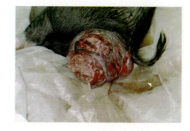

FIGURE 21-3 Rectal prolapses can occur in pigs of any age. They may be caused by stress or constipation or may be secondary to other illness. Immediate veterinary care is required. (*Courtesy of Eric Klaphake, DVM.*)

FIGURE 21-4 The hooves of this pig are severely overgrown, preventing it from walking. The pig was able to move only by *walking* on her knees and scooting her hind limbs at an angle.

FIGURE 21-5 The trims began by using horse hoof nippers to dramatically reduce the length of each hoof before attempting to reshape and realign the hooves with a dremal to allow the pig to walk normally.

FIGURE 21-6 With great skill and the use of a dremel tool, one foot has been restored to normal. Prior to the procedure, injectable anesthesia was administered.

FIGURE 21-7 While anesthetized, the pig's teeth were also examined. Aside from some dental plaque, all is normal.

lateral ear veins. Blood can be drawn from the cranial vena cava. Although this is routinely performed in production pigs, it is not recommended for companion animals.

For procedures not carried out *in the field*, injectable or inhalant (isoflurane) anesthesia is recommended. Pigs can be a challenge under prolonged anesthesia. It is recommended to withhold food 6 to 12 hours prior to surgical anesthesia.

This will reduce pressure from the stomach on the diaphragm, which can affect the respiratory rate. Patients should receive a pre-induction sedation. This may include an anticholinergic (atropine), a tranquilizer (acepromazine), and/or a sedative (diazepam). Pigs are easily stressed, which, in turn, affects their cardiac function and respiration. All pigs under inhalant anesthesia must be intubated with the use of a laryngoscope. Direct visualization of the larynx for intubation can be difficult because of the small mouth, prominent tongue, long soft palate, and ventral slope of the larynx. Pigs are unable to open their mouth very wide, which contributes to the inability to visualize the larynx. It is recommended to use smaller endotracheal tubes for intubation. 1 to 2 ml of 2% lidocaine applied to the arytenoid cartilage can facilitate the ability to intubate the patient. Temperature, oxygen saturation, heart rate, and respiratory rate should be monitored carefully.

Some pigs have an inherited condition that can cause **malignant hyperthermia** (also recognized as porcine stress syndrome). This syndrome is recognized in large-breed pigs but the condition has also affected potbellied pigs. This specific gene is responsible for a muscle disorder and was most likely inherited when out-crossing mini-pigs with larger breeds. Malignant hyperthermia can occur with inhalant agents. This is characterized by a rapid rise in core body temperature, dyspnea, muscle rigidity, and, if prolonged, death. With this condition, respiration rate usually exceeds heart rate. Immediate cessation of the inhalant anesthesia is required, and the patient should be given 100 percent oxygen. The patient needs to be quickly cooled with ice packs and water baths. If an appropriate level of pre-anesthetics is administered, the level may be sufficient to safely stabilize the patient.

A Simple Blood Test has been developed to test for the presence of the specific gene in potbellied pigs that can cause malignant hyperthermia. Pigs testing positive for this gene should not be bred. Through genetic management (not breeding known positive animals) the gene has all but disappeared in hog farm production animals. Knowing the status of a patient pre-anesthesia is also important for the surgical team. The presence of the gene does not contraindicate anesthesia, but it alerts the anesthetist and surgeon to the possibility of occurrence. Further information regarding testing can be obtained from GenAlysis Lab., Inc.

Radiographs of potbellied pigs are useful to reveal the presence of a foreign body or delayed gastric emptying. Gas retention is often associated with these gastric problems. Lateral/oblique and ventro-dorsal views are taken in anesthetized pigs to detect oral or dental disease. A radiograph is useful to detect urolithiasis in potbellied pigs that are showing difficulty with urination. Lameness in potbellied pigs is due to poor nutrition, being overweight, and inadequate hoof trims. Radiographs are useful to detect arthritis and deterioration of the joints associated with the lameness.

Summary

Potbellied or miniature pigs became very popular in the late 1980s. Due to the demand, these pigs were soon out-crossed with other breeds, and as a consequence, many of these *adorable little piglets* matured much larger than promised, often at 350 lbs or more. Disappointed owners were soon faced with the reality of owning a pig and perceptions changed dramatically. As a consequence, there are many pig rescue facilities, many of them full to capacity. Pigs, even though they have been taken in as house pets, are still classified as agricultural animals, and most town ordinances prohibit the keeping of livestock, which contributed even more to pigs being relinquished to shelters. Pigs are highly intelligent and emotional animals. They are herd animals and begin to establish a dominance hierarchy when they are first born. They are born with very sharp canines called needle teeth, which are clipped out when they are one or two days old to prevent injury to the sow and littermates. Pigs are very clean animals and are easily housebroken or trained to use a litterbox if kept indoors. Their natural behaviors of rooting, digging wallows, and challenges for dominance cannot be changed even with their newly given *pet* status. Pigs are omnivores and will eat anything. Obesity is by far the greatest health problem with companion pigs. Pigs require routine vaccinations and regular deworming, hoof and tooth trims, and ear cleaning, and because of their nature and size, almost all procedures require general anesthesia.

*fast*FACTS

MINIATURE PIGS

WEIGHT
- 227–90.9 kg (50–200 lbs)

LIFE SPAN
- 10–20 years average

REPRODUCTION
- **Sexual maturity:** males: 12 weeks (3 months); females: 16–36 weeks (4–6 months)
- **Gestation:** 114 days (3 months, 3 weeks, 3 days)
- **Litter size:** 4–8
- **Weaning age:** 6–8 weeks

VITAL STATISTICS
- **Temperature:** 38.8°C (102°F)
- **Heart rate:** 56–58 beats/min
- **Respiratory rate:** 15–20 breaths/min

DENTAL
- Total 44 teeth
- Dental formula 2 (I 3/3, C 1/1, P 4/4, M 3/3)

VACCINES
- **Erysipelas:** 8–12 weeks, 3 weeks later, annually
- **Leptospirosis:** 8–12 weeks, 3 weeks later, annually
- **Parvovirus (breeder pig):** 5–6 months, 3 weeks later; 3–8 weeks prior to breeding
- *Mycoplasma hyopneumoniae:* 1 week, 3 weeks later, annually

ZOONOTIC POTENTIAL
- Erysipelas
- Coccidia
- Salmonella

Review Questions

1. What are needle teeth and why should they be clipped soon after birth?
2. Why is it important to provide shelter for pigs?
3. What procedures should be included in routine veterinary care?
4. How do the natural behaviors of a pig conflict with human expectations in a companion pig?
5. What vaccinations are recommended for companion pigs?
6. What is prohibited to feed pigs and why?
7. Why should an owner not rub a pig's abdomen?
8. Define the following terms: boar, barrow, gilt, sow.
9. What is pseudorabies?
10. What are the signs of Dippity Pig syndrome?

Case Study I

History: A client telephones requesting a house call for a pet pig. The pig cannot walk, but "drags her hind legs and scoots around on her knees." The owner states that the pig must have been "bitten by a snake or something."

Physical Examination: The house call has been arranged and the veterinarian and her technician arrive to find no one home, but the pig grunts a greeting from the backyard. As they have been called and given permission to do "whatever," the team approach the pig and immediately see the problem. The hooves of all four feet are so overgrown that it has become impossible for the pig to stand.

   a. What will likely be the first procedure performed?
   b. Describe how the second procedure will be achieved.

Case Study II

History: A six-month-old gilt had been admitted for a spay. She has been previously seen as a patient for routine vaccinations and deworming.

Physical Examination: She weighs 67 lbs, and aside from being overweight, the presurgical physical examination indicates that she is in good health. During the operation the patient's vital signs suddenly change dramatically: respiration increases to 62 breaths/min, heart rate is 54 beats/min, and the core body temperature has risen to 105°F.

a. What is the crisis with this patient?

b. What immediate actions should be taken by the anesthetist?

For Further Reference

Blaney, J. (2013). *Diseases/Illnesses—Dippity Pig Syndrome*. California Potbellied Pig Association, Inc. Pleasant Hill, CA

Braun, W., Jr. (1993). Anesthetics and Surgical Techniques Useful in the Potbellied Pig. A Symposium on Potbellied Pigs. *Veterinary Medicine* 88(5):441–447.

Braun, W., Jr. (1993). Helping Your Clients Raise Healthy Potbellied Pigs. A Symposium on Potbellied Pigs. *Veterinary Medicine* 88(5):414–428.

Braun, W., Jr. (1993). When the Patient Is a Pig: Providing Basic Care for Potbellied Pigs. *Veterinary Medicine* 88(5):412.

Ching, F. Ching Farm Sanctuary, Herriman, Utah. Direct communication 2003–2006.

http://www.aphis.usda.gov (accessed June 2006).

http://www.cppa4pigs.org (accessed June 6, 2014).

http://www.cppa4pigs.org (accessed June 26, 2014).

http://www.genalysislaboratory.com (accessed June 1, 2014).

http://www. ncbi.nim.gov (accessed June 7, 2014).

http://www.oaklandzoo.org (accessed June 2006).

http://thepigsite.com (accessed March 6, 2006).

http://www.pigs4ever.com (accessed June 14, 2006).

http://www.trendhunter.com (accessed January 22, 2014).

http://www.twycrosszoo.org (accessed June 7, 2014).

http://www.vspn.org (accessed June 26, 2014).

Ko, J.C.H., Thurmon, J.C., & Benson, J. (1993). Problems Encountered When Anesthetizing Potbellied Pigs. Symposium on Potbellied Pigs. *Veterinary Medicine* 88(5):435–440.

Masson, J.M., & McCarthy, S. (1996). *When Elephants Weep: The Emotional Lives of Animals*. New York: Delta/Dell.

Orr, M.G. (2006). Bird and Exotic All Pets Hospital, Draper, Utah. Direct communication, 1998–2006, regarding recent protocol for injectable anesthetic agents.

UNIT IX

Tsekhmister/Shutterstock

OBJECTIVES

After completing the chapter, the student should be able to

- Describe the correct housing for a backyard chicken flock.
- Provide client education to new chicken owners.
- Provide basic nursing care for chickens.
- Demonstrate restraint techniques for chickens.
- Understand common medical disorders of chickens.

Introduction

Domestication of the chicken (*Gallus* gallus domesticus) can be traced back to 2000 BC. The domestic chicken is a descendant of four species of jungle fowl native to Southeast Asia, with the Red jungle fowl being the most common ancestor. There are over 19 billion chickens worldwide, making the chicken the most common domestic and agricultural animal. Cockfighting, an ancient sport of pitting one bird against another, is believed to have been popular in ancient Greece as long ago as 500 BC. This had a major influence on the domestication and distribution of chickens worldwide. In 1873, the American Poultry Association was organized to help adopt standards of excellence and establish ways to classify the breeds. Poultry is a collective term for farmed domestic birds including turkeys, ducks, and geese.

Two distinct industries have grown from the different types of domestic fowl (poultry). There are purebred chickens that are prized for conformation, plumage, and color. For many people, raising fowl for exhibition has become a hobby and a passion to maintain the purity of many different breeds. The commercial aspect of the poultry industry focuses on chickens bred for meat and egg production. Backyard chicken flocks have become very popular, and the numbers of hobbyists have increased dramatically over the past several years. City regulations in many communities have been amended to allow small flocks, or groups, of chickens to be kept within suburban neighborhoods. Most of the regulations do not allow roosters (male chickens) to live within city limits because of the noise they make. The crowing of roosters is disruptive and loud, and they call not only at dawn but

FIGURE 22-1 A Silkie rooster.

throughout the day, sometimes challenging other neighborhood roosters or calling to their own flock (**FIGURE 22-1**).

With the rise in popularity also comes the downside; animal rescue facilities are seeing an increase in the numbers of backyard chickens because the owners cannot provide for them, and they are relinquished or abandoned. Many times, they are simply turned loose to fend for themselves, creating a public nuisance.

Chicks are most often purchased when they are only a week old and the sex has not been determined. These unsexed chicks are referred to as a **straight run**. As the chicks get a little older, it becomes apparent that there are roosters within the flock. Owners try to relinquish them to animal rescue groups or turn the roosters loose in the neighborhoods because of the regulations prohibiting roosters on the property. Another contributing factor in the increased number of abandoned chickens is that many owners do not want the chickens as they age and the egg count decreases, and they are reluctant to slaughter them for meat.

Chickens can be a great source of enjoyment if properly managed. Many owners think of their chickens as companions and pets. The emotional bond is strong and not unlike the bond between an owner with a more traditional pet (**FIGURE 22-2**). Pet poultry have greater sentimental value than production poultry, and owners are more likely to seek veterinary care.

FIGURE 22-2 A Maran hen.
(© luna4/Shutterstock.com)

Several hundred breeds are available from hatcheries that specialize in chickens for the hobbyist. Choosing the right breed can be difficult, but there are several that have proven to be excellent choices for a backyard poultry flock. The Plymouth Rock is a large but docile and friendly breed and an excellent egg layer (**FIGURE 22-3**). The Rhode Island Red is also an excellent egg layer, and both these breeds lay brown eggs. The Ameraucana and Araucana make good pets and lay medium-sized, blue or green eggs, adding interest and novelty (**FIGURE 22-4**). Leghorns are a little more flighty and noisier but are rated as the top white egg layers.

There Is No Nutritional Difference between a White Egg and a Brown Egg. The preference for brown eggs is consumer driven. When specifically asked why someone prefers brown eggs over white eggs, the responses are usually, "They are fresher," "farm fresh eggs are always brown," or "it means they are free-range, organic, healthier," yet the color of an eggshell actually means none of these things. Eggshell colors are derived from pigments produced by the hen when the shell is formed, and the color of the egg depends on the breed of the chicken (**FIGURE 22-5**).

FIGURE 22-3 A Plymouth Rock hen.

FIGURE 22-4 An Araucana hen.

FIGURE 22-5 Different colors of eggs are produced by the hen. There is no nutritional difference in eggs of different colors.

Ornamental chicken breeds are also very popular as pet chickens because of their smaller size and big personalities, not for their egg-laying ability, or meat production. They are docile and many are unusual looking. A few of these include Silkies (**FIGURE 22-6**), Bantams (**FIGURE 22-7**), and Polish (**FIGURE 22-8**).

FIGURE 22-6 A Silkie hen.

FIGURE 22-7 A Bantam hen.

FIGURE 22-8 A Polish hen with the typical head feathering of this breed.

Environmental factors affect the success or failure of incubating fertilized eggs both naturally, by the laying hen, or in home incubators. Temperature, humidity, and improperly turning the eggs all contribute to hatching failures. Eggs must be turned to prevent the embryo chick from adhering to the shell wall, and hens gently turn their eggs daily.

One to two days prior to hatching, the chick uses an **egg tooth**, a tiny sharp projection on the tip of the upper beak, and begins to peck at the inside of the shell. This is referred to as **pipping**, and the chick creates a hole in the shell, which allows the chick to begin breathing outside air. The chick can be heard peeping while still within the shell. Once the chick is completely free from the shell, it is wet and exhausted from the whole ordeal, which can take up to 24 hours (**FIGURE 22-9**). Over the next several days, the egg tooth will gradually disappear. Some chicks hatch with a **yolk sac**, a small sac attached to the embryo that provides nutrients to the developing chick. A newly hatched chick with a retained egg sac continues to absorb these nutrients from the sac over the next couple of days until it is depleted. Chicks born with egg sac should not be disturbed or handled unnecessarily.

Newly hatched chicks are covered in soft down feathers, but these are insufficient to keep them warm. The hen nestles them under her fluffed-up breast feathers and wings to keep them warm. This period of time is called **brooding**, which lasts an average of three to four weeks, during the time chicks begin growing feathers to replace the down. Keeping chicks warm without the hen requires a heat source to keep the environmental temperature around 110°F. The best source of heat is a 250-watt infrared light. The perimeter temperature of the enclosure should be maintained at around 84°F. The temperature can be decreased by 5°F each week to allow the chicks to adjust and adapt to staying warm by themselves (**FIGURE 22-10**). Chicks need to have an enclosure for the first few weeks to keep them from wandering away from the heat source and dying from

FIGURE 22-9 Pipping occurs when the chick uses its egg tooth to peck at the inside of the shell immediately prior to hatching. Once the chick is completely free of the shell, it is wet and exhausted from the whole ordeal, which can take up to 24 hours. (© Anneka/ Shutterstock.com)

hypothermia. A large rubber stock tank makes an ideal enclosure; the sides are high enough to prevent chicks from jumping out, and it can be easily cleaned. The pen should be cleaned daily and the bedding changed to prevent disease and keep the chicks clean. Aspen shavings are one of the best and most readily available type of bedding because aspen is absorbent and retains warmth. Straw is another commonly used bedding, but is too coarse for small chicks, making it difficult for them to walk, and the stalks have been known to cause eye injuries.

Behavior

Chicken behaviors are not only learned but also instinctive. Thanks to noninvasive techniques, it is known that unhatched chicks can hear and see and that they learn from these experiences to communicate with others once hatched. Chickens are poor fliers but still instinctively roost on perches above the ground to avoid predators.

Flocks of any size maintain a strict **pecking order**; that is, each member of the flock has a specific social status within the flock. The dominant chicken is the caretaker of the flock and warns the others of danger, calls to them when a particularly choice food source is discovered, and often settles squabbles between the others. The pecking order establishes and maintains order within the flock, but a dominant hen or rooster can become overtly aggressive with members of the rest of the flock and people. The dominant chicken will raise the hackle feathers on the neck, stand almost on its toes, demonstrate wing flapping, and beat a subordinate bird with the wings. In some cases, the dominance behavior becomes so aggressive that the dominant hen or rooster plucks out feathers, injures or kills the weaker chicken, and begins to cannibalize the weaker chicken (**FIGURE 22-11**). This display of dominance is sometimes directed toward people, and occasionally, it is more than just a show or a bluff. Chickens have been known to fly at a person, deliver a nasty peck, and rake their spurs on any exposed flesh. Both sexes of chickens have a **spur** on the dorsal side of both legs, but they are more developed in roosters. Spurs are very sharp and capable of inflicting a deep laceration.

FIGURE 22-11 An example of a hen that has been plucked by dominant flock members.

Chickens enjoy taking dust baths, but dusting also keeps the feathers clean and helps with external parasite control. Chickens dig a small depression in the dirt, flop down on one side, and start kicking with their feet to stir up the dust, and then roll over and perform the same ritual on the opposite side.

To Help Control Lice and Mites, many keepers use this behavior, the chicken's urge to bathe in the dust to their advantage by adding **diatomaceous earth** to the area where chickens most often bathe in the dust. Diatomaceous earth is a fine white or gray powder made up of small particles—the skeletons of microscopic diatoms. It is also used as a feed additive and to treat diarrhea.

Chickens scrape their beaks on the ground or a solid surface to clean away debris. It is a side-to-side movement usually followed by a ruffling and resettling of the feathers. Similar to other birds, chickens preen themselves but also fluff up their feathers to set them back in place after being petted or restrained.

Poultry have the ability to perceive vibrations from the ground and in the environment. They have sensory receptors located on the legs and skin. It is thought that this helps warn them of predators. A rooster or a dominant hen will alert the rest of the group with a range of vocalizations to alert them of danger.

Chickens use 30 different sounds to communicate with each other. The sound a mother hen makes to call her chicks to the food source, the **cackle** (a distinctive vocalization) from a hen after laying an egg, and the rooster's calls for the hens to follow him are among those sounds.

Hens do not need roosters to produce eggs. Pullets, young females, reach sexual maturity and begin producing eggs around five to seven months depending on the breed. Increasing the length of the light cycle helps to stimulate egg production.

Some chickens may be slower to mature and begin laying eggs or may refuse to lay their eggs in a nesting box. Placing a dummy egg in the nest box can encourage egg laying within the nesting box. Golf balls, painted rocks, plastic eggs, and small, smooth pieces of wood are all items that can be used. Hens also share nesting boxes, and the visual presence of another hen helps encourage a hen to lay her eggs in the nest box.

Molting is a natural process that occurs during the late summer. During this time chickens gradually lose old feathers, which are replaced with new feathers. A full molt can take up to three months and is dependent on a variety of factors including health status, diet, and seasonal temperatures. During the molt some hens stop laying eggs. Other hens that have been daily egg producers may lay erratically, that is, producing only one egg every few days. This is normal behavior and not a cause for concern.

Housing

Chickens are adaptable but need to be provided with adequate shelter to protect them from the weather and night predators. Chicken houses, or coops, come in a variety of shapes and sizes, and many of them are designed to enhance the landscape. The coop should provide for air circulation and yet be small enough to protect the chickens from the elements. To avoid overcrowding, the recommendation is to provide 1.5 to 2.0 square feet of floor space per adult chicken. Coops should have a variety of nesting boxes to provide a clean area for the hens to lay their eggs and should also be designed for easy access to gather the eggs. The nest boxes should be filled with soft bedding that is at least two to three inches deep to prevent breakage of the eggs. The coop must have enough perches, or roosts, to allow the birds to stay off the floor at night. Roosting at night is a natural behavior to avoid predators. The coop should provide for access to food and water. Many coops have ramps for the chickens to use, and the ramp is lifted and secured at night to further protect the flock from raccoons, foxes, and other invaders. Chickens always *put themselves to bed*, going into the coop at dusk. Coops are often attached to an outside, fenced run (**FIGURE 22-12**).

Diet

Chickens are omnivores and eat a variety of insects, snails, snakes, mice, and greens, including weeds. They may also attack and consume small wild birds that have become trapped in their pens. A variety of commercial seeds mixes (scratch) and pelleted diets are available for poultry. Commercial feeds are formulated to meet the needs of growing or producing chickens and are labeled as starter, grower, or layer feeds. Nutritional requirements vary depending on the age, season, and type of poultry. For example, in colder months and during molting periods, chickens need more protein in their diet, and chicks require small pieces, or *crumble*, which is easier for them to eat. Most commercial

premixed rations do not need supplementation unless the diet is not selected to meet the needs of the chickens. During heavy egg-laying periods, calcium should be supplemented in the form of oyster shells to prevent the eggshells from becoming thin and breaking easily. Grit, small pieces of sand and stone, are essential in the diet of chickens to help with the breakdown of their feed. Grit remains in the gizzard, a grinding muscular organ of the digestive system. Chickens pick up sand and pebbles from their pen throughout the day. Trough feeders are usually used to feed young chicks. Adult chickens can be fed in shallow bowls that vary in size, or the feed can be just scattered on the ground. This gives them the opportunity to engage in natural foraging behavior. Outside, free-range chickens spend most of their days scratching around, looking for food morsels (FIGURE 22-13).

FIGURE 22-13 Chickens can be fed in a pan or the grain can be scattered on the ground.

Stored feed should be kept off the floor in securely lidded containers to keep it dry and away from rodents. If the feed smells rancid or looks moldy, it should be thrown away. Questionable feed can cause health problems, nutritional deficiencies, and in some cases death if consumed. Totally free-range chickens, those allowed access to outside space, will rarely be able to consume a balanced diet. Most people feed table scraps to their flocks, and this can include vegetable peelings, salad ingredients, crushed shells of used eggs, fish and meat scraps, and the shells of peeled shrimp. Food items that should not be fed include onions, avocado or avocado skins, apple cores with seeds (these are toxic to birds), and all spoiled household food. Chickens will come running when presented with this buffet.

A high-quality food can be turned into a mash or gruel with warm water and be fed to a chicken that is recovering from an illness or injury. Offering favorite foods and stirring the mixture in front of the chicken can stimulate the bird's appetite. The addition of Nutrical® paste (a high-calorie dietary paste) to the food or the water provides a debilitated chicken with a concentrated source of calories. When added to the water bowl, a slight swirling of the water makes the paste appear to be a wiggling worm that will tempt most chickens to grab and swallow it; another benefit is that it will not dissolve in water. Commercial avian pellets, soaked and turned into a gruel, also provide balanced nutrition for a recovering bird. Hospitalized chickens that are reluctant to eat can be fed using a large avian feeding tube and syringe to place food and medication directly into the crop.

Water is the most important nutrient for chickens. There is a direct relationship between the amount of water a chicken drinks and the amount of food it consumes. Inadequate amounts of water will affect egg production, growth, and the health of all body systems. It is important that adequate freshwater is provided every day and the water container is cleaned to prevent bacterial growth and disease. In cool weather, an adult chicken consumes an average of 0.05 to 0.08 gallon of water per day. In hot weather, an adult will consume 0.08 to 0.16 gallon per day.

Handing and Restraint

When attempting to capture a chicken, do not chase after it but first try and tempt the bird with food. If it is a social and well-handled bird, it will not be alarmed and will very likely approach. Attempting to chase the chicken (they can run as fast as nine miles per hour) will be futile and is very stressful for the bird and alarming to the entire flock. It is easier to slowly *herd* the chicken into a corner, talking softly. Give the bird a time to assess the situation. When the bird is close to the feet of the captor, the fingers of both hands are spread wide and lowered over the back of the chicken. In most instances, the bird will coopy down, that is, squat on the ground as if presenting to a rooster or another dominant hen. It may then simply be picked up and held with both hands, holding the wings close to the body. Once the chicken is caught, place one hand underneath the chicken and hold both legs between the fingers and place the free arm around the chicken's body (**FIGURE 22-14**). If the chicken becomes anxious and attempts escape, the head can be tucked between the arm and the body of the restrainer. Chickens may also be calmed by tucking the head under a wing. *Rocking*, or swaying the chicken slowly back and forth, will induce a hypnotic state. The chicken can be placed

FIGURE 22-14 Chickens are easily restrained by holding them close to the body and grasping both feet.

FIGURE 22-14 Chickens are easily restrained by holding them close to the body and grasping both feet.

back on the ground, and it will not move until it is gently nudged. If it is impossible to capture a chicken, wait until the flock has roosted for the night. Enter the coop with a flash light and simply collect the desired bird from the roost.

Medical Concerns

Most small backyard chicken flocks are relatively healthy. Chickens hide signs of illness as other flock members may attack a bird when it appears weak. It is not uncommon to see trauma cases from predator attacks. The injuries may involve leg fractures and soft tissue wounds. Leg fractures pose a problem in treatment due to the weight-bearing ratio of a chicken's leg and are complicated further in that these bones are pneumatic and are directly involved in the respiratory system. Lethargy, anorexia, abnormal vocalizations and posture, and atrophy of the pectoral muscles are general signs of illness in a chicken (**FIGURE 22-15**).

A chicken's comb can be also be an indicator of health. The comb is a fleshy crest on top of the head. In most breeds, the color is bright red and stands upright. In some breeds, such as the Silkies, the comb is black or dark blue. If a

FIGURE 22-15 A chicken that is ill will often lay in an abnormal crouched position. (© cynoclub/Shutterstock.com)

chicken is suffering from an illness, the comb appears to have lost color and often becomes limp and falls to one side.

Respiratory Disorders

Newcastle disease and avian influenza are viral diseases carried by wild birds, waterfowl, and poultry originating from diseased flocks. Both diseases cause high mortality. Initial signs may include severe depression, nasal discharge, unusual coloring or swelling in the comb, and a twisted neck. Confirmed cases of either of these diseases must be reported to the state veterinarian and to the Department of Agriculture. This is a zoonotic disease that can cause conjunctivitis and flu-like symptoms in humans.

Mycoplasma causes chronic respiratory disease in poultry. Mild signs include sneezing and head shaking. Oral antibiotics are given to help treat air sacculitis, an infection of the air sacs.

Avian influenza (*bird flu*) is a viral disease transmitted by mosquitoes and through the droppings or nasal secretions of infected wild birds. Birds can suddenly develop influenza and die. Slower progression of the disease can result in diarrhea, depression, discolored combs and wattles (fleshy appendage hanging off the neck), and respiratory signs. Bird flu, the specific H5N1 virus, has zoonotic potential and can cause conditions in humans that vary from conjunctivitis to severe pneumonia. There is a vaccine to protect against avian influenza, but it has been shown that vaccinated birds remain carriers if they are exposed to the virus.

Aspergillosis is a fungal disease that attacks the respiratory tract. The fungus thrives in unsanitary environments. The fungal spores are inhaled and settle in the bird's air sacs. Respiratory signs vary and are nonspecific depending on the degree of infection. Antifungal therapy is administered and can be effective in treating aspergillosis in most cases. With severe infections, lesions are created in the air sacs and can result in the death of the bird.

Digestive Disorders

Thrush (candidiasis) is a yeast infection caused by unsanitary conditions and careless storage of food. The infection grows inside the crop and the mouth and involves the upper digestive system. Birds become lethargic, with ruffled feathers; the vent is inflamed; and birds have chronic diarrhea. Treatment involves cleaning the environment and treating the affected bird with antifungal medications.

Two strains of Salmonella are seen in poultry. *Salmonella pullorum* is associated with high mortality in birds that are three weeks old or younger. Chickens are the most commonly affected species by this bacterium, but it can be seen in other domesticated fowl, parrots, and doves. Transmission occurs orally or via the yolk of the egg laid by a positive hen. Death rates increase under stressful conditions or in already health-compromised chickens. Common signs include decreased appetite, lethargy, white diarrhea (white pasting around vent, commonly called Pasty Butt), gasping, and possible lameness. Diagnostic testing in a clinical setting for individual birds includes culturing the droppings on a selective culture media or performing a simple blood agglutination test on a drop of blood. Clinics may find these test cards available through their County Extension agent as a negative test is required to participate in any exhibition. Antibiotic therapy is recommended for individual chickens or euthanasia if a flock is involved. Most birds that have recovered from any of the Salmonella subspecies are resistant to the effects of being re-infected but can become asymptomatic carriers. Vaccines are not used to prevent or treat this disease because vaccines can interfere with the elimination of carrier birds and testing for the disease.

Nondigestive Disorders

Marek's disease (MD) has become one of the most commonly diagnosed diseases in backyard chickens. This disease is caused by a herpes virus that causes tumors in various parts of the bird. Silkies and Sebrights are particularly susceptible to MD. Once birds become infected with the disease and survive, they become lifetime carriers. The disease is highly contagious and is spread through feather dander. There are three forms of this disease, defined by where the tumors are found: ocular, visceral, or neural. Birds often die of secondary problems brought on by the weakened immune system. Paralysis, one wing drooping, head tilt, and changes in the droppings are signs of MD. A vaccine is available for poultry, and it is administered to chicks while they are still in the egg or up to one day after hatching. There is no cure for this disease, only supportive treatment.

Fowl pox is a viral disease that is spread by biting insects and by direct contact with infected poultry. The virus enters through the comb and wattles and from scratches or scrapes. Scabs form in the mouth and throat. Most birds improve without treatment, but a few may suffer from secondary infections. Most birds recover and become immune for the rest of their lives. A vaccine is available, and it is recommended to be given every few years to maintain a high level of antibody protection.

There are several vaccinations for chickens. Backyard hobbyists often see no need to vaccinate their small, isolated flock while other clients may request complete vaccinations. The most accurate sources of information about vaccines and recommendations for a specific area are the state veterinarian's office or the local county agricultural extension office. Recommendations often change when disease outbreaks are reported.

Disorders of the reproductive tract can occur in hens with high egg production. Metritis, an inflammation of the oviduct, is often caused by a bacterial infection. It can affect eggshell formation, uterine contractions, and growth in the chick. Metritis can be responsible for other conditions such as egg binding and egg yolk coelomitis.

Egg binding occurs when the hen is unable to deliver the egg, and this condition can rapidly become life threatening. Nutritional deficiency, obesity, young hens, and an excessively large egg are common causes of this condition. Signs of egg binding include walking with legs extended laterally, squatting and straining, frequent passing of wet droppings, and ataxia. Initial treatment may start with applying a warm, damp cloth to the vent and a water-based lubricant around the vent opening with a cotton-tipped applicator. If the hen is still unable to lay the egg, veterinary assistance is required. Veterinary treatment includes injections of calcium gluconate to elevate the blood calcium level and **oxytocin**, a hormone that stimulates uterine contractions. The hen should be placed in warmed incubator with a damp towel and observed carefully. If she is still unable to deliver the egg, further, more invasive assistance is required. The veterinarian may attempt a manual delivery of the whole egg or implode the egg if it is visible in the cloaca. Implosion of the egg necessitates inserting a large bore needle (16 to 18 gauge) into the visible portion of the eggshell and carefully withdrawing the contents into a syringe. This collapses the egg, and the shell may be withdrawn with forceps. If a radiograph reveals that the egg is well up within the oviduct, anesthesia and surgical removal may be necessary.

Egg yolk coelomitis is caused by the presence of yolk within the abdominal cavity. The cause is usually from an egg that has broken within the oviduct. This creates an inflammatory response in the abdominal lining (peritoneum) and leads to a bacterial infection and can result in death of the hen.

Inadequate housing and poor nutrition can cause various leg or feet problems in poultry. Splayed leg in young birds is a condition where the legs fan out laterally (splay) and the bird is unable to stand up straight. This results from a young bird being housed on a slick or hard surface. Genetic deformities of the leg and toes are associated with poor nutrition from the hen. Trauma to the leg or toes can be a result of poor housing, leg bands, or predators (**FIGURE 22-16**).

FIGURE 22-16 Frost bite on a chicken's toes due to inadequate housing.

Parasites

Lice and mites are frequently seen ectoparasites in backyard chickens. Paleness of the comb, which suggests anemia, and a decreased pectoral mass with a prominent keel bone are indicators of possible parasitism. The Northern Fowl mite (*Ornithonyssus* sp.) appear as black or gray colonies on the face, comb, and wattle. These mites invade the base of the feather shaft and cause scaling and anemia. It is difficult to eradicate the mites because they live in the wood of the coop. The Red Roost mite (*Dermanyssus gallinae*) attacks birds at night and infests the cracks or joints of roosts and nests. This mite is the most common and difficult to eradicate from the coop. Heavy infestation of mites have the potential of killing a young bird as they, like all ectoparasites, are blood feeders and cause anemia. The Scaly Leg mite (*Knemidocoptes*) burrows between the scales of the leg and foot and causes problems that can make it difficult for the chicken to walk. Damp conditions, contaminated bedding, and inadequate ventilation can predispose a chicken to this mite. The damaged scales should be carefully removed following a warm water soak with Epsom salts. Ivermectin injections kill the adult Scaly Leg mite (**FIGURE 22-17**).

FIGURE 22-17 An example of Scaly Leg, which is caused by a mite.

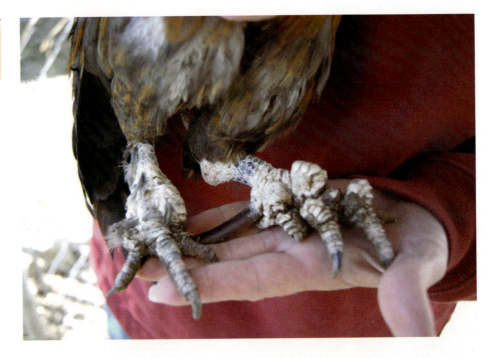

When examining a bird for the presence of lice, carefully check beneath the wings, on the head and legs, and around the vent. The eggs of lice (nits) appear as white clusters attached to the feathers. Lice live their entire life cycle on the chicken. Commercial powder can be used to dust the bird to treat against lice and mites.

Protozoan parasites in poultry commonly cause coccidiosis and blackhead. The protozoan, coccidia, is transferred from chicken to chicken through droppings. Large populations predispose the flock to coccidia. A diseased chicken will be hunched over, have fluffed feathers, have diarrhea, and be reluctant to move. Coccidia can cause high mortality in chickens.

Histomonas meleagridis is the protozoan responsible for blackhead disease. This protozoan causes a 10 to 15 percent mortality in chickens. Transmission occurs through a nematode (roundworm), *Heterakis gallinarum*, or through a **paratenic** host, a host that is an incidental substitute for the usual host, such as the earthworm. The parasite attacks the liver and cecum of the bird. Clinical signs are a blue or black discoloration of the skin on the head (cyanosis) and yellow, watery droppings. Currently, there is no medication that has shown to have a positive outcome in treating the disease. A roundworm commonly found in chickens (Ascaridia) is seen in backyard chickens that are raised in a densely populated environment where the birds are overcrowded. The roundworm lives in the intestinal tract and causes weight loss and diarrhea. Fenbendazole has shown to be an effective treatment against most roundworm infections.

Clinical Procedures

The techniques used in collecting diagnostic samples from poultry are the same used with the pet parrot. To collect blood from a chicken, a 23- to 25-gauge needle on a 3-ml syringe is used depending on the size of chicken. The maximum amount of blood that can be safely collected is one percent of the bird's bodyweight. The brachial vein (wing vein) is a good site for blood collection. The jugular vein and the medial metatarsal (leg vein) can also be used as collection sites.

Oral antibiotic therapy is preferred to injectable antibiotics. Owners placing the antibiotic in drinking water need to be cautioned when utilizing this route. A debilitated chicken may not be drinking and most medications will start to destabilize chemically after a short period of time in water. Placing the antibiotic in a syringe and delivering the dose directly into the mouth is more effective; however, there is a risk of aspiration since the glottis is just caudal to the tongue. Clients should be reminded to give oral medications slowly, place the medication in the lower beak, and allow the patient to swallow to prevent aspiration.

Dehydration can further complicate any medical issue. The hydration status can be assessed by slightly pinching loose skin just behind the neck to see if it falls back into place quickly. To correct mild dehydration, Gatorade® or Pedialyte® can be diluted 50/50 with water. Place this solution in a syringe and slowly drip it into the mouth by gently opening the beak and placing the drops on the inside of the lower beak. This method will prevent aspiration. The advantage of using either of these is that they both contain electrolytes.

Inhalant anesthesia is recommended over injectable anesthesia in performing surgical procedures in chickens. Anesthesia can be induced by masking the chicken using either isoflurane or sevoflurane. Injectable anesthetics such as ketamine, diazepam, xylazine, and propofol have been used and have been shown to be effective for surgical anesthesia in chickens. Accessing a vein in a chicken can be difficult, and the vein can easily collapse during anesthesia administration. Recovery from injectable anesthetics is prolonged. The recovering patient should be wrapped in a towel and held, or placed in a prewarmed incubator and supported sternally with the use of rolled towels. Anesthesia is required when attempting fracture repair, tumor removal, or soft tissue repair.

Radiographs, as with all species, assist with diagnosis of a disease, fracture, and other abnormalities. The use of digital radiography has provided the veterinarian with the ability to easily visualize bone and soft tissue problems. The hypnotizing technique used for restraint can assist with taking radiographs without anesthesia. If positioning the chicken requires anesthesia, masking down the bird with gas anesthesia is recommended. Ventro-dorsal and lateral views are taken of the whole body of a chicken. Lateral and anterior/posterior views of the wings and legs are taken to detect fractures and bone disease.

Rigid endoscopy techniques are utilized to visualize internal organs, perform biopsies of tissue, collect samples for culture, and remove foreign bodies. During endoscopic procedures, the chicken is anesthetized with an inhalant anesthesia.

Summary

An increased interest in backyard chickens has brought more chicken patients into the veterinary hospital. Preventative health care is important in keeping the chicken healthy. Most medical disorders are a reflection of poor husbandry and malnutrition. Common signs of illness in a chicken are poor feathering, weight loss (prominent keel bone), faded wattle and comb coloring, lethargy, change in droppings consistency, and nasal discharge. Mites, lice, and coccidia are common parasites found in chickens. The treatment for most parasitic diseases involves not only treating the patient but also ridding the environment of the parasite. Clinical procedures in a chicken are similar to those in other avian patients.

*fast*FACTS

BACKYARD CHICKENS

ADULT WEIGHT
- **Ornamental breeds:** 4.5–7 lbs average
- **Egg-layer breeds:** 6–9.5 lbs average
- **Meat breeds:** 8.5 – 11 lbs average

LIFE SPAN
- 8 years

REPRODUCTION
- **Sexual maturity:** 18–24 weeks
- An egg is normally produced every 24–26 hours
- **Incubation:** 21 days to hatch

VITAL STATISTICS
- **Temperature:** 102–104°F
- **Heart rate:** 250–300 beats/min

- **Respiratory rate:** 15–30 breaths/min (hens normally have a higher respiratory rate than roosters)

ZOONOTIC POTENTIAL
Bacterial
- Salmonella
- Mycobacteria
- Campylobacter
- Colibacter (*E. coli*)
- Pasteurella

Viral
- Newcastle disease
- Bird flu (H5N1 virus) *potential*

Parasitic
- Cryptosporidia
- Coccidia

Review Questions

1. Describe a brooding hen and the reason for this behavior.
2. What should the average environmental temperature be for young chicks?
3. What is the purpose of the pecking order?
4. List several food supplements that can be feed to chickens.
5. Describe one method of increasing the nutritional value of food to recovering chicken.
6. List several signs that would indicate an ill chicken.
7. Where are lice commonly found on a chicken?
8. Why is Newcastle disease a reportable disease?
9. What is the recommended route for administering antibiotics to chickens?
10. Why is it necessary for chickens to be able to roost?

Case Study I

History: A chicken is presented at the veterinary hospital with scaling around the face and neck area. The client's chief concern is that the chicken has not been very active lately and the comb does not seem as red and flops to one side.

Physical Examination: The patient is very subdued and severely underweight with a prominent keel bone. The technician performing the initial examination also notes evidence of diarrhea around the vent.

   a. What are some of the causes that have contributed to the condition of this bird?

   b. What further questions need to be asked of the client to aid in a diagnosis by the veterinarian?

Case Study II

History: A young pullet is struggling to walk and appears to be dragging the right wing. The client reports that he has a dozen birds the same age and feeds them "chicken scratch." They all seem fine except, he adds as an afterthought, they seem to have diarrhea.

Physical Examination: The patient is unable to walk without support. The right wing is weak and held lower than the left.

   a. What disease would be an initial rule-out for the veterinarian?

   b. What diagnostic tests are likely to be recommended?

For Further Reference

Backyard Chicken: Chicken Behavior. http://www .youtube.com (accessed 2014).

Barber, J. (2012). *The Chicken: A Natural History*, Race Point Publishing. New York, New York.

Clouse, M.B. *Caring for a Sick or Injured Chicken.* http://www.brittonclouse.com (accessed 2014).

Flora, J.H. (1977). *ABC of Poultry Raising: A Complete Guide for the Beginner or Expert, 2nd Edition.* New York: Dover Publications.

Frame, D.D. (2010). *Basics for Raising Backyard Chickens.* USU Extension. AG/Poultry/2010-02pr.

Frame, D.D. (2012). *Causes and Control of Selected Diseases Related to Backyard Chicken Flocks in Utah.* USU Extension. AG/Poultry/2012-04pr.

Harris, D.J. (2011, November). *Backyard Chickens* (pp. 73–77). NAVC Clinician's Brief.

Harris, D.J. (2012). *Causes and Control of Selected Diseases Related to Backyard Chicken Flocks in Utah.* USU Extension. AG/Poultry/2012-04pr.

http://www.avianweb.com (accessed January 15, 2014).

http://www.backyardchickens.com (accessed March 21, 2014).

http://www.cdc.gov (accessed March 21, 2014).

http://www.edis.ifa.ufl.edu (accessed March 21, 2014).

http://www.mypetchicken.com (accessed January 18, 2014).

http://www.thepoultrysite.com (accessed 2014).

Salmonella Pullorum, Pullorum Disease, 'Bacillary White Diarrhoea'. The Poultry Site. http://www.thepoultrysite (accessed 2014).

Stone, N. (2007, July). *Common Diseases of Backyard Poultry.* State of Victoria Government Agriculture. AG1012.

Professional Organizations and Associations

American Veterinary Medical Association

1931 N. Meacham Road, Suite 100
Schaumburg, IL 60173-4360
www.avma.org

ASPCA Animal Poison Control Center

1717 S. Philo Road, Suite 36
Urbana, IL 61802
www.aspca.org/pet-care/animal-poison-control
(888)426-4435
(24/7 Animal Poison Control Center hotline)
Emergency contact, fee may apply.

Association of Avian Veterinarians

P O Box 9
Teaneck, NJ 07666
720-458-4111
www.aav.org

Association of Exotic Mammal Veterinarians (AEMV)

618 Church Street, Suite 220
Nashville, TN 37219
www.aemv.org
Memberships open to veterinarians, veterinary staff, and veterinary students.

Association of Reptilian and Amphibian Veterinarians

810 East 10th, P O Box 1897
Lawrence, KS 66044
(800)627-0326
arav.allenpress.com/arav/
Quarterly Bulletin, Membership Directory, Annual Conference and Proceedings
Memberships open to veterinarians, veterinary technicians, and veterinary students.

Association of Veterinary Technician Educators

11428 38th Street South
Horace, ND 58047
www.avte.net

Association of Zoo Veterinary Technicians

www.azvt.org

Avian Biotech International

1684 Metropolitan Circle
Tallahassee, FL 32308-3731
(800)514-9672
www.avianbiotech.com
DNA sexing and avian disease diagnostics.

Centers for Disease Control and Prevention

1600 Clifton Road
Atlanta, GA 30320-4027
800-CDC-INFO (800-232-4636)
www.cdc.gov

Diagnostic Center for Population and Animal Health

P O Box 30076
Lancing, MI 48909-7576
www.animalhealth.msu.edu
Provides *diagnostic assistance to veterinarians and animal owners.* This is information regarding emerging animal and public health concerns.

Journal of Exotic Pet Medicine

www.elsevier.com
Quarterly publication by subscription.

National Association of Veterinary Technicians in America (NAVTA)

P O Box 1227
Albert Lea, MN 56007
www.navta.net
Sponsors continuing education, quarterly *NAVTA Journal*, job search, careers, opportunities for veterinary technicians and veterinary assistants.

North American Potbellied Pig Association

15525 E. Via Del Palo
Gilbert, AZ 85298
(480)899-8941
www.petpigs.com

North American Veterinary Community Conference

NAVC Headquarters
5003 SW 41st Blvd.
Gainesville, FL 32608-4930
www.navc.com
Annual Conference (January), Orlando, FL, Proceedings
Clinician's brief, regional conferences, online courses.

Western Veterinary Conference

2425 E Oquendo Road
Las Vegas, NV 89120
702-739-6698
www.wvc.org
Annual Conference, Proceedings (February)
Request conference brochure.

Zoological Education Network

P O Box 19357
West Palm Beach, FL 33416
(800)946-4782
www.exoticdvm.com
Publish *Exotic DVM* magazine (bi-monthly), Annual Conference: International Conference on Exotics (ICE), and Proceedings.
Contact for subscription and conference details.

Aberrant Behavior Syndrome (ABS): Aggressive behavior of alpacas that is associated with bottle-fed crias and inappropriate human behavior.

Acroceridae: A genus of flies, commonly called *small-headed flies*; the larva are parasites of spiders, invading the book lungs and feeding on internal organs.

aculeus: The stinger of a scorpion.

acute: Something short, sharp, and sudden, as in pain. Opposite to chronic, which is continual, re-occurring.

adenoma: Tumor of a gland.

air sacculitis: An inflammation of the air sacs in a bird.

alfalfa: A legume plant grown and harvested for hay and used in the formulation of pelleted animal feeds.

alopecia: Loss of hair.

altricial: Animals born with their eyes and ears closed and with no visible hair growth. They are entirely dependent on maternal care for survival.

ambulate: Being able to move around, the ability to walk.

amplexis: In amphibians, the time when eggs are passed by the female and fertilized externally by the male. The clasping of the male to the female.

anaphylactic (shock): An exaggerated allergic response severe enough to potentially cause death. Also see **anaphylaxis**.

Anaphylaxis: A severe, potentially life-threatening allergic reaction.

anogenital distance: The visual difference between the anus and the genitalia.

anointing: Refers to the hypersalivation of hedgehogs. Hedgehogs create a foamy saliva and deposit it on their quills in response to a new scent or taste.

anorexia: A lack of appetite, refusal of food. Appetite is psychological whereas hunger is physiological.

anthelmintic: A deworming product.

anthrax: An infectious, fatal disease of warm-blooded animals, especially of cattle and sheep, caused by the bacterium *Bacillus anthracis*.

anthropomorphic: Interpreting animal behaviors in terms of human emotions, thoughts, and behaviors.

antihistamine: A drug that is used to relieve the symptoms of an allergic response.

anting: Another term for anointing in hedgehogs.

anura: Amphibians without tails: frogs and toads.

apnea: A temporary suspension of breathing.

aquatic: Living in water.

arachnid: A member of a class of invertebrates that includes spiders, scorpions, ticks, and mites. A group of arthropods with eight legs.

arboreal: Tree-living.

arthropod: An invertebrate animal with hard segmented bodies and jointed legs.

arthrosclerosis: Hardening of the arteries.

ascites: An abnormal accumulation of fluid in the abdomen.

ascorbic acid: Vitamin C, found in many fruits and vegetables, especially citrus fruits, necessary in the prevention of scurvy.

aspiration: Inhalation of fluids or solids into the lungs.

asymmetrical: Not identical on both sides of a central line.

asymptomatic: An animal showing no signs of having or carrying a disease or condition.

ataxia: A lack of muscle coordination.

atrophy: Wasting or decreasing, as in the size of an organ or a tissue.

auscultating: Listening for sounds within the body with the use of a stethoscope.

autonomy: The ability to voluntarily release a limb or appendage.

aviary: A place where birds are housed, enclosed outdoor housing for birds, or a bird-breeding facility.

aviculture: The breeding and keeping of birds.

barrow: A castrated adult male pig.

Betadine: A dilute iodine solution used for flushing wounds.

blood feather: A new and growing feather with a blood supply.

boar: Term for a male cavy, hedgehog, or pig.

boggling: A term used to describe protrusion of the eyeballs in a contented rat.

bolt hole: The entrance to a rabbit warren or a hide box provided for a small animal.

book lungs: Folded, accordion-like respiratory organs where gas exchange takes place.

bowel: Lower part of the large intestine.

brachyodont: Teeth that have long, closed roots.

bristle feathers: Sensory feathers around the eyes and nares of birds.

bronchodilator: A drug that causes the bronchi of the lungs to expand, increasing air flow.

brooding: The period of time when a hen sits on the eggs keeping them warm until they hatch; the incubation period provided by the hen.

browse (n.): Leaves, shoots, and other vegetation consumed by animals.

browsing (v.): Consuming these items.

Bruce effect: Abortion of embryos in recently bred mice in order to re-breed with a newly introduced male.

bruxism: Grinding of the molar teeth.

buccal: Referring to the cheek side of the oral cavity, the side of the tooth that is next to the cheek.

buck: A male rabbit.

cackle: The distinctive vocalization made by a hen when an egg is laid.

calcium gluconate: A liquid form of calcium used to replenish metabolic deficiencies.

Campylobacter: A bacteria found in the oral, gastrointestinal, and reproductive tracts of several animal species.

cannulation: Inserting a tube (cannula) into a duct or body cavity.

canopy: Uppermost branches and overlapping foliage of closely growing trees.

captive bred (CB): Species that have been bred in captivity. The parents may or may not have been wild caught.

carapace: The upper part of a chelonian shell.

carcinoma: Tumor that is cancerous.

cardiocentesis: Puncture of a chamber of the heart for a diagnosis test or therapy.

carrion: Dead animal flesh eaten by other animals.

caseous: Something that resembles cheese or curd.

caudal: Toward the tail.

caudata: Amphibians with tails: salamanders and newts.

cavy: Correct name for a guinea pig.

cecotrophs: Soft fecal pellets eaten directly from the anus by rabbits and rodents. In the rabbit, they are encased in a mucous membrane and also called *night feces*.

cecum: The first part of the large intestine that forms a pouch and acts as a fermenting chamber.

cephalothorax: The fused head and thorax of some species, especially arachnids.

cestodes: A class of parasites that includes tapeworms.

chelating agent: A substance used to remove heavy metals from the body by chemically binding with the metals.

chelicerae: The mouth parts of an arachnid.

cheliped: The large front claws of a crab.

cheyletiella: Also known as *walking dandruff*; a mite that infests rabbits; has zoonotic potential.

chitin: The hard substance that forms the exoskeleton.

choana: An opening (slit) in the roof of the mouth; the nasopharynx in birds.

choke: An obstruction in the esophagus.

cloaca: The organ where the urinary tract, GI tract, and reproductive functions meet in birds and reptiles.

cock fighting: An ancient sport of pitting one rooster against another in a fight, usually to death. It is illegal in the United States.

cognition: Possessing awareness, an ability to understand.

colony (colonies): A group of the same species of animals living together in a particular area.

comb: the fleshy tissue on the head of many species of poultry. Also see **wattles**.

congenital: A condition that is present at birth.

coop: Housing for poultry, a chicken coop.

coopy down: A behavior of chickens to squat on the ground as if presenting to a rooster or another dominant hen.

coprophagic: Animals that consume fecal material. It is necessary in some species for digestion and ingestion of some vitamins, beneficial bacteria, and protein.

crepuscular: Animals that are active at dawn and at dusk.

cria: The young of an alpaca or a llama. (There is no specific term for an adult other than male or female; a neutered male is called a gelding.)

crop milk: A liquid produced in the crop by both male and female pigeons and doves to feed their young.

crustaceans: Aquatic-dwelling arthropods.

crystalluria: A condition of having crystals in the urine.

cunniculture: Raising rabbits for meat and other by-products.

cuterebra: The parasitic larvae of large flies that burrow under the skin of the host animal.

cuttlebone: The endoskeleton of a cuttlefish.

cyanosis: A bluish coloration of the skin and mucous membranes due to a decrease in the level of blood oxygen.

dam: The mother of any animal.

decapod: A crustacean with ten legs.

deciduous: One of the two sets of teeth that mammals possess; also called *baby teeth*, they are shed and replaced by permanent, adult teeth.

deglove: The traumatic removal of all layers of skin, like removing a glove.

dermal bone: Bone that originates directly from the skin; bony structure that makes up the skeletal structure of a chelonian shell.

dermatophytosis: A condition caused by a dermatophyte; a fungal infection.

desiccation: A condition of complete dryness; an absence of moisture or body fluid.

detritus: The debris discarded and disintegrating material dropped by animals while feeding.

dewclaw: A rudimentary first digit on the inner (medial) side of the leg.

dewlap: The loose skin under the throat and neck of a female rabbit and some reptiles.

diatomaceous earth: A fine white or gray powder made up of small particles—the skeletons of microscopic diatoms—and used as a feed additive or as a dust bath for poultry; has insecticidal properties.

Dippity Pig Syndrome: A syndrome that causes the back of a potbellied pig to become very sensitive to touch and frequently causes the skin to bleed. Etiology unknown.

dissociative agent: A drug that temporarily depresses neuronal function, producing total or partial loss of sensation with or without the loss of consciousness.

diuretic: A drug that increases urination and elimination of excess body fluid.

diurnal: Active during the day.

doe: A female rabbit.

down: Feathers that are close to the body and provide warmth.

drilling: A territorial or aggressive sound produced by cavies by rapidly chattering their teeth.

dry bite: A bite by a venomous animal, tarantula or snake, without the injection of venom.

dry heaves: Nonproductive retching in an attempt to vomit.

duodenal: Pertaining to the beginning portion of the small intestine.

dwarf: An animal or a plant much smaller than the average of its kind or species.

dysecdysis: Difficulty during the shedding process.

dyspnea: Difficult or labored breathing.

dystocia: Difficulty during the birthing process and usually requiring human assistance.

ecchymosis: Hemorrhagic spots on the skin.

ecdysis: The normal shedding of skin in a healthy reptile.

ecosystem: A community of organisms and nonliving elements interacting in a specific area, for example, deserts, rainforests, and wetlands.

ectotherm: An animal that cannot produce its own body heat and is dependent on an external heat source to regulate metabolic function and normal activity. Also termed *cold-blooded*.

eczema: A superficial skin condition causing redness and itching; can also develop into oozing, crusty sores.

edema: An abnormal accumulation of fluid in the intercellular spaces and body cavities.

egg tooth: a tiny sharp projection on the tip of the upper beak that a chick uses to break open the shell and allow it to hatch. It disappears shortly after hatching.

egg yolk peritonitis: Inflammation and infection caused by egg yolk in the abdominal cavity.

elodontoma: Tumor-like growth that develops due to an overgrowth of molar teeth roots.

en utero: Within the uterus.

endemic: A disease that is known to be in a specific area or animal population.

endoskeleton: A framework of bones that provides structure and support to the body.

endotherm: An animal capable of generating and regulating its own body temperature. Also called *warm-blooded*.

enrichment: Items added to an animal's habitat to provide stimulation and encourage normal behaviors.

enteritis: An inflammation of the intestines.

enterotoxemia: A condition of having bacterial toxins in the blood that originated from the intestines.

epigastric furrow: The location of the reproductive organs of a female tarantula.

epiglottis: The thin elastic cartilaginous structure located at the root of the tongue that folds over the glottis to prevent food and liquid from entering the trachea during the act of swallowing.

episodically: Occurring sporadically.

epizootic: An epidemic of animal disease within the same area and at the same time.

erythema: An abnormal redness of the skin.

erysipelas: An infectious bacterial disease in pigs that can cause swollen joints, lameness, systemic infection, and skin lesions.

estivate: A dormant state during the heat of the summer.

estrus: The period during the reproductive cycle when a female is receptive to the male.

ethology: The scientific study of animal behavior.

eugenol: The chemical extract from clove oil, used to anesthetize amphibians.

eutherian: Mammals that develop a placenta that nourishes the developing young, connecting them to the uterus until the time of birth.

exophthalmosis: An abnormal protrusion of the eyeball, partially or completely, from the eye socket.

external coaptation device: Device used on the outside of the limb to stabilize fractures, for example, splints, rods, casts.

extra label: Using a medication for species outside of the label's recommended use.

exudate: Fluid that has leaked from blood or lymph vessels; usually a result of inflammation or trauma.

falconry: A sport using birds of prey to hunt small game. It includes falcons, hawks, eagles, and owls.

feral: Animals that were once domesticated but have returned to living in a wild state.

ferreting: Capturing wild rabbits with the use of a ferret that is placed into the warren.

filoplumes: Small, hair-like feathers with barbs along the shaft that help guide the bird in flight and reposition the feathers.

fitch: A European polecat that is a close relative to the domestic ferret. Also describes the coat color of some ferrets.

fixed formula: Refers to diets that are constant; the ingredients and percentages do not change with market availability and price.

flock: Collective term for a group of birds, for example, flock of chickens, poultry flock.

fly strike: An area on an animal that has been bitten repeatedly by small flies causing irritation and inflammation often resulting in exudates and scabbing.

fomite: Any inanimate item that can transmit disease-causing organisms.

forage: Grazed foodstuffs; food found by the animal; animals turned out to forage for food.

foraging: Actively searching for food.

fowl cholera: A contagious disease of chickens and other fowl caused by the Pasteurella bacteria.

fowl pox: A disease of poultry caused by the avian poxvirus. Transmitted by mosquitoes and by direct contact with infected birds.

fowl: Collective term for domesticated members of the genus *Gallus*, the pheasant family.

free choice: Referring to unhindered access to food at all times.

free range: Allowed to graze or forage freely rather than being confined to a feedlot or a small enclosure.

fry: The young of fish.

fur slip: The ability of the chinchilla to release patches of fur to evade capture.

fuzzy: A young rat or mouse with a light covering of hair.

Gallus: The genus of birds that includes domestic fowl and birds of the pheasant family.

gang: The collective term for mice.

gaping: Frequent opening of the mouth or holding the mouth open.

gastric: Relating to, or associated with, the stomach.

gelding: A neutered male alpaca or llama.

genome: The complete set of hereditary factors encoded in DNA.

gestation: The period of time between conception and birth.

giardia: A zoonotic protozoal intestinal parasite common to many species.

gizzard: The grinding muscular organ of the digestive system of birds.

glider mills: A term that refers to the indiscriminate breeding of sugar gliders for profit.

glottis: The opening at the upper part of the larynx between the vocal cords.

gout: A condition caused by excess urates being deposited in the intestinal tract and joints.

grit: Small pieces of sand and stone ingested by poultry, essential in the diet to help with the breakdown of feed in the gizzard.

guanaco: A non-domesticated South American camelid related to alpacas and llamas.

gut flora: The normal bacteria found in the digestive tract.

gut motility: The normal movement of the intestinal tract during digestion.

gut stasis: The cessation of normal gut activity.

gut load: The practice of feeding prey animals and insects vitamin- and mineral-enriched food prior to feeding the prey to another animal.

guttural: A sound coming from the back of the throat.

gymnophiona: The limbless amphibians, caecilians.

hair follicle: The skin layer (epidermis) that encloses hair roots. The follicle is the site of growing hair.

hallucinogen: A mind-altering chemical that produces vivid but unreal mental images and alters perception.

Harderian gland: A gland found behind the eyes of many rodents, including hamsters and gerbils; produces a lipid-based substance; and when over stimulated, is the cause of *red tears*.

heaving: Abdominal muscle movements with each breath.

hematoma: A circumscribed collection of blood, usually clotted, in a tissue or an organ; caused by a break in a blood vessel.

hematuria: The presence of blood in the urine.

hemipenes: The reproductive organs in some male reptiles.

hepatic lipidosis: Fat infiltration into liver cells; also known as *fatty liver disease*.

herbivore: An animal that eats only plant material.

herpetology: The study of reptiles and amphibians.

hind-gut fermenter: A term given to a species with a large cecum where food undergoes further digestion through bacterial action.

hob: A male ferret; also sometimes referred to as a *dog ferret*.

hoglet: An immature hedgehog.

hopper: A juvenile mouse or rat pup.

huacaya: A breed of alpaca known for its short, crimped fiber.

humming: The soft sound alpacas and llamas make when communicating.

hutch: The outdoor cage or housing for a rabbit.

hydrometer: An instrument that measures the specific gravity of a liquid.

hygrometer: An instrument that measures environmental humidity.

hyper: A prefix denoting an excess amount.

hypercalcinurea: Having an excessive amount of calcium in the urine.

hyperestrogenism: A condition of unspayed female ferrets that results in an overproduction of estrogen.

hyperkalemic: Relating to an abnormally high concentration of potassium in the blood.

hyperkeratosis: An abnormal overgrowth of keratin.

hypo: A prefix denoting a lower-than-normal amount.

hypsodontic: Describes an animal with teeth that are open rooted and grow continually.

immuno-compromised: A patient without a strong immune system, one that has been weakened due to age or illness.

in-breeding: The mating of closely related animals.

incus: The anvil-shaped small bone or ossicle in the middle ear. It connects the malleus to the stapes. Also known as the *anvil*.

induced ovulator: A species that requires the stimulation of mating before eggs are released from the ovaries.

inflammatory bowel disease (IBD): Chronic condition of the GI tract.

inguinal: The uppermost groin area between the rear legs. Area described by both internal and external anatomy, for example, inguinal canal.

insectivore: An animal that feeds primarily on insects.

insulinoma: A tumor of pancreatic cells.

invertebrate: An animal without a backbone (vertebral column).

isotonic: A fluid with the same tension as normal body fluid, for example, 0.9% saline solution.

Jacobson's organ: An olfactory organ located in the roof of the mouth.

jill: A female ferret.

joey: The young of a marsupial animal.

keratin: A dense, hard protein layer that forms hair, nails, horn, and spines

kindle: The process of giving birth or being *in kindle* (pregnant) for a rabbit.

kit: An immature rabbit or ferret.

Kurloff body: A cytoplasmic inclusion normally found in the monocytes of guinea pigs. The function is unknown.

lactation: The period of time when a dam is producing milk.

lagomorph: The order of mammals that includes rabbits, hares, and pikas.

larval stage: In amphibians and reptiles, the stage between the embryo (egg) and adult.

lavage: Washing, especially of a hollow organ, such as the stomach or lower bowel.

lay-breeders: People who breed animals for their own interest or as a hobby.

lumpy jaw: A sub-mandibular abscess in a guinea pig. The common name for cervical lymphadenitis.

lymphoma: An invasive tumor that involves the lymphatic system.

maggot: The larva of many insects, especially those that feed on decaying flesh.

malignant: Neoplasms that usually have rapid and invasive growth and spread to other areas of the body; cancerous growths.

malignant hyperthermia: A rapid and potentially lethal rise in core body temperature seen in pigs and attributed to anesthetic agents. Genetic in origin and pigs may be tested for the presence of this gene prior to anesthesia.

malleus: The largest of the three bones found in the middle ear of mammals. Also called the *hammer*.

malocclusion: A poor alignment of the biting or grinding surface of teeth.

mandible: The lower jaw or mouth part.

Marek's disease: A herpes virus that causes tumors in various parts of a chicken.

mastitis: Inflammation of the mammary glands.

melena: Meaning black or with a very dark pigment. Stool (feces) that is black with digested blood.

metabolic bone disease (MBD): A condition in which bone is destroyed and becomes rubbery due to a calcium/phosphorous nutritional imbalance.

metastasis: The spread of a disease to multiple organs; usually cancer.

metatheria: Animals born in an embryonic state that make their way to either a pouch or a teat attachment to complete their development; marsupials.

metritis: Inflammation of the uterus.

microhematocrit: A method of determining packed cell volume with the use of a centrifuge.

microhematocrit tube: Small glass vial used to obtain a minimal about of blood for sampling.

molt (molting): Discarding and replacing an exoskeleton or term used when birds molt old feathers and replace them with new.

monkey pox: Caused by an orthopoxvirus and closely related to the human smallpox virus, Variola.

monogamous: Having one mate for life.

monogastric: An animal with a single stomach; a non-ruminant.

mosaic: Refers to the patterned coats of some chinchilla.

MS-222 (tricaine methanesulfonate): An anesthetic agent that is dissolved in water and used in fish and amphibian anesthesia.

mucoid: Relating to or resembling mucus.

Mycoplasma: Bacteria that cause respiratory disease.

nasal septum: The dividing bone and cartilage that separates the nasal cavities.

nebulizer: A device that issues a spray, usually medicated water.

needle teeth: The small, sharp canines that piglets are born with. They are usually removed when the piglets are a day old.

nematodes: A class of parasites that includes roundworms.

neonate: A newborn of any species.

neoplasia: A new (neo) and abnormal growth; a tumor that may be either benign, harmless, or malignant, cancer causing.

Newcastle's disease: A highly contagious viral disease (paramyxovirus) of poultry with high mortality. Zoonotic potential.

nocturnal: Animals that are active at night.

nomenclature: The scientific naming of species by genus and species.

NSAID: A non-steroidal anti-inflammatory drug.

nuptial pads: A thick padding on the feet of some male amphibians that assist in clasping a female during amplexis.

nutritional secondary hyperthyroidism: Another name for metabolic bone disease.

obligate nasal breather: An animal that cannot breathe through its mouth.

omnivore: An animal that eats both plants and animals.

OOP: An acronym for *out of pouch*, a method of estimating the age of a young sugar glider.

os marsupalia: The bones that are unique to marsupials and form part of the pelvic girdle.

osmosis: The exchange of fluid through a semipermeable membrane to establish equal concentrations of a fluid on each side of the membrane.

osteodystrophy: Decalcification of bone due primarily to dietary insufficiency.

OTC: The abbreviation for a drug available without a prescription and bought *over the counter*.

otitis interna: An inflammation of the inner ear.

oviparous: Reproduction by laying eggs that are incubated and hatch outside of the body.

oxytocin: A hormone that stimulates uterine contractions.

palatable (palatability): A food item that has a pleasant taste.

paleontology: The study of fossils.

pandemic: A widespread epidemic of a disease, potentially with worldwide implication.

paratenic: An incidental host that is not necessary for the life cycle of a parasite.

parasite: An organism that lives on or within another species without benefit to the host.

parotid glands: The largest pair of salivary glands, located near the ear.

parturition: The process of giving birth.

passerines: Small perching birds, including songbirds, with three toes forward and one toe pointing backward.

pasty butt: Breeders' term commonly used to describe a white pasting of fecal material caked around the vent of a chicken due to a contagious strain of Salmonella.

patagium: The flap of fur-covered skin on each side of a sugar glider connected from carpus to tarsus.

pathogen: Any disease-causing organism.

pecking order: The social structure of a chicken flock.

pectines: The feathery, toothed appendages on the ventral side of some arthropods.

pedipalps: The most anterior legs of a scorpion or a tarantula, modified and much larger than other pairs of legs.

perianal: The area around or in approximation to the anus.

perineum: The area between the tail and the genitals; the anogenital space.

photo-period: The natural, seasonal range of daylight hours.

piloerection: When hairs are erected, lifted upright, in pain, fear, or aggression.

pinky: A newborn rat or mouse without hair.

pinna: The fleshy, cartilaginous part of the outer ear.

pipping: The term used when a chick begins to break the shell from the inside; beginning of hatching process.

pith (pithing): Destruction of brain and spinal tissue; a method used to obliterate the senses before the dissection of living tissue, especially amphibians.

placenta: An organ that forms with pregnancy and lines the uterine wall. It provides for diffusion of nutrients and circulating blood between the dam and the fetus, and the elimination of waste. The connection is through the umbilical cord.

plastron: The ventral part of a chelonian shell.

plummage: The complete feathering of a bird.

plantar: The sole or bottom of the foot.

pododermatitis: An inflammation on the surface of the foot, with or without a bacterial infection.

poikilotherms: Animals that are totally dependent on their environment to regulate body temperature and metabolic activity.

polydipsia: Excessive or abnormal thirst.

polyestrus: Species that have two or more breeding cycles in a given period of time, for example, *seasonally polyestrus*.

polyphyodontic: Teeth that are shed and replaced throughout life.

polyuria: Excessive production and excretion of urine.

porphyrin: A red discharge from the Harderian gland, located behind the eyes of many rodents.

postmortem: After death.

postpartum: The period of time shortly after giving birth.

POTZ: Refers to the *Preferred Optimum Temperature Zone* giving reptiles the ability to digest prey, fight off infection, and perform day-to-day functions.

poultry: Collective term for farmed domesticated chickens, ducks, geese, and turkeys.

powder down: Feathers that produce a fine white powder that provides waterproofing.

precocial: Animals that are fully developed at birth with their eyes open and a full coat of hair and are able to function with a degree of independence.

preening: To smooth or clean the feathers with the beak or bill.

prehensile: Refers to a tail adapted for grasping or wrapping around objects.

probiotics: Cultured beneficial bacteria fed to restore normal gut flora.

prophylactic: A treatment or medication used to prevent a disease instead of targeting a disease that is already present.

protozoan: A single-celled organism.

proventriculus: The true stomach in birds that contains digestive enzymes that starts to break down food.

pseudo rabies: A strain of herpes virus that causes nervous system signs and intense itching; primarily a disease of pigs. Also called *mad itch* and Aujeszky's disease.

psittacines: Birds with a hooked beak; the parrots with two toes forward and two pointing back.

polytetrafluoroethylene (PTFE): The chemical that is released when nonstick cookware becomes overheated. Extremely toxic to birds.

pullets: Young hens that have not yet started to lay eggs.

pumice: A fine volcanic dust provided for chinchillas to roll in and maintain their fur.

purulent: Containing, discharging, or causing the production of pus.

pyuria: The presence of pus in the urine.

quick: The area of the nail bed supplied with blood and nerve endings.

ration: A fixed amount of a specific food item allowed per day.

rectrices: The tail feathers of a bird, used to balance and help control flight.

red leg: A disease of amphibians characterized by the leg tissue becoming red and ulcerated. It is usually a result of contaminated water and bacterial invasion.

red sneezes/red tears: The discharge of excessive amounts of porphyrin, giving the appearance of blood coming from the eyes or being sneezed out.

reminges: The primary flight feathers of a bird.

rhinitis: Inflammation of the mucous membranes of the nose.

rickets: A condition/disease of growing animals due to a dietary deficiency of phosphorous or vitamin D that affects bone growth and normal development.

roost: A perch for chickens that is off the floor and away from predators; also used to describe when the birds are sleeping for the night, *gone to roost*.

rooster: A mature male chicken.

rubber jaw: A consequence of calcium depletion in the bone of the mandible; the jaw bones become soft and rubbery.

scratch: Commercially produced poultry feed.

scurvy: A disease caused by deficiency of vitamin C; characterized by spongy and bleeding gums, bleeding under the skin, and extreme weakness.

scutes: The individual sections of the plastron.

seasonal feeders: Animals with a diet that varies depending on natural food availability and abundance.

sebaceous glands: The secreting glands of the skin usually in close association with hair follicles.

selective breeding: Choosing specific breeding pairs that may produce the desired traits in their offspring.

septicemia: A bacterial infection of the blood.

sexually dimorphic: Species with visual differences between the sexes: color, shape, size, having distinct male and female characteristics.

shelf-life: The amount of time from production to use without losing any of the benefits of a product; expiration date; *use by* date.

slam feeding: Feeding an excessive amount of prey items at one time in an attempt to promote rapid growth.

slobbers: In rabbits, a moist dermatitis from a chronically wet dewlap.

snuffles: A bacterial disease of rabbits caused by *Pasteurella multocida* resulting in heavy nasal and ocular discharge.

softbills: Birds that do not eat seeds but feed primarily on insects and soft fruit.

sore hocks: Another name for pododermatitis with ulcers and lesions of the hock.

sow: A mature female cavy, hedgehog or pig.

specific gravity: The weight of a substance compared to the weight of a liquid, usually water.

spermathacae: A sperm packet that can be stored by a female tarantula until eggs are produced or is discarded during a molt.

spiderlings: Immature tarantulas.

spinnerettes: The small organs on the dorsal/ventral surface of spiders and tarantulas that produce silk.

spur: The sharp horn-like talon on the caudal leg of **chickens**; more pronounced in roosters and used for fighting and self-defense.

squab: A young pigeon or dove that has not yet left the nest or loft.

stapes: The innermost of a chain of three small bones in the middle ear of mammals; also known as the *stirrup*.

straight run: A group of newly hatched chicks that haven't been sexed.

stomatitis: Inflammation of the mouth.

stridulation: The chirping noise made when two body parts, usually appendages, are rubbed together.

substrate: Any material used on the floor of an enclosure or animal habitat.

suri: A breed of alpaca known for its long, straight fiber.

syndactylism: The condition of having two or more fused digits as in the chameleons and sugar gliders.

syndrome: A group of clinical signs, which taken together, are suggestive of a specific medical condition or disease.

tadpoles: The limbless, aquatic larva of a frog or toad.

telson: The tip of the scorpion's tail where the stinger (aceuleus) is located.

terrestrial: Animals that live on land.

thrush (Candidiasis): a fungal infection caused by unsanitary conditions and careless storage of food.

timothy: A type of grass grown for grazing animals and also dried for hay.

torpor: An inactive state; a physiological mechanism for coping with extremes of heat or a scarcity of food.

torticollis: Also called *wry neck*. The muscles of the neck contract, causing the head to tilt and twist to one side.

trematodes: A class of parasites that includes flukes.

trichobezoar: A hairball in the stomach.

tularemia: A zoonotic disease caused by a contagious Pasteurella strain, *Francisella tularensis*.

turbid: Describes a cloudy appearance in a urine sample.

tusks: Incisor teeth of pigs that are open rooted and grow continually; especially large in boars and used for fighting.

umbilicus: The navel; the area of attachment of the umbilical cord to the fetus prior to birth.

United States Department of Agriculture (USDA): Responsible for developing and executing federal government policies on farming, agriculture, forestry, and food.

urethral cone: Describes the external genitalia of both male and female chinchillas and degus.

urolith: A calculus or stone in the urine or urinary tract.

urolithiasis: The condition of having uroliths or bladder stones.

uropygial gland: The gland located at the base of the tail in some species of birds that produces an oil used in preening.

urticating: An irritating barbed hair of tarantulas that infiltrates the skin of humans or other animals, causing a condition, urticaria, which is similar to hives.

vector: An animal carrier of disease, usually an insect, which transfers pathogens via a bite.

ventriculus: Commonly called the gizzard. The strong, tough muscles of the ventriculus grind the food before it enters the small intestine.

vicuna: A camelid of the Andes related to the alpaca, llama, and guanaco.

vivarium: An enclosed habitat that contains both living plants and animals.

viviparous: Refers to animals that give birth to live young.

volplane: The ability to glide through the air.

vulva: The external female genitalia of mammals.

warren: Connecting underground tunnels dug and inhabited by rabbits.

wattles: Fleshy appendage under the lower beak of poultry.

weeping: A distinctive vocalization of the degu that may continue off and on for several hours.

Whitten effect: First observed in mice, when female estrus cycles become synchronized with the introduction of a new male.

whorls: An area where the hair forms a circle rather than laying flat.

Yersinia pestis: Bacteria that are found in the gut of fleas and thrive in the blood of rats. It is the causative agent of bubonic plague.

yolk sac: A small sac attached to the embryo that provides nutrients to the developing chick via the egg yolk

zoologist: One who studies the branch of biology that deals with animals and animal life.

zoonotic: Used to describe diseases that are transmitted directly from animals to humans. The causative agent may be bacteria, virus, fungi, protozoa, or parasites. Common routes of transmission include inhalation, direct contact, and exposure to urine and fecal material.

INDEX